ENVIRONMENTALLY SUSTAINABLE VITICULTURE

Practices and Practicality

ENVIRONMENTALLY SUSTAINABLE VITICULTURE

Practices and Practicality

Edited by
Chris Gerling

APPLE
ACADEMIC
PRESS

Apple Academic Press Inc. | Apple Academic Press Inc.
3333 Mistwell Crescent | 9 Spinnaker Way
Oakville, ON L6L 0A2 | Waretown, NJ 08758
Canada | USA

©2015 by Apple Academic Press, Inc.

First issued in paperback 2021

Exclusive worldwide distribution by CRC Press, a member of Taylor & Francis Group

No claim to original U.S. Government works

ISBN 13: 978-1-77463-386-1 (pbk)
ISBN 13: 978-1-77188-112-8 (hbk)

Library of Congress Control Number: 2014953722

Library and Archives Canada Cataloguing in Publication

Environmentally sustainable viticulture: practices and practicality/edited by Chris Gerling.

Includes bibliographical references and index.
ISBN 978-1-77188-112-8 (bound)
1. Viticulture. 2. Viticulture--Environmental aspects. 3. Sustainable agriculture. I. Gerling, Chris, editor

SB388.E58 2015   634.8'8   C2014-907066-7

Apple Academic Press also publishes its books in a variety of electronic formats. Some content that appears in print may not be available in electronic format. For information about Apple Academic Press products, visit our website at **www.appleacademicpress.com** and the CRC Press website at **www.crcpress.com**

ABOUT THE EDITOR

CHRIS GERLING

Chris Gerling is part of the Cornell Enology Extension Laboratory (CEEL), which conducts applied research trials, industry workshops and custom analysis. He is the project manager of the Vinification and Brewing Laboratory at the New York State Agricultural Experiment Station in Geneva, New York, and he is Cornell University's extension associate for enology in New York State. Formerly a commercial winemaker, he now engages with the farm-based beverage industry to maximize the quality and sustainability of wine, spirits, and ciders. He serves as a liaison between the research and commercial sectors to help ensure that research is industry-driven wherever possible and that the industry learns of and benefits from research results.

CONTENTS

ACKNOWLEDGMENT AND
HOW TO CITE

The editor and publisher thank each of the authors who contributed to this book. The chapters in this book were previously published in various places in various formats. To cite the work contained in this book and to view the individual permissions, please refer to the citation at the beginning of each chapter. Each chapter was read individually and carefully selected by the editor; the result is a valuable compendium that provides an informative overview of the best sustainable practices in the wine industry. The chapters included examine the following topics:

- Chapter 1 offers a good overview of the basic questions and research areas that form the foundation of the other articles collected here.
- Chapter 2 uses Italian vineyards as a case study to offer significant information to the rest of the world, suggesting sustainability initiatives and practices that work and don't work.
- Talk of sustainability is common, but it's difficult to know if we're comparing apples to apples and oranges to oranges. Chapter 3 offers suggestions for comparison standards that can be applied internationally.
- New Zealand is another important wine region, and Chapter 4 focuses on factors that influence the adoptions of environmental practices. The question is asked: would it more practical to focus on more pragmatic concerns, such as finances and operations, rather than the need for these innovations from a purely environmental perspective? Apparently, viticulture can be as shortsighted as any other industry when it comes to assessing the long-term value of sustainability practices.
- Chapter 5 focuses on water use, one of the obvious targets for sustainable viticulture. From a purely practical standpoint, water scarcity is a growing problem in many wine-growing regions of the world, forcing viticulture to come to terms with this element of sustainability, like it or not.
- Like water use, land use is a challenge viticulture faces. Chapter 6 suggests several useful considerations for sustainable vineyards, including herbicide

and fertilizer use, as well as other land management practices such as adjusting planting to topography.

- Chapter 7 argues that spray techniques are relevant for sustainability practices, since the more efficient the spray, the fewer chemicals that need to be applied. This is similar to the rational for reducing the amount of plastic in water bottles: it does not remove the problem, only makes it a little smaller in a world where the demand for water bottles—and pesticides—is not going to go away.

- Chapter 8 points out the range of climates—and thus environmental factors—that sustainable viticulture must consider. The article offers a nice, practical discussion of various sustainability factors, from water use to isotopic signatures, from stress responses to canopy temperature.

- Sustainable viticulture will only become an across-board reality when consumers demand it. The local foods movement is relevant to viticulture. Chapter 9 focuses on Australian vineyards, but some of the conclusions are helpful to the entire industry.

- "Organic," like "local," is a word with growing consumer popularity. If, as Chapter 10 indicates, organic can be achieved alongside all the other wine qualities consumers value, it could prove to be a boon to sustainable viticulture.

- Vineyards don't grow in vacuums. They are part of larger ecosystems, affected by other factors in those ecosystems, such as natural pest control and pollination. Chapter 11 argues that protecting these ecosystems will benefit viticulture.

LIST OF CONTRIBUTORS

A.-F. Adam-Blondon
INRA-URGV 2, rue Gaston Crémieux CP 5708 F-91057, Evry, France

C. Anderson
CSIRO Plant Industry, PO Box 350, Glen Osmond SA 5064, Australia

Jaume Arnó
Department of Agricultural and Forest Engineering, Research Group in AgroICT & Precision Agriculture, Universitat de Lleida, Rovira Roure 191, 25198 Lleida, Catalonia, Spain

A. Bouquet
UMR BEPC, Campus Agro-M/INRA, 2 place Viala, 34060, Montpellier, France

Ettore Capri
Istituto di Chimica Agraria ed Ambientale, Università Cattolica del Sacro Cuore di Piacenza, 29122 Piacenza, Italy

Leonardo Casini
GESAAF Department, University of Florence, Piazz.le Cascine 18, Firenze, Italy

Alessio Cavicchi
Department of Education, Cultural Heritage and Tourism, University of Macerata, P.le Bertelli, 1, 62100, Macerata, Italy

Cassandra Collins
School of Agriculture, Food and Wine, University of Adelaide, Waite Campus, Private Mail Bag 1, Glen Osmond SA 5064, Australia

Chiara Corbo
Istituto di Chimica Agraria ed Ambientale, Università Cattolica del Sacro Cuore di Piacenza, 29122 Piacenza, Italy

Moya Costello
Southern Cross University, Lismore, New South Wales, Australia

Ross Cullen
Commerce Faculty, Lincoln University, Lincoln, PO Box 85084, Christchurch 7647, New Zealand

I. B. Dry
CSIRO Plant Industry, PO Box 350, Glen Osmond SA 5064, Australia

Alexandre Escolà
Department of Agricultural and Forest Engineering, Research Group in AgroICT & Precision Agriculture, Universitat de Lleida, Rovira Roure 191, 25198 Lleida, Catalonia, Spain

Steve Evans
Flinders University, Adelaide, South Australia, Australia

A. Feechan
CSIRO Plant Industry, PO Box 350, Glen Osmond SA 5064, Australia

J. Flexas
Grup de Recerca en Biologia de les Plantes en Condicions Mediterrànies, Departament de Biologia (UIB-IMEDEA), Universitat de les Illes Balears, Carretera de Valldemossa Km 7.5, 07122 Palma de Mallorca, Balears, Spain

Sharon L. Forbes
Commerce Faculty, Lincoln University, Lincoln, PO Box 85084, Christchurch 7647, New Zealand

Davide Gaeta
Dipartimento di Economia Aziendale, Università degli Studi di Verona, Verona, Italy

Montserrat Gallart
Department of Agri Food Engineering and Biotechnology, Universitat Politècnica de Catalunya, Esteve Terradas 8, Campus del Baix Llobregat D4, 08860 Castelledfels, Barcelona, Spain

A. Gallé
Grup de Recerca en Biologia de les Plantes en Condicions Mediterrànies, Departament de Biologia (UIB-IMEDEA), Universitat de les Illes Balears, Carretera de Valldemossa Km 7.5, 07122 Palma de Mallorca, Balears, Spain

J. Galmés
Grup de Recerca en Biologia de les Plantes en Condicions Mediterrànies, Departament de Biologia (UIB-IMEDEA), Universitat de les Illes Balears, Carretera de Valldemossa Km 7.5, 07122 Palma de Mallorca, Balears, Spain

Emilio Gil
Department of Agri Food Engineering and Biotechnology, Universitat Politècnica de Catalunya, Esteve Terradas 8, Campus del Baix Llobregat D4, 08860 Castelledfels, Barcelona, Spain

Russell Greenberg
Migratory Bird Center, Smithsonian Conservation Biology Institute, National Zoological Park, Washington, DC, United States of America

Rachel Grout
Commerce Faculty, Lincoln University, Lincoln, PO Box 85084, Christchurch 7647, New Zealand

J. Gulías
Grup de Recerca en Biologia de les Plantes en Condicions Mediterrànies, Departament de Biologia (UIB-IMEDEA), Universitat de les Illes Balears, Carretera de Valldemossa Km 7.5, 07122 Palma de Mallorca, Balears, Spain

Diego Ivan
Department of Biology, University of Padova, Padova, Italy

Julie A. Jedlicka
Department of Environmental Studies, University of California Santa Cruz, Santa Cruz, California, United States of America

A. M. Jermakow
CSIRO Plant Industry, PO Box 350, Glen Osmond SA 5064, Australia

Cate Jerram
Adelaide Business School—School of Marketing and Management, University of Adelaide, 09.03 Nexus 10, Adelaide SA 5064, Australia

Markus Keller
Department of Horticulture and Landscape Architecture, Irrigated Agriculture Research and Extension Center, Washington State University, Prosser, WA 99350, USA

Lucrezia Lamastra
Istituto di Chimica Agraria ed Ambientale, Università Cattolica del Sacro Cuore di Piacenza, 29122 Piacenza, Italy

Monica Laureati
Department of Food, Environmental and Nutritional Sciences (DeFENS), Università degli Studi di Milano, Milano, Italy

Deborah K. Letourneau
Department of Environmental Studies, University of California Santa Cruz, Santa Cruz, California, United States of America

Jordi Llop
Department of Agri Food Engineering and Biotechnology, Universitat Politècnica de Catalunya, Esteve Terradas 8, Campus del Baix Llobregat D4, 08860 Castelledfels, Barcelona, Spain

Jordi Llorens
Department of Agri Food Engineering and Biotechnology, Universitat Politècnica de Catalunya, Esteve Terradas 8, Campus del Baix Llobregat D4, 08860 Castelledfels, Barcelona, Spain

Lorenzo Marini
DAFNAE - Department of Agronomy, Food, Natural Resources, Animalsandthe Environment, University of Padova, Legnaro (Padova), Italy

H. Medrano
Grup de Recerca en Biologia de les Plantes en Condicions Mediterrànies, Departament de Biologia (UIB-IMEDEA), Universitat de les Illes Balears, Carretera de Valldemossa Km 7.5, 07122 Palma de Mallorca, Balears, Spain

Andrew Metcalfe
Mathematical Science School, University of Adelaide, North Terrace Campus, 6.43 Ingkarni Wardli, Adelaide SA 5005, Australia

Juri Nascimbene
Department of Biology, University of Padova, Padova, Italy

Ella Pagliarini
Department of Food, Environmental and Nutritional Sciences (DeFENS), Università degli Studi di Milano, Milano, Italy

A. Pou
Grup de Recerca en Biologia de les Plantes en Condicions Mediterrànies, Departament de Biologia (UIB-IMEDEA), Universitat de les Illes Balears, Carretera de Valldemossa Km 7.5, 07122 Palma de Mallorca, Balears, Spain

M. Ribas-Carbo
Grup de Recerca en Biologia de les Plantes en Condicions Mediterrànies, Departament de Biologia (UIB-IMEDEA), Universitat de les Illes Balears, Carretera de Valldemossa Km 7.5, 07122 Palma de Mallorca, Balears, Spain

Joan R. Rosell-Polo
Department of Agricultural and Forest Engineering, Research Group in AgroICT & Precision Agriculture, Universitat de Lleida, Rovira Roure 191, 25198 Lleida, Catalonia, Spain

Irina Santiago-Brown
School of Agriculture, Food and Wine, University of Adelaide, Waite Campus, Private Mail Bag 1, Glen Osmond SA 5064, Australia

Cristina Santini
Faculty of Agriculture, Università San Raffaele, Via Val Cannuta 247, Roma, Italy

Ricardo Sanz
Department of Agricultural and Forest Engineering, Research Group in AgroICT & Precision Agriculture, Universitat de Lleida, Rovira Roure 191, 25198 Lleida, Catalonia, Spain

H. R. Schultz
Institut für Weinbau und Rebenzüchtung, Forschungsanstalt Geisenheim, von-Lade Str. 1, D-65366, Germany and Fachhochschule Wiesbaden, Fachbereich Geisenheim, von-Lade Str. 1, D-65366 Geisenheim, Germany

M. Stoll
Institut für Weinbau und Rebenzüchtung, Forschungsanstalt Geisenheim, von-Lade Str. 1, D-65366, Germany

M. R. Thomas
CSIRO Plant Industry, PO Box 350, Glen Osmond SA 5064, Australia

M. Tomàs
Grup de Recerca en Biologia de les Plantes en Condicions Mediterrànies, Departament de Biologia (UIB-IMEDEA), Universitat de les Illes Balears, Carretera de Valldemossa Km 7.5, 07122 Palma de Mallorca, Balears, Spain

Michela Zottini
Department of Biology, University of Padova, Padova, Italy

INTRODUCTION

Viticulture is necessarily a long-term commitment. As perennial crops, grapevines must be nurtured over years and even decades. Decisions made this season will impact the seasons to come, and any short-term gains must be weighed against the longer-term consequences.

Sustainable practices are therefore essential for any vineyard, because concerns about pruning and crop loads pale in comparison to those regarding whether or not there will be enough water, whether the soil will remain fertile enough to support vines, and whether the climate will even allow grapes to survive in the near future. Thoughtful and well-researched investigations into the best courses of action, the best ways to assess sustainability programs and the best ways to consider viticulture in a larger environmental context—research like that included in this compendium—are all necessary to keep the industry moving in the right direction and to make sure that grapes planted today are growing well for years to come.

Chris Gerling

Sustainability plays a key role in the wine industry, as shown by the attention paid at several levels by the academia, institutions and associations. Nevertheless, the principle itself of sustainability opens a wide debate and it significantly affects firms in all their activities. Using a systematic literature review, Chapter 1, by Santini and colleagues, highlights some of the questions that academics must face when they approach the issue of sustainability with a specific focus on the wine industry. In particular the paper aims to: highlight where research is going and what has already been done; define the contribution of background research in explaining the determinants of sustainable orientation in the wine industry; and understand the role of research (and academics' social responsibility) for the diffusion of a sustainable orientation within the wine industry. The purpose of this paper is to provide a detailed overview of the main research contributions to the issue of sustainability in the wine industry.

The Italian wine industry is strongly committed to sustainability: the stakeholders' interest in the topic is constantly growing and a wide number of sustainability programs have been launched in recent years, by both private businesses and consortiums. The launch of these initiatives has signaled the commitment of farmers and wine producers to the implementation of sustainability principles in viticulture and wine production, which is a positive signal. Unfortunately, however, the varied design of the sustainability initiatives and the differences in the objectives, methodologies, and proposed tools risks to create confusion, and undermine the positive aspects of these initiatives. In order to bring some clarity to this topic, in Chapter 2, Corbo and colleagues present a comparison of the most important sustainability programs in the Italian wine sector, with the overall objective of highlighting the opportunity to create synergies between the initiatives and define a common sustainability strategy for the Italian wine sector.

Chapter 3, by Santiago-Brown and colleagues, documents and compares the most prominent sustainability assessment programs for individual organisations in viticulture worldwide. Certification and engagement processes for membership uptake; benefits; motives; inhibiting factors; and desirable reporting system features of viticultural sustainability programs, are all considered. Case-study results are derived from nine sustainability programs; 14 focus groups with 83 CEOs, Chief Viticulturists or Winemakers from wine grape production organizations from five countries (Australia, Chile, New Zealand, South Africa and the United States); 12 semi-structured interviews with managers either currently or formerly in charge of the sustainability programs; researcher observations; and analysis of documents. Programs were categorized by their distinct program assessment methods: process-based, best practice-based, indicator-based and criterion-based. We found that programs have been created to increase growers' sustainability, mainly through the direct and indirect education they receive and promote, and the economic benefit to their business caused by overall improvement of their operations. The main finding from this study is that the success of each of these programs is largely due to the people driving the programs (program managers, innovative growers and/

or early adopters) and the way these people communicate and engage with their stakeholders and peers.

The Greening Waipara Project developed and introduced a number of ecologically and environmentally-focused practices to the Waipara vineyards and wineries of North Canterbury, New Zealand. Chapter 4, by Forbes and colleagues, describes the practices that were introduced to the Waipara wine industry as part of the Greening Waipara Project and evaluates the adoption of these environmental innovations by wine businesses. In addition, this paper examines the sustainability of these practices in terms of business costs and benefits. Data for the evaluation was obtained from a survey of vineyards and wineries in the Waipara region. Results reveal that adoption of the environmental innovations is relatively low and varies across wine growing properties. Furthermore, the costs associated with the innovations tend to outweigh the benefits gained by the businesses.

Improving water use efficiency (WUE) in grapevines is essential for vineyard sustainability under the increasing aridity induced by global climate change. WUE reflects the ratio between the carbon assimilated by photosynthesis and the water lost in transpiration. Maintaining stomata partially closed by regulated deficit irrigation or partial root drying represents an opportunity to increase WUE, although at the expense of decreased photosynthesis and, potentially, decreased yield. It would be even better to achieve increases in WUE by improving photosynthesis without increasing water loses. Although this is not yet possible, it could potentially be achieved by genetic engineering. Chapter 5, by Flexas and colleagues, presents current knowledge and relevant results that aim to improve WUE in grapevines by biotechnology and genetic engineering. The expected benefits of these manipulations on WUE of grapevines under water stress conditions are modelled. There are two main possible approaches to achieve this goal: (i) to improve CO_2 diffusion to the sites of carboxylation without increasing stomatal conductance; and (ii) to improve the carboxylation efficiency of Ribulose-1,5-bisphosphate carboxylase/oxygenase (Rubisco). The first goal could be attained by increasing mesophyll conductance to CO_2, which partly depends on aquaporins. The second approach could be achieved by replacing Rubisco from grapevine with Rubiscos from other C3 species with higher specificity for CO_2. In summary, the physiological bases and future prospects for improving grape yield and WUE under drought are established.

Vineyards are amongst the most intensive forms of agriculture often resulting in simplified landscapes where semi-natural vegetation is restricted to small scattered patches. However, a tendency toward a more sustainable management is stimulating research on biodiversity in these poorly investigated agro-ecosystems. The main aim of Nascimbene and colleagues in Chapter 6 was to test the effect on plant diversity of management intensity and topography in vineyards located in a homogenous intensive hilly landscape. Specifically, this study evaluated the role of slope, mowing and herbicide treatments frequency, and nitrogen supply in shaping plant diversity and composition of life-history traits. The study was carried out in 25 vineyards located in the area of the Conegliano-Valdobbiadene DOCG (Veneto, NE Italy). In each vineyard, 10 plots were placed and the abundance of all vascular plants was recorded in each plot. Linear multiple regression was used to test the effect of management and topography on plant diversity. Management intensity and topography were both relevant drivers of plant species diversity patterns in our vineyards. The two most important factors were slope and mowing frequency that respectively yielded positive and negative effects on plant diversity. A significant interaction between these two factors was also demonstrated, warning against the detrimental effects of increasing mowing intensity on steep slope where plant communities are more diverse. The response of plant communities to mowing frequency is mediated by a process of selection of resistant growth forms, such in the case of rosulate and reptant species. The other two management-related factors tested in this study, number of herbicide treatments and N fertilization, were less influential. In general, our study corroborates the idea that some simple changes in farming activities, which are compatible with grape production, should be encouraged for improving the natural and cultural value of the landscape by maintaining and improving wild plant diversity.

Spraying techniques have been undergoing continuous evolution in recent decades. Chapter 7, by Gil and colleagues, presents part of the research work carried out in Spain in the field of sensors for characterizing vineyard canopies and monitoring spray drift in order to improve vineyard spraying and make it more sustainable. Some methods and geostatistical procedures for mapping vineyard parameters are proposed, and the development of a variable rate sprayer is described. All these technologies are interesting in terms of adjusting the amount of pesticides applied to the target canopy.

The rapidly increasing world population and the scarcity of suitable land for agricultural food production together with a changing climate will ultimately put pressure on grape-producing areas for the use of land and the input of resources. For most grape-producing areas, the predicted developments in climate will be identical to becoming more marginal for quality production and/or to be forced to improve resource management. This will have a pronounced impact on grapevine physiology, biochemistry and ultimately production methods. Research in the entire area of stress physiology, from the gene to the whole plant and vineyard level (including soils) will need to be expanded to aid in the mitigation of arising problems. In Chapter 8, Schultz and Stoll elaborate on some key issues in environmental stress physiology such as efficient water use to illustrate some of the challenges, current limitations and future possibilities of certain experimental techniques and/or data interpretations. Key regulatory mechanisms in the control of stomatal conductance are treated in some detail and several future research directions are outlined. Diverse physiological aspects such as the functional role of aquaporins, the importance of mesophyll conductance in leaf physiology, night-time water use and respiration under environmental constraints are discussed. New developments for improved resource management (mainly water) such as the use of remote sensing and thermal imagery technologies are also reviewed. Specific cases where our experimental systems are limited or where research has been largely discontinued (i.e. stomatal patchiness) are treated and some promising new developments, such as the use of coupled structural functional models to assess for environmental stress effects on a whole-plant or canopy level are outlined. Finally, the status quo and research challenges around the 'CO_2-problem' are presented, an area which is highly significant for the study of 'the future' of the grape and wine industry, but where substantial financial commitment is needed.

Grapevine reproductive development extends over two seasons, and the genotypic expression of yield potential and fruit composition is subject to environmental impacts, which include viticultural manipulations, throughout this period. In Chapter 9, Keller reviews current knowledge on yield formation and fruit composition and attempts to identify challenges, opportunities and priorities for research and practice. The present analysis of published information gives a critical appraisal of recent advances concerning variables, especially as they relate to global climate change, that influence yield formation and fruit composition at harvest. Exciting discoveries in fundamental research on the one hand and an increasing focus on outcomes and knowledge transfer on the other are enabling the development and implementation of practical recommendations that will impact grape production in the future. Future research should aim to minimise seasonal variation and optimise the profitable and sustainable production of high-quality fruit for specific uses in the face of climate change, water and labour shortages, shifting consumer preferences and global competition. Better control of product quantity and quality, and differentiation to meet consumer demands and market preferences will enhance the competitiveness and sustainability of the global grape and wine industries.

The Eurasian winegrape *Vitis vinifera* has little or no genetic resistance to the major fungal pathogens, powdery mildew (*Erysiphe necator*) and downy mildew (*Plasmopora viticola*). These pathogens were first introduced into French vineyards from North America in the 1800s before spreading to all major grape producing regions of the world. As a result, grape production is highly dependent on the use of fungicides. With the increasing financial and environmental costs of chemical application and the emergence of fungicide-resistant strains, the introduction of natural genetic resistance against these fungal pathogens is a high priority for viticultural industries worldwide. In Chapter 10, Dry and colleagues utilize a number of different molecular approaches to increase our understanding of the basis of resistance to these important major fungal pathogens and to identify potential new sources of genetic resistance. This review will outline the progress and the potential of each of these different molecular strategies to the generation of fungal-resistant grapevine germplasm.

Bibere vinum suae regionis, to drink wine from one's own region, attempts to match the neologism 'locavore', local eater, with one for wine. In Chapter 11, Costello and Evans compare drinking in two regions: the surrounds of Adelaide, South Australia, an area of international repute for wine-making, and the subtropical Northern Rivers, on the far north coast of New South Wales—not a diverse wine-growing area because of high rainfall and humidity that produce grape-destroying mildew/fungus, but bordering a number of 'new' wine areas. Issues under consideration include distribution and access, choice and cost. We also survey the reasons for consuming wine in particular, and consuming it locally, including sustaining economies, environments, societies, cultures and identities, and investigate the idea of the local per se.

In recent years, produce obtained from organic farming methods (i.e., a system that minimizes pollution and avoids the use of synthetic fertilizers and pesticides) has rapidly increased in developed countries. This may be explained by the fact that organic food meets the standard requirements for quality and healthiness. Among organic products, wine has greatly attracted the interest of the consumers. In Chapter 12, by Pagliarini, trained assessors and regular wine consumers were respectively required to identify the sensory properties (e.g., odor, taste, flavor, and mouthfeel sensations) and to evaluate the hedonic dimension of red wines deriving from organically and conventionally grown grapes. Results showed differences related mainly to taste (sour and bitter) and mouthfeel (astringent) sensations, with odor and flavor playing a minor role. However, these differences did not influence liking, as organic and conventional wines were hedonically comparable. Interestingly, 61% of respondents would be willing to pay more for organically produced wines, which suggests that environmentally sustainable practices related to wine quality have good market prospects.

Insectivorous Western Bluebirds (*Sialia mexicana*) occupy vineyard nest boxes established by California winegrape growers who want to encourage avian conservation. Experimentally, the provision of available nest sites serves as an alternative to exclosure methods for isolating the potential ecosystem services provided by foraging birds. In Chapter 13, Jedlicka and colleagues compared the abundance and species richness of avian foragers and removal rates of sentinel prey in treatments with songbird nest boxes and controls without nest boxes. The average species richness of avian insectivores increased by over 50 percent compared to controls. Insectivorous bird density nearly quadrupled, primarily due to a tenfold increase in Western Bluebird abundance. In contrast, there was no significant difference in the abundance of omnivorous or granivorous bird species some of which opportunistically forage on grapes. In a sentinel prey experiment, 2.4 times more live beet armyworms (*Spodoptera exigua*) were removed in the nest box treatment than in the control. As an estimate of the maximum foraging services provided by insectivorous birds, we found that larval removal rates measured immediately below occupied boxes averaged 3.5 times greater than in the control. Consequently the presence of Western Bluebirds in vineyard nest boxes strengthened ecosystem services to winegrape growers, illustrating a benefit of agroecological conservation practices. Predator addition and sentinel prey experiments lack some disadvantages of predator exclusion experiments and were robust methodologies for detecting ecosystem services.

PART I

OVERVIEW

CHAPTER 1

SUSTAINABILITY IN THE WINE INDUSTRY: KEY QUESTIONS AND RESEARCH TRENDS

CRISTINA SANTINI, ALESSIO CAVICCHI, AND LEONARDO CASINI

1.1 INTRODUCTION

The wine industry is definitely engaged in sustainability. The emerging interest in sustainability is confirmed by a growing body of academic literature as well as by the rise of new academic journals and scientific communities. Also, the industry has shown an involvement in sustainability in general; people in the wine industry wonder about the effectiveness of sustainable practices and under what conditions it pays to be oriented towards sustainability. Talking about sustainability opens up a multitude of research issues, especially in wine, where being sustainable is often misunderstood with being organic or biodynamic. This paper investigates background research on sustainability in wine; it outlines what are the main challenges that scholars must face when they deal with this research issue. After having provided a description of research trends, the paper will highlight the determinants of a firm's orientation towards sustainability and the role that research has in promoting sustainability.

1.2 SO MANY GREEN NUANCES

The word "sustainability" has so many definitions that it holds a shadow of ambiguity (Warner 2007). Sustainability can be seen as a concept based on various principles (de Bruyn and van Drunen 2004): economic principles (maximising welfare and improving efficiency), ecological principles (living within carrying capacities and conservation of resources) and equity-principles that concern intragenerational and intergenerational equity (the disparity of wealth among different regions of the world or among generations).

(Ohmart 2008) gives an idea of how complex it is to be sustainable in agriculture: "sustainability involves everything you do on the farm, including economics, environmental impacts of everything done on the farm and all aspects of human resources, including not only you and your family but your employees and the surrounding community" (Ohmart 2008): 7.

Nevertheless, there is no univocal sustainable behaviour and some companies should be considered more sustainable than others. (Isaak 2002) distinguishes between green and green–green businesses; green–green businesses are green oriented since their start up, whilst green businesses become green after that manager—who are not inspired by ethical issues—have intuited the benefits (in terms of marketing, corporate image positive feedbacks or cost savings) that being "sustainable" might create for the company.

In a recent study, Szolnoki (2013) points to the idea that wineries have of sustainability: misunderstandings and differences in the approach between countries and wineries emerge.

Some countries are "greener" than others according to the degree of companies' sustainable behaviour. Globally spoken, California holds a leading position among the most sustainable agricultural producing countries: Warner (2007) describes the efforts spent by the Californian wine grape industry for reaching and educating growers about quality issues and sustainability; the availability of place-based networks of production has facilitated social learning among grape growers. (Warner 2007) says: "More than any other group of California growers, winegrape growers are operationally defining sustainability as agricultural enterprise viability, environmental quality and product quality" (p.143). In Northern California there are about 40 industry organizations advocating sustainability in addition to several associations that support organic viticulture at a national level (see among others: Washington State Association of Wine grape Growers; Oregon Wine Advisory Board; New York Wine & Grape Foundation; Penn State Cooperation Extension; Wine Council of Ontario). A case worth to be mentioned is the one of Lodi region in California (Ohmart 2008) that effectively shows how a local economic system could respond to the call for sustainability: after having released a workbook programme that encompasses all the sustainable practices in winemaking, results have been monitored in order to assess how principles have been implemented and to examine action plans carried by companies and associations. The great success of the Lodi programme relies on the active involvement of growers that is the result of a successful combination of workshops, a proactive behaviour of associations and effective communication flows.

Starting from the behaviour adopted by wineries, some scholars (Casini et al. 2010) have proposed a model that would help to classify wineries' orientation in terms of sustainability. In the model, "devoted" wineries have a strong orientation towards sustainability that is emphasised in customer communication; those companies must invest in customers and employees training and education; furthermore devoted wineries must ensure an alignment between their corporate and managerial visions. Another category of wineries, the so called "unexploiters", stands half the way between devoted and "laggards" wineries, or those who would never adopt sustainable practices. Unexploiters usually decide to adopt sustainable practices, but do not inform other people (clients, first of all) about their decision. Consequently the benefits that might be gained through a sustainable orientation are limited. At the opposite of unexploiters stand opportunists, wineries that do not have a particular interest in sustainability, but tend to heavily highlight the few sustainable practices introduced.

As it can be guessed, companies can choose among various alternatives: it is not only a matter of being green or not, but they can also choose among a multitude of "green nuances".

We conceive sustainability as a behaviour adopted to respond to stimuli, whether they are external or internal to the firm. This perspective introduces three elements into the discussion: firstly, the presence and the type of stimuli or drivers; secondly, the degree of responsiveness that characterises the organization; and thirdly, a firm's motivations.

In other words, we can say that orientation towards sustainability depends on how the following questions can be answered: Who cares about sustainability issues? How much do I and my organization care about sustainability? Why should we care about sustainability?

This paper aims to systemically analyse the main academic contributions to the issue of sustainability in the wine industry in order to outline insights that can depict strategic, managerial, consumer and organizational implications, and to highlight what are the main challenges that scholars must face when they get into this research issue. After having provided a description of where the research is going, the paper will explain the determinants of a orientation towards sustainability among firms and it will outline the role of research in promoting sustainability.

1.3 KEY DRIVERS OF SUSTAINABILITY

An analysis of the drivers of sustainability is, in our opinion, the first step to understand the relationship between firms and sustainability. This section introduces the issue of drivers of sustainability, by highlighting the findings emerging from background research (the role of institutions and associations, the role of top management and entrepreneurs, etc.) and the role played by drivers in defining wineries' orientation towards sustainability. We assume that an exploration of the incidence of perceived stimuli on companies' choices represents a way for explaining firm's behaviour; by conceiving sustainability as a behaviour adopted by firms in responding to selected stimuli, we focus on the impact that forces (external or internal to the firm) have on a firm's strategy; being sustainable represents one of the strategic choices that firms can make. The presence of drivers affecting a firm's orientation towards sustainability partially explains the differences in the overall degree of sustainability at a firm or at a country level. It is almost impossible to define a general ranking for estimating a country's overall orientation towards sustainability: the numerous indexes available simply confirm the differences among countries, but it is extremely hard to classify them, because of the differences in index composition and in the trait of sustainability under observation. A focus on the key drivers of sustainability offers a balanced solution to this problem. Both academics and practitioners increasingly emphasize the issue of drivers. The company Accenture has elaborated a list of six key drivers of sustainability, that "are not only reshaping the way businesses and governments operate, but also redefining the value they deliver" (from corporate website). The list includes: consumer demand for sustainable products and services; stakeholder influence; resource depletion; employee engagement; capital market scrutiny; regulatory requirements.

Also background research (Dillon and Fischer 1992; Lawrence and Morell 1995; Winn 1995; Bansal and Roth 2000; Davidson and Worrell 2001; Marshall et al. 2005; Gabzdylova et al. 2009) has highlighted the role of drivers—whether they are conceived as internal/external or internal/institutional—to describe a firm's adoption of a sustainable behaviour.

Internal drivers are all those drivers that take place within the firm: they are ethical motives inspiring top management and entrepreneurs as well as strategic intentions based on the recognition of an advantage that might arise from sustainability. External drivers, instead, take place in the firm's external environment.

1.4 INSTITUTIONS, ASSOCIATIONS, REGULATORS AND MARKET DEMAND

External drivers happen outside of the firm and include pressures arising from institutions, customers, communities, associations, environmental groups, activists, regulators and competitors.

Background research has highlighted the role played by industry associations in creating "sustainable awareness" among grapegrowers and wineries Broome and Warner (2008; Silverman et al. 2005; Warner 2007).

A key factor of success in spreading sustainable practices is local players' networking capacity. In some specific areas, such as California, agroecological partnerships have fostered the adoption of sustainable agricultural practices (Swezey and Broome 2000; Dlott 2004) and they have proactively spread a green orientation among wineries (Broome and Warner 2008).

Environmental concerns have progressively found a diffusion among wineries and became strongly related to corporate image. New Zealand is heavily investing in environmental issues: "The New Zealand wine industry aims to be the first in the world to be 100% sustainable. The Sustainable Winegrowers New Zealand (SWNZ) programme introduced in 1995 is a framework of industry standards set up to achieve this by vintage 2012" (from the website: http://www.newzealand.com).

Also corporate activism should be considered, as shown by the efforts spent by individual companies for promoting practices that would reduce gas emission and waste. The case of The Wine Group, in the US, highlights the consideration that large companies give to environmental issues: in 2008 The Wine Group has launched a website (http://www.betterwinesbetterworld.com) to document how "Bag in Box" can help in reducing emissions and waste (http://www.winebusiness.com). Both the New World and the Old World face similar environmental challenges but they strongly differ in terms of fertiliser usage, that is significantly lower in Europe (http://www.eea.europa.eu).

The development of specific programmes for sustainable winegrowing has fostered the adoption of "ground to bottle" practices for producing grapes and wine (Broome and Warner 2008). This is highlighted by the willingness that institutions and organizations show in providing long term financial support to sustainability programmes and training activities: (Warner 2007) underlines the need for continuous investments in reinforcing a commitment to sustainability.

Institutions and regulators have a prime role in enhancing wineries' interest towards sustainability through funding the adoption of specific practices and education programmes (Swinbank 2009).

A orientation towards sustainability among competitors can foster a me-too mechanism with the result of spreading sustainable practices in the competitive environment: after that Mondavi has introduced the flange-type bottle with a C-cap on the market (Murphy 2000), other wineries in the market have shared—consciously or unconsciously—the same principles that have inspired Mondavi before the product launch.

Consumers' involvement in sustainability is also reshaping wineries' interest toward this issue, as described by (Bisson et al. 2002): "As consumers become more aware of the vulnerability of our global environment, the demand for sound agricultural production practices is increasing. In the future, the perception of the producer as a conscientious environmental steward will be an important influence on the consumer's purchasing decision. This is due in part to the fact that the typical wine consumer is well educated and affluent" (p.698). Consumers' pressure has created a market for wines inspired by environmental issues, such as organic or biodynamic wines (Forbes et al. 2009): in some countries, such as the UK, organic wine moved from a niche to a mainstream position (Sharples 2000).

1.5 ENTREPRENEURS AND TOP MANAGEMENT

Most of the research has focused on explaining the role of external drivers in enhancing a sustainability orientation within firms, but less research has been done about internal drivers.

A consistent body of research can be found in the general management and business strategy literature, that analyses the role of people involved within the organization in promoting a sustainability orientation: various issues have been investigated such as the role of top management's values in determining sustainability orientation (Berry and Rondinelli 1998; Quazi 2003), entrepreneurial commitment to sustainability (Shaltegger 2002) or management practices and principles reshaped by a sustainability orientation (Atkin et al. 2012; Warner 2007).

In some cases, niche research fields have emerged by providing a "sustainability" perspective to diffused and internationally adopted research approaches: this is the case of Ecopreneurship (ecological entrepreneurship) or the Natural Resource Based View, a version of the Resource Based View of the Firm approach mainly based on environmental issues.

Ecopreneurship is a term that has been introduced in early 1990s (Bennett 1991; Berle 1991; Blue 1990) and that renames a growing body of literature that investigates most of the critical questions in entrepreneurship from an ecological and environmental perspective. The works by Walley and Taylor (2002), Shaltegger (2002) and Schaper (2002) provide a comprehensive overview of this research field. From this research the prominent role that personality traits can have on the degree of a sustainability orientation within firms emerges: for instance, Regouin (2003) has highlighted that reasons behind a firm's conversion to organic farming depend on personal traits such as curiosity, flexibility, risk propensity and creativity in exploring innovative marketing approaches. Although there is a growing body of academic literature that is exploring the "internal" drivers towards sustainability, only a few studies have been done on wine.

1.6 SUSTAINABILITY AND STRATEGY

(Bonn and Fisher 2011) say that sustainability is often a missing ingredient in strategy: there is a great debate on corporate social responsibility, corporate environmentalism, sustainable practices adoption, green marketing, green corporate image, etc., but the issue of sustainability is not considered as priority in strategy making.

Research in wine has focused on the relationship between a sustainability orientation and competitive advantage. It has been shown how being organic contributes to an effective differentiation (Bernabeu et al. 2008): Delmas et al. (2008) explore the case of a winery in California (the Ceago winery, owned by Fetzer), that has chosen to produce organic wine to differentiate its product from the mass; (Pugh and Fletcher 2002) examine how a wine multinational corporation (the BRL Hardy) focuses on a specific and different market segment through one of its controlled brands (Banrock Station) that supplies organic wine to the market.

Gilinsky and Netwon (2012) provides useful insights for understanding if incorporating an Environmental Management System (EMS) into business models positively or negatively affects wineries' performance. From the research the relevance that EMS has in pursuing a differentiation strategy for some of the wineries who employ EMS has emerged. The literature shows that little attention has been paid to the benefits that implementing EMS might have for wineries (Forbes and De Silva 2012).

1.7 THE ROLE OF RESEARCH

Next to the wine industry, also research in the wine business is going green. Research in the field of sustainability in wine has been fostered by the growing interest of the industry and by the active role of institutions—that funds specific research programmes—associations or individual companies. Supporting research has resulted in a renewed interest in sustainability with the final result of promoting further research. When observing some cases—such as the Washington State Wine Industry—we can say that university research has fostered the development of the wine industry (Stewart 2009). (Ohmart 2008) suggests that a successful diffusion of sustainable practices among grapegrowers depends on two factors: rigorous science and its effective delivery to grapegrowers, two issues that partially explain the differences in terms of penetration and diffusion of sustainable practices in viticulture.

In their analysis of the history of winemaking in California, Guthey and Whiteman (2009) say that funded university research has contributed to shape Californian wine production thanks to the useful inputs provided for developing winemaking practices and understanding human environment relationships.

The field of sustainability in the wine industry appears as a breeding ground for the development of academics and university collaborations: (Lee 2000) provides a general framework that can be used for describing the benefits arising from the relationship between academics and industry. In general it can be said that collaboration between research institutions and the industry (1) may be helpful in solving technical problems, (2) may facilitate the access to useful findings; and (3) may make the implementation of innovation easier. It is not surprising that industry heavily supports research in some countries: we can cite among others the cases of the Wine and Food Institute in California cofounded by the Robert Mondavi Winery and the Anheuser-Busch Foundation and Ronald and Diane Miller of Silverado Vineyards (http://www.winespectator.com). Another case worth to be mentioned is the Australian Wine Research Institute, that has actively promoted research in the field of wine in general and has had a relevant role in spreading a sustainable culture among wineries. Research has been stimulated in new world countries and not only in California or Australia, as the case of Vinnova from Chile shows. Great efforts have been spent for codifying research insights and facilitating knowledge dissemination and accessibility: some countries, such as New Zealand and Chile, have developed Codes of Sustainability, to promote the adoption of sustainable practices among wineries.

Conducting research on sustainability has some social implications and researchers who are working in this field have a social responsibility: with their work, researchers can foster the adoption of sustainable practices among wineries at different levels and they can indirectly contribute to the growth of the overall welfare of people living in a certain area.

1.8 RESEARCH ORIENTATIONS: A SELECTIVE SYSTEMATIC LITERATURE REVIEW

1.8.1 METHODOLOGY

Where is research going and what has been done? In order to answer this specific question we have carried out a systematic literature review by analysing academic databases and some wine academic journals. In particular we have performed a keywords based research in the following academic search engines: ISI Web of KnowledgeSM, Scopus SciVerse® and EBSCO (that contains Econlit, Business Source Premiere and Greenfile databases).

In our research we did not want to use generic "scientific" search engines (i.e., Google Scholars or Mendeley) and to perform a search on specific academic databases that are widely diffused among scholars.

The keywords used, combined with the word "wine" are: green, organic, sustainable, sustainability, biodynamic, ecopreneurship, environment. We have also selected some academic journals specialised in wine, and we have checked the presence of articles that examine the issue of sustainability in wine; the journals selected are: the *International Journal of Wine Business Research*; the Australian journal of *Grape and Wine Research*; the *Journal of Wine Research, Enometrica* and the *Journal of Wine Economics*. We have decided not to focus on analyses of practices: there is a wide literature on environmental and organic practices in the wine industry, but it mainly focuses on winemaking and agronomic aspects and we are interested in management, strategic and marketing. Background research has provided us useful inputs for performing our systematic literature review; in particular the works by (Lobb 2005) and (Thieme 2007) have been helpful for designing our methodology. The work by (Hart

1998) has been extremely useful for understanding how to analyse results. After having verified their contents, the articles have been included in a database that has been created for sorting and analysing results. We have then classified the articles collected into four main categories that have been built on the basis of major JEL classifications.

1.8.2 TOPICS, GEOGRAPHIC AREA AND RESEARCH TECHNIQUES EMPLOYED

The main four categories corresponding to our classification of the main research bodies are (Table 1): (1) strategy; (2) entrepreneurial and top management behaviour; (3) consumer behaviour; and (4) supply chain management and certification. It is important to observe the differences emerging from the geographic area of research in which research has been carried out.

The category "strategy" includes all those articles that deal with the issues of business strategy and sustainability. It is a matter of fact that research on strategy is mainly performed in the New World Countries: Chile, New Zealand, US, Australia and Argentina lead the way to understand the links between wine and sustainability in a strategic orientation. Research techniques employed are often qualitative and case study research is frequently performed. One of the reasons could be the necessity to explore the main drivers of pressure towards sustainability taking into account the motivations and opinions of different wineries' stakeholders. In fact, according to Flint (2009), in order to conduct such exploratory research, an appropriate methodology such as grounded theory is necessary that has been used to reveal how social actors interpret and act within their environments. In other papers, the aim is to enlighten an entire sector at national or regional level and for this reason a multidisciplinary case study approach is employed (Guthey and Whiteman 2009; Cederberg et al. 2009). The topics investigated in this category are diverse, but two trends stand out: at a country level the analysis is carried out to understand the boundaries of emerging organic wine industry and the implications to promote place branding activities; at firm level the interest is for internal and external pressures towards sustainable and environmental practices.

TABLE 1: Topics, geographical coverage and techniques used in selected literature

Topics	Author(s)	Geographical coverage	Study tipology	Technique	Sustainability aspects
Strategy	Atkin et al. (2012)	US	Quantitative	Survey	Links between environmental strategy and performance
	Forbes and De Silva (2012)	New Zealand	Quantitative	survey	Environmental Management System
	Cederberg et al. (2009)	Chile	Qualitative	Case study at country level	Potentiality of industry organic wine
	Flint and Golicic (2009)	New Zealand	Qualitative	In-depth Interviews (Grounded Theory)	Drivers of wine industry sustainability
	Guthey and Whiteman (2009)	US (California)	Qualitative	Case study	Firm-ecology relationships
	Gabzdylova et al. (2009)	New Zealand	Mixed	Interviews	Internal and external drivers of sustainability
	Pullman et al. (2010)	US	Mixed	Interviews	Sustainability practices
	Bonn and Fisher (2011)	Australia	Qualitative	Case study	Sustainability as a business strategy
	Sinha and Akoorie (2010)	New Zealand	Quantitative	Multivariate Analysis	Environmental practices
	Sampedro et al. (2010)	Spain	Qualitative	Interviews	Environment as a business strategy
	Warner (2007)	California	Qualitative	Interviews and Focus Groups	Links between sustainability and place-based branding
	Preston (2008)	France and Australia	Qualitative	Case study	Change in supply chain practices
	Novaes Zilber et al. (2010)	Argentina	Qualitative	Case study	Potentiality of industry organic wine
	Poitras and Getz (2006)	Canada	Qualitative	Case study	Host community perspective

TABLE 1: *Cont.*

Topics	Author(s)	Geographical coverage	Study tipology	Technique	Sustainability aspects
Entrepreneurial and Top Management Behaviour	Marshall et al. (2010)	US and New Zealand	Quantitative	Multivariate Analysis	Motivations for improving environmental performance
	Marshall et al. (2005)	US	Qualitative	Focus groups and Interviews	Environmental behavior drivers
	Cordano et al. (2010)	US	Quantitative	Multivariate Analysis	Drivers of adoption of voluntary EMP
	Silverman et al. (2005)	US	Quantitative	Multivariate Analysis	Drivers to improve environmental performance
Consumer Behaviour	Brugarolas et al. (2010)	Spain	Quantitative	Contingent Valuation	Organic wine
	Forbes et al. (2009)	New Zealand	Quantitative	Descriptive Analysis	Green production practices in vineyards
	Mann et al. (2012)	EU- Switzerland	Quantitative	Survey based on interviews	Determinants organic wine consumption
	Bernabeu et al. (2007)	Spain	Quantitative	Conjoint Analysis	Organic wine
	Thogersen (2002)	Denmark	Quantitative	Multivariate Analysis	Organic wine
	Krystallis et al. (2006)	Greece	Quantitative	Factor Analysis	Organic wine
	Loureiro (2003)	US (Colorado)	Quantitative	Probit Model	Environmental friendly label
	Blondel and Javaheri (2004)	France	Quantitative	Experimental procedure	Organic wine

TABLE 1: *Cont.*

Topics	Author(s)	Geographical coverage	Study tipology	Technique	Sustainability aspects
	Fotopoulos et al. (2003)	Greece	Mixed	Means-end chain analysis	Organic wine
	Barber (2010)	USA	Quantitative	Multivariate Analysis	Environmental friendly labels
	Barber et al. (2010)	USA	Quantitative	Multivariate Analysis (Factor, Discriminant)	Environmental friendly labels
	Bernabeu et al. (2008)	Spain	Quantitative	Multivariate Analysis	Organic wine
Supply chain management and certification	Desta (2008)	California	Quantitative	Cross sectional survey	Code of sustainable winegrowing practices
	Ohmart (2008)	California	Mixed	Context analysis based on secondary data	Code of sustainable winegrowing practices
	McManus (2008)	Australia	Qualitative	Case study	Environmental sustainability
	Ardente et al. (2006)	Italy	Qualitative	Case study	Estimation of direct and indirect env impact with POEMS methodology and simplified LCA
	Colman and Paster (2009)	Global	Qualitative	LCA Analysis	Impact on environment based on a carbon calculator model
	Marchettini et al. (2003)	Italy	Qualitative	Emergy analysis	Ecological performance of wine production
	Cholette and Venkat (2009)	US	Quantitative	LCA analysis	Employ CargoScope tool to analyze the carbon and energy profiles of wine distribution

Even more concentrated is the research investigating "entrepreneurial and top management" drivers for the adoption or improvement of environmental behaviour: Marshall, Cordano and Silverman use the Theory of Planned Behaviour and the Theory of Reasoned Action mainly in US (California) and New Zealand wineries. Results are not univocal because the weight of internal and external pressures, attitudes and subjective norms can vary among cases and the papers give evidence of interactions among considered variables.

Contrary to the other research fields, the consumer behaviour field is investigated worldwide. Europe seems to be focusing more on consumers' perception of—and willingness to pay for—organic wine, while the New World research is oriented to a more complex issue such as the environmental friendly label or a more general topic as green production practices. This is the category where quantitative analysis and statistical techniques are more used and developed.

Finally, research in the field of supply chain management and certification aims to give an overview of various attempts to implement codes of sustainable winegrowing practices and to reduce the impact of environment based activities on carbon emission; these studies are carried out both in the New and Old World. It is worth to emphasize the various methods employed to analyse impacts and efficiency of practices on the environment; not a single technique or tool seems to have been recognized worldwide as a standard for such measurement and thus more research is needed.

A brief final note is about the kind of journal and the year of publication: only 5 of the papers collected have been published before 2005. This highlights how "young" this field of study is. Particularly, the field of strategy seems to be the newest one.

An analysis of journals reveals a multidisciplinary interest in sustainability and wine: journals such as *Renewable Agriculture and Food Systems, Journal of Cleaner Production, E:Co Emergence: Complexity and Organization, International Journal of Sustainable Development & World Ecology* devote specific attention to the various facets of sustainability; on the other side the *Journal of Wine Research* and the *International Journal of Wine Business Research* have a wine sector focus. Then we can find another kind of reviews with a general focus on the agri-food sector (*British Food Journal, Food Quality and Preference* and *Acta Agriculturae*

Scandinavica Section B-Soil And Plant Science, Journal of Rural Studies)
or new journals with specific topics on firms and sustainability such as
Business Strategy and the Environment.

1.9 CONCLUSIONS

We have seen how a tight relationship between academics and industry
can provide benefits to the wine industry and can improve its overall
orientation towards sustainability: research can help winegrowers in the
adoption of sustainable practices and can provide answers to some managerial issues.

Scholars suggest to focus research on a few critical aspects, such as the
reconfiguring and understanding of economic performance and the creation of the conditions for incremental adjustment and multidisciplinary
learning to happen (Guthey and Whiteman 2009).

Research has a social responsibility in the development of a sustainability orientation in the wine business: once spread, research results can
motivate wineries to adopt a sustainable behaviour and create a sustainability awareness among industry and consumers.

The main challenge is "to change perceptions and mind-sets, among
actors and across all sectors of society, from the over-riding goal of increasing productive capacity to one of increasing adaptive capacity, from
the view of humanity as independent of nature to one of human and nature
as coevolving in a dynamic fashion with the biosphere" (Folke, 2002, in
Guthey and Whiteman, 2009); research plays a key role in the achievement of this goal, and by helping managers and people during the learning process and the adaptation of the organization to the evolving social
conditions.

Some scholars perceive the role played by the role of drivers in the defining a sustainability orientation as critical: "We encourage further intra-in-
dustry, as well as inter-industry, research in order to better understand when
internal and external drivers are most critical, and perhaps at times, less
critical, in ushering in environmental stewardship" (Marshall et al. 2005).

One of the key emerging research questions to focus on is: "under what conditions sustainability happens". The wine industry is particularly suitable for research on sustainability, as it has been shown by the analysis of the literature we have performed.

Anyway, although sustainability issues are affecting the wine industry all over the world, research does not show how to keep the path of such a diffusion and it is much more intensive in some countries rather than others, as it has emerged from the analysis provided. It can be said that research is more concentrated and focused on sustainability in those countries where the pressure of drivers is stronger.

The originality of our paper relies in being the first classification about research on sustainability and wine. Our paper aimed to identify the main methodologies and research techniques used, as well as the main problems observed by scholars. Further investigations to highlight any relationship between university research and the pressure of key drivers should be carried out.

REFERENCES

1. Better Wine. http://www.betterwinesbetterworld.com. Accessed 10 September 2012
2. New Zealand Industry Sustainability. http://www.newzealand.com
3. Wine Business. http://www.winebusiness.com
4. EEA. http://www.eea.europa.eu
5. Ardente F, Beccali G, Cellura M, Marvuglia A (2006) POEMS: A case study of an italian wine-producing firm. Environ Manage 38(3):350-364
6. Atkin T, Gilinsky A, Newton SK (2012) Environmental strategy: does it lead to competitive advantage in the US wine industry? Int J Wine Bus Res 24(2):115-133
7. Bansal P, Roth K (2000) Why companies go green: a model of ecological responsiveness. Academy of management journal 43(4):717-736
8. Barber N (2010) "Green" wine packaging: targeting environmental consumers. Int J Wine Bus Res 22(4):423-444
9. Barber N, Taylor DC, Deale CS (2010) Wine Tourism, Environmental Concerns, and Purchase Intention. J Trav Tour Manage 27(2):146-165
10. Bennett SJ (1991) Ecopreneuring: The Complete Guide to Small Business Opportunities from the Environmental Revolution. New York: Wiley.
11. Berle G (1991) The Green Entrepreneur: Business Opportunities that Can Save the Earth and Make You Money. Blue Ridge Summit, PA: Liberty Hall Press.

12. Bernabéu R, Martínez-Carrasco L, Brugarolas M, Díaz M (2007) Differentiation strategies of quality red wine in castilla-la mancha (spain); Estrategias de diferenciación del vino tinto de calidad en Castilla-la Mancha (España). Agrociencia 41(5):583-595

13. Bernabeu R, Brugarolas M, Martinez-Carrasco L, Diaz M (2008) Wine origin and organic elaboration, differentiating strategies in traditional producing countries. Brit Food J 110(2):174-188

14. Berry MA, Rondinelli DA (1998) Proactive environmental management: a new industrial revolution. Acad Manage Exe 12(2):38-50

15. Bisson L, Waterhouse AL, Ebeler SE, Walker A, Lapsley JT (2002) The present and future of the international wine industry. Nature 418(8):696-699

16. Blondel S, Javaheri M (2004) Valueing organic farming: an experimental study of the consumer. Proceedings of the XVTH International Symposium on Horticultural Economics and Management, 2004.

17. Blue J (1990) Ecopreneuring: Managing For Results. London: Scott Foresman.

18. Bonn I, Fisher J (2011) Sustainability: The missing ingredient in strategy. J Bus Strat 32(1):5-14

19. Broome J, Warner K (2008) Agro-environmental partnerships facilitate sustainable wine-grape production and assessment. Calif Agr 62:133-141

20. Brugarolas M, Martinez-Carrasco L, Bernabeu R, Martinez-Poveda A (2010) A contingent valuation analysis to determine profitability of establishing local organic wine markets in Spain. Ren Agr Food Syst 25:35-44

21. Bruyn SD, Drunen MV (2004) Sustainability and Indicators in Amazonia: conceptual framework for use in Amazonia. Report number: W-99/37. Institute for Environmental Studies..

22. Casini L, Corsi A, Cavicchi A, Santini C (2010) Hopelessly devoted to sustainability: marketing challenges to face in the wine business. In: Proceedings of the 119th EAAE Seminar 'Sustainability in the Food Sector: Rethinking the Relationship between the Agro-Food System and the Natural, Social, Economic and Institutional Environments. Capri, Italy. June, 30th –July, 2nd, 2010

23. Cederberg P, Gustafsson JG, Martensson A (2009) Potential for organic Chilean wine. Acta Agr Scan - B – Soil and Plant Science 59(1):19-32

24. Cholette S, Venkat K (2009) The energy and carbon intensity of wine distribution: A study of logistical options for delivering wine to consumers. J Cl Prod 17(16):1401-1413

25. Colman T, Paster P (2009) Red, White, and 'Green': The Cost of Greenhouse Gas Emissions in the Global Wine Trade. J Wine Res 20(1):15-26

26. Cordano M, Marshall RS, Silverman M (2010) How do Small and Medium Enterprises Go "Green"? A Study of Environmental Management Programs in the US Wine Industry. J Bus Eth 92(3):463-478

27. Davidson WN, Worrell DL (2001) Regulatory pressure and environmental management infrastructure and practices. Bus Soc 40:315-342

28. Delmas MA, Doctori-Blas V, Shuster K (2008) Ceago vinegardens: How green is your wine? Environmental differentiation strategy through Eco-labels. AAWE Working Paper n. 14

29. Desta A (2008) Conventional versus Environmentally-Sensitive Wines: The Status of Wine Production Strategies in California North Coast Counties. J Bus Pub Aff 2(1):1-17

30. Dillon P, Fischer K (1992) Environmental Management in Corporations: Methods and Motivations. Tufts University, Boston (MA): Center for Environmental Management.

31. Dlott J (2004) California Wine Community Sustainability Report. San Francisco: California Sustainable Winegrowing Alliance.

32. Flint DJ, Golicic SL (2009) Searching for competitive advantage through sustainability: A qualitative study in the New Zealand wine industry. Int J Phy Dist Log Manage 39(10):841-860

33. Forbes SL, De Silva T (2012) Analysis of environmental management systems in New Zealand wineries. Int J Wine Bus Res 24(2):98-114

34. Forbes SL, Cohen DA, Cullen R, Wratten SD, Fountain J (2009) Consumer attitudes regarding environmentally sustainable wine: An exploratory study of the New Zealand marketplace. J Clea Prod 17(13):1195-1199

35. Fotopoulos C, Krystallis A, Ness M (2003) Wine produced by organic grapes in Greece: using means end chains analysis to reveal organic buyers' purchasing motives in comparison to the non-buyers. Food Qual Prefer 24(7):549-566

36. Gabzdylova B, Raffensperger JF, Castka P (2009) Sustainability in the New Zealand wine industry: drivers, stakeholders and practices. J Clea Prod 17:992-998

37. Guthey GT, Whiteman G (2009) Social and ecological transitions: Winemaking in California. E:CO Emer: Compl Org 11(3):37-48

38. Hart C (1998) Doing a literature review. London: Sage Publication.

39. Isaak R (2002) The making of the ecopreneur. Greener Management International 38:81-91

40. Krystallis A, Fotopoulos C, Zotos Y (2006) Organic consumers' profile and their willingness to pay (WTP) for selected organic food products in Greece. J Int Cons Market 19(1):81-106

41. Lawrence AT, Morell D (1995) Leading-edge environmental management: motivation, opportunity, resources and processes. Res Corp Soc Perf Pol, Supp 1:99-126

42. Lee YS (2000) The Sustainability of University-Industry Research Collaboration: An Empirical Assessment. J Tech Trans 25(2):111-133

43. Lobb A (2005) Consumer trust, risk and food safety: A review', Food Economic. Acta Agr Scand Section C 2(1):3-12

44. Loureiro ML (2003) Rethinking New Wines: Implications of Local and Environmentally Friendly Labels. Food Pol 28:5-6

45. Mann S, Ferjani A, Reissig L (2012) What matters to consumers of organic wine? Brit Food J 114(2):272-284

46. Marchettini N, Panzieri M, Niccolucci V, Bastianoni S, Dorsa S (2003) Sustainability indicators for environmental performance and sustainability assessment of the productions of four fine Italian wines. Int J Sust Dev World Eco 10(3):275-282

47. Marshall RS, Cordano M, Silverman M (2005) Exploring individual and institutional drivers of proactive environmentalism in the US wine industry. Bus Str Env 14:92-109

48. Marshall RS, Akoorie MEM, Hamann R, Sinha P (2010) Environmental practices in the wine industry: An empirical application of the theory of reasoned action and stakeholder theory in the United States and New Zealand. J World Bus 45(4):405-414

49. McManus P (2008) Mines, wines and thoroughbreds: Towards regional sustainability in the upper hunter, Australia. Reg Stud 42(9):1275-1290

50. Murphy J (2000) Understanding creative marketing, Practical Winery. Available at: http://www.practicalwinery.com/julaug00p42.htm

51. Novaes Zilber S, Friel D, Nascimento LF M d (2010) Organic wine production: the case of Bodega Colomé in Argentina. Int J Wine Bus Res 22(2):164-177

52. Ohmart C (2008) Innovative outreach increases adoption of sustainable winegrowing practices in Lodi region. Cal Agric 62(4):142-147

53. Poitras L, Getz D (2006) Sustainable Wine Tourism: The Host Community Perspective. J Sust Tour 14(5):425-448

54. Preston D (2008) Viticulture and Winemaking in Contemporary Rural Change: Experience from Southern France and Eastern Australia. J Wine Res 19(3):159-173

55. Pugh M, Fletcher R (2002) Green international wine marketing. Austr Market J 10(3):76-85

56. Pullman ME, Maloni MJ, Dillard J (2010) Sustainability Practices in Food Supply Chains: How is Wine Different? J Wine Res 21(1):35-56

57. Quazi AM (2003) Identifying the determinants of corporate managers' perceived social obligations. Manag Decis 41(9):822-831

58. Regouin E (2003) To convert or not to convert to organic farming. In: Organic agriculture–Sustainability, markets and policies. Proceedings of an OECD workshop, September 2002. Washington, DC. pp 227-235

59. Sampedro EL, Sánchez MBG, López JCY, González ER (2010) The environment as a critical success factor in the wine industry: Implications for management control systems. J Wine Res 21(2):179-195

60. Schaltegger S (2002) A Framework for Ecopreneurship Leading Bioneers and Environmental Managers to Ecopreneurship. Greener Management International 38:45-58

61. Schaper M (2002) Introduction: the essence of ecopreneurship. Greener Management International 2002(38):26-30

62. Sharples L (2000) Organic wines – the UK market: a shift from 'niche market' to 'mainstream' position? Int J Wine Market 12(1):30-41

63. Silverman M, Marshall RS, Cordano M (2005) The greening of the California wine industry: Implications for regulators and industry associations. J Wine Res 16(2):151-169

64. Sinha P, Akoorie MEM (2010) Sustainable Environmental Practices in the New Zealand Wine Industry: An Analysis of Perceived Institutional Pressures and the Role of Exports. J Asia-Pac Bus 11(1):50-74

65. Stewart C (2009) The science of wine: Washington State University scientists and the development of the Washington wine industry, 1937–1992. Doctoral Thesis, Washington State University.

66. Swezey SL, Broome JC (2000) Growth Predicted in Biologically Integrated and Organic Farming Systems in California. Cal Agric 54(4):26-35

67. Swinbank A (2009) Sustainable Bioenergy Production and Trade. Issue Paper No. 17, ICTSD Programme on Agricultural Trade and Sustainable Development. University of Reading.
68. Szolnoki G (2013) A cross-national comparison of sustainability in the wine industry. J Clean Prod 53:243-251
69. Available online 5 April 2013
70. Thieme J (2007) The World's Top Innovation Management Scholars and Their Social Capital. J Prod Inn Manage 24:214-229
71. Thogersen J (2002) Direct experience and the strength of the personal norm - Behavior relationship. Psych Market 19(10):881-893
72. Walley EE, Taylor DW (2002) Opportunists, Champions, Mavericks...? Greener Management International 2002(38):31-43
73. Warner KD (2007) The quality of sustainability: Agroecological partnerships and the geographic branding of California winegrapes. J Rural Stud 23:142-155
74. Winn MI (1995) Corporate leadership and policies for the natural environment. sustaining the natural environment: Empirical studies on the interface between nature and organizations. In: Research in Corporate Social Performance and Policy ed. J.E. Post. Greenwich, London: JAI Press. pp 127-161

CHAPTER 2

FROM ENVIRONMENTAL TO SUSTAINABILITY PROGRAMS: A REVIEW OF SUSTAINABILITY INITIATIVES IN THE ITALIAN WINE SECTOR

CHIARA CORBO, LUCREZIA LAMASTRA, AND ETTORE CAPRI

2.1 INTRODUCTION

Although the wine industry can be seen as less "dirty" than other sectors, as for example the chemical one [1,2], wine producers and vine growers have been increasingly engaged in sustainability driven by different forces, first of all the environmental concerns. The wine industry, indeed, has to face a number of environmental issues and challenges. The literature reports several environmental sustainable practices and these aspects are often mentioned as relevant: soil management, water management, wastewater, biodiversity, solid waste energy use, air quality, and agrochemical use [3]. Producers have to limit the use of chemicals, promoting their sustainable use in order to preserve and enhance the level of biodiversity and soil fertility. Water must also be managed responsibly by minimising con-

From Environmental to Sustainability Programs: A Review of Sustainability Initiatives in the Italian Wine Sector. © 2014 by the authors; licensee MDPI, Basel, Switzerland. Sustainability, 6,4 (2014). doi:10.3390/su6042133. Licensed under Creative Commons Attribution 3.0 Unported License, http://creativecommons.org/licenses/by/3.0/.

sumption and reducing run-off of contaminated wastewater. Furthermore, wineries must manage the landscape, to protect the health and safety of workers, as well as minimize its impact on the community (from chemical spray drifts, odor, and noises) [1,2].

The global wine industry also faces institutional and stakeholders pressures. The pressures from governments and environmental groups, the growing interest from consumers for green products and the higher commitment to export in countries with a strong attention for "sustainable products" are among the "institutional drivers" to sustainability [4,5,6]. Finally, managers' personal values, entrepreneurs' personal motivations, and employees' environmental attitudes can be considered as important drivers to guide the wine industry towards sustainability, given the fact that the sector is mainly made by small-medium companies, and there is a frequent coincidence between the ownership and the management [6]. Moreover, in the wine sector, preserving the environment is a sort of "natural instinct" for winemakers, concerned in maintaining proper environmental conditions and in preserving the natural resources in order to maintain the productivity of the land, not only for the present business but also to the future generations of winemakers that will manage the farm [7].

The term "sustainability" also has to be interpreted from a social and economic point of view: only an equal consideration of the ecological, economic and social dimensions of sustainability can lead to the achievement of (among the others) "changing unsustainable patterns of production and consumption and protecting and managing the natural resource base of economic and social development" [8]. Therefore, the three interdependent and mutually reinforcing pillars of sustainability must always be jointly considered in order to define viticulture as "sustainable", promoting aspects such as, the health and safety of workers, the Company's contribution to the rural and local development, and the economic viability and profitability of the measures taken.

The importance of considering all the dimensions is clearly laid out in the definition of "sustainable viticulture" given by the "Organisation Internationale de la Vigne et du Vin" in its Resolution of the Comité Scientifique et Technique (CST) 1/2004 addressing the issue of sustainability in the production of grapes, wines, spirits, and other vine products [9], and laying the foundations for the guidelines for the production, processing and packaging of products further released in 2008 [10]. Sustainable viticulture, hence, has been defined as a "global strategy on the scale of the grape production and processing systems, incorporating at the same time the economic sustainability of structures and territories, producing quality products, considering requirements of precision in sustainable viticulture, risks to the environment, product safety and consumer health, and valuing of heritage, historical, cultural, ecological, and aesthetic aspects".

At the International level, Countries in the "New Wine World" have been the pioneers in introducing sustainability in the wine industry (vine growing and wine production): In 1992, the Lodi Winegrape Commission from California launched an Integrated Pest Management program that is considered the "foundation" of sustainability winegrowing programs. Since then, several guidelines and programs for sustainable winegrowing and production have been defined by institutions and organizations around the world: for example, the "California Sustainable Winegrowing Alliance" in California, the "Wine Sustainable Policy" in New Zealand and the "Integrated Production of Wine Scheme" in South Africa [5,11]. However, it is important to highlight that although the development of several sustainability initiatives and the establishment of a number of certification schemes—a univocal definition of "sustainable viticulture" still does not exist, and nor do internationally recognized sustainability indicators. Furthermore, the integration of social and economic aspects of sustainability still seems quite insufficient [12].

Italy has now also stepped up to the challenge. Although the widest number of publications and projects regarding sustainable wine have been produced in countries other than Italy [13] the quantity of academic literature dedicated to the topic of sustainability in the Italian wine sector is increasing [14], with particular focus on environmental sustainability indicators [15,16,17] greenhouse gas emissions and the use of Life Cycle Assessment methodology [18,19,20]. Sustainability has often also been a dominant element for international wine conferences and events recently held in Italy. For example, the issue of sustainability in the wine sector was included among the topics of the international event "Vinitaly 2013" (the major Italian fair in the wine sector) and 2014's edition will host a special area dedicated to organic wine. In May 2013, at the International Congress "Enoforum 2013", several sessions were dedicated to sustainability, and in November 2013, the International Congress "Sustainable Viticulture and Wine Production: steps ahead toward a global and local cross-fertilization" was hosted in the framework of the XXV International Enological and Bottling Equipment Exhibition ("Salone Internazionale Macchine per Enologia e Imbottigliamento"—SIMEI). Apart from the interest from the research world, the most consistent sign of interest for sustainability in the wine sector in Italy can be considered the wide range of sustainability programs launched in recent years by private producers and consortiums. It looks like a "wave" of sustainability is overwhelming the entire sector and if, on one hand, this is a positive signal of the concern regarding the issue of sustainability in viticulture, on the other hand confusion can arise among vine growers and wine producers. The large number of different strategies, guidelines and practices, indeed, make their comparison extremely complicated, and there is the risk that farmers and producers do not have a clear understanding of the opportunities and benefits deriving from the implementation of a certain sustainability program. The lack of clarity can also affect consumers, due to the growing range of wines sporting "sustainability-sounding" names and adjectives (such as sustainable, organic, natural, free, eco-friendly, etc.), but adequate explanations do not always accompany these names.

In this paper, the authors report on a comparative and qualitative analysis of several sustainability programs currently being used in the Italian wine industry. Due to the lack of similar studies in the Italian context, the authors tried to use a new approach, as pragmatic as possible, basing the evaluation of the sustainability programs on parameters, such as the presence of certain elements in the program (sustainability protocol, practical tools for the sustainability improvement, certifications, and labels), the scientific consistency and originality, marketing and communication issues. The information for this research was obtained through interviews with the managers and supervisors of the selected programs. The overall objective was to underline the similarities and differences between programs, and suggest a new approach built on the cooperation and integration between the numerous initiatives in order to create a unique framework for sustainable vitiviniculture in Italy.

This paper begins with the presentation of the objective, and the research method is also outlined, followed by the presentation of results. Further discussions are presented in the final section.

2.2 OBJECTIVE

This study reviews the most relevant sustainability programs in the Italian wine industry, in order to underline the main features of each program and highlight how they differ from each other in terms of objectives, assessment methodologies and tools, innovative and scientific features, as well as level of completeness and transparency. Rather than stating the validity or the efficacy of the considered sustainability programs, this work aims to undertake an evaluation that can generate the information needed to build a tailored sustainability framework for the Italian wine sector, an approach that could bring uniformity to the sector, while still taking into consideration the specific needs of each company.

2.3 METHODOLOGY

In the first phase, a literature review was conducted in order to know if similar analyses (about the comparison of sustainability initiatives in the wine sector) had been previously produced. The existence of some papers about the cross-cultural comparison of sustainability practices in the wine sector [5] or the use of Environmental Management Systems in a specific area (namely New Zealand) [21] have been found, but unfortunately not so many examples of papers specifically referring to this topic (particularly regarding Italy) have been detected at the moment the authors were writing this article. Hence, due to the lack of similar analysis, the authors developed a pragmatic approach that was followed for the review and analysis of the selected programs.

2.3.1 PROGRAMS SELECTION

The sustainability programs analyzed were chosen based on an analysis of the "state of the art" of sustainability initiatives in the wine sector, meaning the most "relevant" initiatives at the moment existing in Italy according to the number of wineries joining the programs, the participation in congresses, and the number of informative articles in sector magazines and journals.

It is important to underline that nowadays a wide variety of sustainability initiatives have been implemented by single wine makers, but the focus of this paper is on membership programs. This choice has been driven first of all by the final objective of this work, which is to highlight the necessity of creating a unique sustainability framework that could fit the widest possible number of Italian wineries. Secondly, the importance of Institutions and Industry Associations in influencing the "sustainable awareness" has been highlighted by several researchers: it is enough to think of California's experience, where the creation of strong local organizations have brought about the successful "collaborative sustainability" [14,22]. Corporate Activism can also be seen as an important driver for sustainability in the wine sector. Collaborative working can be a tool to enhance the spread of knowledge, as well as to reduce the risks and minimize the costs deriving from the implementation of sustainability initiatives. Being part of a "sustainability network" is, for the single winery, an opportunity to enhance and reinforce their relationship with Institutions, sponsors, consumers' associations and society. Finally, for a large part of consumers to know that the winery is part of a sustainability network that is managed by a credible group, a Trade Association or an Institution can be interpreted as a guarantee of truthfulness. The following sustainability initiatives have been selected for the analysis:

- Tergeo
- Magis
- SOStain
- V.I.V.A. ("Valutazione dell'Impatto della Vitivinicoltura sull'Ambiente") Sustainable Wine
- ECO-Prowine
- Ita.Ca/Gea.Vite

- Vino Libero
- New Green Revolution (Montefalco 2015)
- Organic wine
- VinNatur

The choice of including organic wine was made because—although not clearly claimed as "sustainable"—organic production can be interpreted by consumers as "environmental friendliness". Minimization of inputs in the vineyard and the cellar organic practices confers the idea of an environmentally friendly wine, both from the consumer's and the producer's point of view [23,24,25,26].

The association between the characteristics of sustainability and organic has often been made by researchers analyzing environmental protection initiatives or management systems in the wine sector (e.g., [1,27,28]) In the review of studies about the perception of organic food presented by Schleenbecker and Hamm it was noticed that, beside aspects as nutrition and animal welfare, the belief that these products are less damaging for the environment can drive the choice of organic food [29]. Organic food associations themselves claim that their products are more respectful of the environment (for example, the official website of the Italian Association for Organic Farming reports that organic farming is able to offer solutions for a more "environmentally friendly" agriculture [30]) For this reason, we decided to include organic wine in our analysis.

For similar reasons (confusions between the words "sustainable" and "natural") we also decided to include in the analysis an association of "natural wine" producers (VinNatur), as a representative of the Natural Wine movement and philosophy.

2.3.2 DATA COLLECTION

Data and information, collected between May and November 2013, were firstly taken from the website and informative material (brochures, advertisements, press releases, etc.) available for each program. At a later stage, a representative of each program was contacted (informally, or via email/professional networks) and interviews were held over the phone and/or via email exchanges in order to gain additional information and verify that previously collected. Interviewees were also asked to confirm the correctness of the information collected from websites, informative materials, and congresses and conferences. The interview started by presenting the scope of the research, then questions were asked in order to obtain the following information:

- General purposes and objective of the program/initiative;
- Implementation status of the program;
- Presence or absence of specific elements (listed in Section 3.3.1 "Glossary");
- Which elements of the so-called "three pillars" of sustainability are accounted for in the program;
- "Boundaries" of the program: is sustainability only considered in the vineyard, or are the cellar practices also analyzed?
- Names of the subjects designing and promoting the program;
- Presence of elements related to transparency and communication, such as emission of a sustainability report for participants of the program, availability (to consumers) of the evaluating system used, the presence of a website, and a sustainability labeling scheme;
- Presence of a third party verifier for certification or validation of the winery's results.

The same questions were posed during each interview. After this phase of interviews, a specific "card" was filled in for each program, listing all the information collected and it was sent to the interviewee in order to verify its correctness [31]. The last step was to create a comprehensive document listing all the information collected for each program in order to facilitate the comparison of all the information.

2.3.3 ANALYSIS

In order to conduct the analysis, two preliminary steps were defined, namely: (1) The creation of a "Glossary" and (2) The definition of the analysis methodology.

2.3.3.1 GLOSSARY

The word "sustainability" has so many definitions that some authors speak about a wide variety of "green nuances" [14]: there is no single sustainable behavior, nor is there a single definition of elements included in the various sustainability programs. Hence, the first step was to establish a common terminological base, a "glossary" according to which the analysis could be conducted.

The following words were defined:

- *Sustainability Protocol* in this context, defines the "document" that states the requirements a producer has to satisfy and/or the management and behavioral standards to be followed in order to reach sustainability goals and to be admitted to join the program. It should not be confused with technical disciplinary or assessment manuals (since their scope is to provide practical guidelines and tools to improve sustainability) [32].
- *Management tools* are defined as practical tools provided by the program for sustainability assessment and improvement. Examples of management tools are the self-assessment questionnaires (also called "checklist"), the technical/scientific tools (e.g., a web-platform for sustainability assessment), the guidelines (more detailed if compared with the Protocol), the training tools (workshops, seminars, manuals, etc.) and the specific indicators to be used for the evaluations (for example: a program requiring companies to assess social sustainability by means of the calculation of indicators as the number of women employed in the company or the employees' turn-over).
- *Calculators* are used to express a concise and comparative measure of sustainability performances. Hence, they are used to explicate complex results in a single measure (e.g., Carbon Footprint, Water Footprint, or a final indicator of sustainability).
- *Validation* is the procedure through which the compliance with the program's rules, the transparency and accuracy of information, the achievement of a minimum level of sustainability are checked. These checks can be conducted by the program's staff, or by a third party.
- *Certification* is the outcome of a formal process by which an independent and accredited Body declares that a product or a system is in compliance with a specific standard (rules or regulation) provided by an International Body (e.g., ISO standards).

2.3.3.2 DEFINITION OF THE ANALYSIS METHODOLOGY

The "analysis" consists in the examination of the gathered information and the coherence between what was stated by the interviewed persons and the information collected in the previous phase. When something was not clear enough, explicit explanations were sought from the program's responsible person(s). The analysis was then developed according to the following steps:

- Definition of the implementation status of each program (definition phase, pilot phase, operating).
- Detection of the elements. The presence of the series of elements defined in the program (Protocol, Management Tools, Calculators, Label, Validation, Certification) was detected.
- Analysis of completeness. For each program analyzed it was asked and reported which "sustainability pillars" were taken into account (if only the Environmental or also the Economic and Social aspects were considered) and—within each area—which elements were taken into account, namely [33]:

  1. Environmental pillar: Air, Water, Soil, Biodiversity, Energy management, Packaging, Transports, Pesticides and Fertilizers, Waste Management, Landscape, Raw Materials (this last one—"Raw Materials"—means that the program evaluates the attention from producers dedicated to the sustainability of the materials they purchase for their operations).
  2. Economic: Direct Economic Impacts, Indirect Economic Impacts, Evaluation of the territorial resources raw material and labor force), responsibility towards workers (health, safety, training, etc.).
  3. Social: Responsibility towards residents and inhabitants (people living nearby the farm), responsibility toward the local community, responsibility toward consumers.

- Analysis of the boundaries (if the sustainability program covers operations in the Vineyard, in the Cellar, or both).
- Analysis of the consistency in terms of science and innovation (Academic bodies and Institutions involved in the project; innovation level).
- Analysis of the transparency and communication aspects (availability and clarity of information available to the consumers and stakeholders, in the form of sustainability reports, labeling, websites, availability of the evaluation system used by the program to assess the sustainability of the winery).
- Analysis of the "sustainability label", its content and the provided information.
- Analysis of the "verification" type. A distinction between in/out values and "gradual" vales has been made. Indeed, observing the majority of the sustainability programs, it was seen that in order to be accepted in a certain program, or to have access to a label or certification, some programs ask their wineries to comply with certain rules, whereas other times it is a matter of respecting values or thresholds. Therefore, the analysis was conducted according to this distinction.
- Analysis of an independent third party involvement (for validation or certification, if provided).

TABLE 1: Name and descriptions of the Italian sustainability initiatives analyzed in the study.

Name	Description
Tergeo [34]	An initiative of the UIV—Unione Italiana Vini, an Italian wine trade association.
	Objective: to support the environmental, social and economic sustainability in the Italian wine sector, enhancing the "knowledge and technology transfer" from Companies and Researchers to farmers and wine producers. It acts like a "collector" of initiatives proposed by Companies, Universities and Research Centers dedicated to the promotion of sustainability. Tergeo works mainly with two instruments: the "Matrix" and the "Applications". The Matrix is an assessment tool proposed by the Commitee to assess a company's "sustainability positioning". "Applications" are tools proposed by the Partners (Research Centers, Universities, Company but not wineries) that can help wine producers to be more sustainable. These tools are submitted for evaluation by Tergeo Scientific Committee, that is composed by distinguished academic Professors and Researchers and experts of the wine sector; they evaluate the proposal and, if it is accepted, the "Application" is proposed to the members (producers and farmers) of the Association.
	Participation: 170 wineries and 9 main companies operating in the agricultural sector.
Magis [35]	A sustainability program initiated by Bayer CropScience in cooperation with the University of Milan.
	Objective: to promote sustainability in viticulture and minimize environmental impact by using precision viticulture techniques. Monitoring vineyards and distributing fertilizers or agrochemicals in a more precise way to enable reduced interventions in the field, as well as reducing the amount of wastes and the overall environmental impact.
	Participation: approximately 106 wineries.
SOStain [36]	A sustainability program promoted by the Observatory for Productivity and Efficient use of Resources in Agriculture - OPERA of the Università Cattolica del Sacro Cuore, in cooperation with other Italian universities and research centers.
	Objective: to promote environmental, social and economic sustainability in Sicily. The program is characterized by the "cycle of continuos improvement", an iterative process through which each winery can assess, monitor and improve its sustainability performances.
	Participants: 2 big Sicilian wineries. The project is open to all the wineries in the region.

TABLE 1: *Cont.*

Name	Description
V.I.V.A. Sustainable Wine [37]	It is a project launched in 2011 by the Italian Ministry for the Environment, Land and Sea in order to evaluate the wine-sector sustainability performance, based on Water & Carbon Footprint calculation, with the participation of some large Italian wine–producing companies, Universities and Research institutes.
	Objective: to establish a common methodology for the environmental, social and economic sustainability assessment in the wine sector using 4 indicators (Air, Water, Vineyard, Territory) and to propose a label and a smart-phone enabling the final consumers to recognize producers committed to the Project.
	Participation: 9 pilot companies.
ECO-Prowine [38]	It is a European project funded under the framework CIP - EcoInnovation. 6 European countries are participating (including Italy).
	Objective: to promote sustainability in the wine sector through the use of LCA methodology and to create a label for European sustainable viticulture. Social and economic aspects are also taken into account, the latest through the use of Life Cycle Costing methodology.
	Participation: 105 pilot wineries in Italy, Spain, Portugal, Bulgary, Greece, Austria.
Ita.Ca/Gea.Vite [39,40]	Two sustainability initiatives promoted by an Italian Agronomic Institute (Studio Sata)
	Objective: Ita.Ca (Italian Wine Carbon Calculator) is a tool to calculate the greenhouse gas emissions, specifically set for the Italian wine sector, and built upon indications from the International Wine Carbon Calculator (IWCC). Gea.Vite is a program to assess the efficiency and sustainability of the winery. It is composed by several indicators and tools (Ita.Ca is one of the tools in the program).
	Participation: 47 wineries.
Vino Libero [41]	A program initiated by an Italian wine entrepreneur, who started with his products and then involved other wineries.
	The program aims mainly at promoting the production of wine free from chemical fertilizers, weed killers and excess sulfites.
	Participation: 12 producers, 62 restaurants, 75 winehouses.
New Green Revolution (Montefalco 2015) [42]	Started in 2009, the project has been developed by the Associazione Grandi Cru of the Italian wine region "Montefalco Sagrantino".
	Objective: creation of a environmental, social and economic sustainability protocol specifically designed for the Region (a territorial model of sustainable development).
	Participation: 7 wineries (all located in the Montefalco area).

TABLE 1: *Cont.*

Name	Description
Organic wine [43]	A wine can be defined "Organic" when it is produced according to the Regulation of the European Commission (EC) no. 203/2012, that is: (in the vineyard) produced from "organic" grapes; (in the cellar) produced using only products and processes authorized by the Regulation (EC) No 203/2012.
	Until 2012, there were no EU rules or definition of "organic wine". Only grapes could be certified organic and only the mention "wine made from organic grapes" was allowed. In February 2012, new EU rules have been agreed. The new regulation has identified oenological techniques and substances to be authorized for organic wine, including a maximum sulphite content (set at 100 mg per liter for red wine and 150 mg/L for white/rosé).
VinNatur [44]	VinNatur is a consortium of wineries (across all Europeo) producing so-called "natural wine". At the present time, an official or legal definition for "natural wine" does not exist; however, there are many unofficial codes of practices or definitions released by several associations of natural wine producers. Objective: promoting a wine that is produced with the lowest possible number of human interventions in the vineyard and in the cellar, in order to enhance the link between the territory of origin, the final product and its taste. In general, "natural wines" are produced from organically or bio-dynamically grown grapes. Grapes must be hand-picked, and no sugars, foreign yeasts and bacteria must be used. The use of sulphites must be strictly limited, and no heavy manipulation are permitted (micro-oxygenation, reverse osmosis, spinning cone, cryoextraction. Participation: 96 producers in Italy (and 66 accross Europe). pursued particularly limiting the quantity of sulfites and chemicals.
	Participation: 96 producers in Italy (and 66 accross Europe).

2.4 PROGRAM DESCRIPTION

Table 1 summarizes the main features of the selected programs.

2.5 RESULTS

2.5.1 IMPLEMENTATION STATUS

The analysis was always conducted taking into account the current implementation status of the different programs, as showed in Table 2, according to the following definitions:

- *Definition phase*: project defined in all its components but still not tested on companies.
- *Pilot phase:* project already defined but in a testing phase on pilot companies.
- *Operating*: project fully working, already tested on companies and completely defined in all its components.

This type of information was requested in order to ensure the highest grade of objectivity of the analysis. Indeed, not all the projects are in the same phase of implementation and particularly for some elements—for example, the presence of a label to certify the adhesion of a winery to the program or the requirement for a sustainability report—we received answers during the interviews such as "this element is not present at the moment, but it is expected to be inserted later in the program". Hence, we decided to conduct the analysis on the base of the current status of each program, listed in Table 2.

TABLE 2: Implementation status of the analyzed sustainability programs.

Programs	Status		
	Definition	Pilot	Operating
Tergeo		•	
Magis			•
SOStain			•
V.I.V.A.		•	
ECO-Prowine	•		
Ita.Ca/Gea.Vite			•
Vino Libero			•
New Green Revolution			•
Organic Wine			•
VinNatur			•

Notes: The black spot indicates the "Status" (Definition, Pilot or Operating) of each program. Please note that this "Status" refers to the moment the authors are writing the article.

2.5.2 ELEMENT DETECTION

According to the "Glossary" defined at the beginning of the analysis, a "sustainability program" can be characterized by the presence of some main elements, namely a protocol, the management tools and a specific calculator (e.g., Carbon Footprint). Some programs have, as a final output, a label of sustainability, which is a useful way of informing consumers and stakeholders about the sustainability of a specific product and the commitment of the winery participating in the program. Finally, a program may also require the participating companies to undergo a verification (made by the program's staff or a third party) or a certification. Results of the analysis are shown in the Table 3.

TABLE 3: Results of evaluation of the elements of sustainability programs *.

Programs	Protocol	Management Tools	Calculators	Label	Verification	Certification
Tergeo		•			•	
Magis	•	•		•	••	
SOStain	•	•	•		•	
V.I.V.A.		•	•	•	••	
ECO-Prowine		•	•	•	•	
Ita.Ca/Gea.Vite	•	•	•		•	
Vino Libero	•	•		•	•	
New Green Revolution	•	•	•	•	••	
Organic Wine	•			•		•
VinNatur	•					

Notes: The black spot shows the presence of the element in the program (blank cellars mean that the element is not considered by the program). Only for validation, the single dot means that the evaluation has been provided by the program's staff; double dots indicate that the validation has been provided by the program's staff and a third party.

2.5.3 COMPLETENESS

As highlighted in the "Introduction", sustainability is often defined as a three-dimensional concept, composed by the environmental, social and economic "pillars" [8]. Although this concept is widely recognized, the fields of application of sustainability are very different: indeed, numerous different sustainable practices are reported in the literature, addressing, for example, soil management, water management, wastewater, biodiversity, solid waste, energy use, air quality, and the use of agrochemicals [5]. This part of the analysis aimed to understand which "pillars" are taken into consideration by each program, as well as which elements within each pillar.

Given that the elements that are useful to assess sustainability in its three dimensions are a wide number, those considered as representative of the pillars were chosen by analyzing the main indicators nowadays used by companies and organizations when assessing and communicating their sustainability performances. Furthermore, a distinction was made between those elements considered "directly" or "indirectly". A certain aspect of sustainability, for example, the impact on water or biodiversity, can be evaluated directly if the winery is asked to assess it through the application of specific indicators, or if a specific questionnaire is designed to extract this information. Alternatively, an element can be considered in an indirect way when the company is not asked to assess it, but the program implicitly considers the aspect. For example, a sustainability program focusing on precision farming, allowing farmers to reduce treatments in the vineyard can, under certain conditions, make them save money and farm in a more effective manner. Here, the economic sustainability is not a direct objective of the program but is without any doubt indirectly considered.

The results, shown in the Table 4, are clear: the majority of the programs take into account the three dimensions of sustainability, but, in doing this, they use a different level of specification. The most complete programs seem to be Tergeo, V.I.V.A., New Green Revolution 2015 and SOStain (the two last, however, are characterized by a strong "territorial vocation"). The VinNatur Consortium and Organic Wine take into account mainly the environmental aspect. The social pillar is also considered but referring particularly to food safety (they aim to provide consumers with a wine with a very limited content of sulphites and chemical residues).

TABLE 4: Completeness of programs according to the environmental, economic and social aspects of sustainability included in the program itself *.

Programs	ENVIRONMENTAL											ECONOMIC			SOCIAL			
	A	W	S	B	E	Pk	T	PF	Ws	L	RM	Dir	In	Loc	Em	R	Cm	Cn
Tergeo	•		*															
Matrix* Applications	*	*	*	*	*	*	*	*	*									
Magis	•	•	•	•	•	•	•	•										
SOStain	•	•	•	•	•	•	•	•	•	•								
V.I.V.A.	•	•	•	•	•	•	•	•	•	•	*	*	*		*	*	*	•
ECO-Prowine	•	•	•	•	•	•	•	•	•	•	•	*	•	•	•	•	•	•
Ita.Ca.Gea.Vite	•	•	•	•	•	•	•	•	•	•	•	•	•	•	•	•	•	•
Vino Libero	•	•	•	•	•	•	•	•										
New Green Revolution	•	•	•	•	•	•	•			•	•	•	•	•	•	•	•	•
Organic Wine	•	•	•	•	•	•												•
VinNatur	•																	

* The black point means that the aspect is completely assessed. The grey point indicates that the aspect is not considered in a direct way (e.g., a program promoting less use of chemicals could indirectly lead to an increased economic efficiency, consisting in cost reductions).

Legend: A: Air; W: Water (impact on its quality and quantity); S: Soil (erosion, quality etc.); B: Biodiversity; E: Energy management; Pk: Packaging; T: Transport; Pf: Pesticides and fertilizers; Ws: Waste management; L: Landscape; Rm: Raw materials (sustainability along the supply chain); Dir: Direct economic impacts; In: Indirect economic impacts (the company is committed in activities having a positive return at the local level (e.g., eno-tourism, R&D activities implemented at a local level, etc.); Loc: Local (Employees/Raw Materials preferably from the local territory); Em: Employees (the company is responsible towards its workers (health, safety, training, working conditions, etc.); R: Residents, Inhabitants (the company is responsible towards people living near the farm); Cm: Community (the company is responsible towards the local community); Cn: Consumers (quality, health, transparency, communication etc.

2.5.4 BOUNDARIES

The boundary analysis aimed at highlighting the extent of the sustainability assessment proposed by each program in terms of "physical" boundaries (vineyard and winery).

As shown in Table 5, almost all programs aimed to assess and improve the sustainability performances both in the field and the vineyard, except for Magis that—at least for the moment—is mainly focused on "precision agriculture" (hence, on the field). Of course, these results for boundaries should be analyzed in comparison with the elements previously considered in the "Completeness" analysis, in order to avoid interpreting all the programs as being at the same completeness level just because operations in both the vineyard and the winery are considered.

TABLE 5: "Physical boundaries" of analyzed projects.

Programs	Vineyard	Winery	Notes
Tergeo	•	•	Winery considered only by "applications" (not in the matrix).
Magis	•		
SOStain	•	•	
V.I.V.A.	•	•	
ECO-Prowine	•	•	
Ita.Ca/Gea.Vite	•	•	
Vino Libero	•	•	
New Green Revolution	•	•	
Organic wine	•	•	Winery considered only for what concerns the regulation of sulphites, chemicals and additives (food safety).
VinNatur	•	•	

Notes: The black spot indicates the areas (Vineyard/Winery) considered within each program. The grey spot means that the program provides "sustainability indications" only in relation to certain parts of the area.

TABLE 6: Academic Bodies/Research Centers and Institutions involved in the projects.

Programs	University/Research Center	Institutions
Tergeo	University of Milan;	
	University Cattolica del Sacro Cuore, PiacenzaUniversity of Verona;	
	University of Naples "Federico II";	
	University of Padova	Regional Delegate;
	Ministry for Rural Policies	
Magis	University of Milan;	
	University of Turin (Department of Agrarian, Forest and Food Sciences);	
	University of Florence (Department of Economics, Engineering, Agrarian and Forest Sciences and Technologies);	
	University of Bari (Institute of Sciences of Food Production—National Research Council)	
SOStain	University Cattolica del Sacro Cuore, Piacenza (OPERA)	
	University of Milan;	
	University of Palermo	Ministry of Rural Policies (sponsorship)
V.I.V.A.	University Cattolica del Sacro Cuore (OPERA);	
	University of Turin (Agroinnova);	
	University of Perugia (Res.Cent. on Biomasses)	Ministry for the Environment, Land and Sea
ECO-Prowine	University Cattolica del Sacro Cuore, Piacenza (OPERA);	
	Research Centers: IPVE (Portugal), Aeiforia (University Cattolica), CIRCE (University of Zaragoza)	European Community (framework Competitiveness and Innovation Framework Programme-EcoInnovation)
Ita.Ca/Gea. Vite	University of Milan	
Vino Libero	University of Turin	Piedmont Region (Framework: European Agricultural Fund for the Rural Development)
New Green Revolution	University of Milan;	

TABLE 6: *Cont.*

Programs	University/Research Center	Institutions
	Parco Tecnologico Agroalimentare Umbria 3A- Agrifood Technological Park of the Umbria Region);	
	ConfAgricoltura Umbria	PSR/Innovation;
Umbria Region		
Organic wine		European Community Regulation No. 203/2012.
VinNatur	Experimental Center for Sustainable Viticulture	

2.5.5 CONSISTENCY (SCIENCE AND INNOVATION)

A criteria, (certainly not exhaustive, but definitely relevant), to examine the scientific nature of a program is the presence of members belonging to Universities, Research Centers, and the scientific community. All such members can participate in the program as promoter subjects or as part of the scientific/technical committee. In the table, the scientific subject within each program has been listed.

In this analysis, consistency is also given by the "grade" of innovation and originality of programs. Given that each program always start from existing regulations and legislations (especially regarding the aspects related to food safety, hygiene and safety at work), it is possible to state that not all the programs have the same grade of originality for dealing with sustainability. Table 6 and Table 7 show the results for both these aspects.

Looking at Table 6, it is possible to notice how different the "innovation proposal" of each program is. Some programs are a "synthesis" of existing protocols, rules or "good practices" already in practice, but usually ensuring the final consumer a higher level of quality for wines produced by companies joining the program. The innovation grade is higher for those programs that propose a detailed sustainability assessment through check-lists, questionnaires, on-line tools, etc. Finally, the innovation is at a maximum level for initiatives introducing completely new methodologies, equipment, indicators, and management tools to assess and enhance sustainability performance.

TABLE 7: Main innovations generated by each program.

Programs	Programs' Innovations
Tergeo	It is the first "platform" aiming to collect sustainability tools and initiatives. It is able to link requests from winemakers and farmers with the solutions and products proposed by companies, Universities, etc.
	Applications: tools innovative by definition, they are all validated according to a scientific process.
	"Tergeo" matrix for the evaluation of sustainability positioning.
Magis	Focus on "precision farming" and innovation in the sector.
	The Platform enables the continuous monitoring by researchers and improvement for farmers.
	Label and validation by a Third Party.
SOStain	First complete sustainability program for wineries in Sicily (it assesses environmental, social and economic sustainability along all the chain). Strong focus on a specific territory.
	Transparency towards consumers (companies are obliged to edit a sustainability report).
V.I.V.A.	Use of completely new indicators, or existing ones but adapted to the wine sector.
	High transparency (disciplinary is public) and involvement of a Third Party for the disciplinary evaluation.
	Communication: innovative way of communicating with the final consumer through a specific label, QR (Quick Response) Code and smart-phone app.
	Procedure to obtain the label: it is necessary to be verified by a third party that can be chosen by the single winery.
ECO-Prowine	Application of LCA-LCC methodology specifically adapted to the wine sector
	Creation of a specific "label" for sustainable wine that can be recognized across all Europe as a standard in the sector.
	Statistical approach to obtain the influencing factors in the impact indicators.
	Program including European different countries.
Ita.Ca/Gea. Vite	Ita.Ca: first GHG emissions calculator for the Italian wine sector.
	Integration of social and economic sustainability with elements related to an effective management of the winery.
Vino Libero	The Disciplinary aims to combine, in a harmonic manner and as an "improved summary", the requirements from national and regional guidelines for organic and integrated production.
	Ability of the program to involve—a part from wineries - also wine houses, restaurants, etc.

TABLE 7: *Cont.*

Programs	Programs' Innovations
New Green Revolution	Strong focus on local territory (Montefalco area, in Umbria Region).
	Application and adjustment of GeaVite's tools to a specific area.
	Realization of a new machine that can enhance sustainability in farming.
	Involvement of a Third Party for the validation of the protocol and of companies
Organic wine	Innovative when it was at first proposed (proposing an "alternative" way of farming), nowadays is only the "protocol" that can give rise to a certification.
VinNatur	It is the first consortium aiming to join all the producers of natural wine across Europe and promote research in this sector.

2.5.6 TRANSPARENCY AND COMMUNICATION

Elements considered in this part of the analysis were the following:

- Reporting: does the program ask wineries to edit and publish a report about sustainability performance and improvement?
- Availability of the evaluation system: is the "evaluation system" available to the final consumers? Are consumers allowed to know how the program assesses a company's sustainability?
- Web: does the program communicate to consumers via a dedicated website? Is information clear and transparent?
- Label: is a "sustainability label" provided as the final output of the program, in order to allow the final consumer to recognize products and companies committed in a sustainability-improving path?

Results are shown in Table 8.

Regarding the "Report", only SOStain specifically asks its wineries to edit a sustainability report, presenting the main results of the implemented program.

The issue of the "Evaluation System" is quite controversial, due to the fact that the majority of programs allow the final consumers to know how the sustainability assessment is conducted (i.e., which parameters were evaluated), but not all programs offer the same level of transparency (in terms of provided information). In this sense, the highest values of transparency seemed to be reached by V.I.V.A. (its manuals are publicly available) and ECO-Prowine (the assessment manual for wineries is available on the website of the project). The Vino Libero disciplinary is available on the website of the project, as well as the one for VinNatur: the consumer can know the "topics" considered by the programs but no detailed information is provided. Particularly for Vino Libero, the disciplinary is available on the website, nevertheless is not really clear how a company's compliance with it is assessed (except for the content of sulphites).

TABLE 8: Results for the analysis regarding the transparency and accessibility of information given to final consumers.

Programs	Report	Evaluation System	Web	Label
Tergeo			•	
Magis			•	•
SOStain	•		•	
V.I.V.A.		•	•	•
ECO-Prowine		•	•	•
Ita.Ca/Gea.Vite		•	•	
Vino Libero		•	•	•
New Green Revolution			•	•
Organic Wine		•	•	•
VinNatur		•	•	

Notes: The black spot indicates the presence of the communication and informational tools within each program: the report, (the availability of) the evaluation system, the specific web site and the label); The grey spot means that the element is only partially used in the program.

The regulation for Organic Wine is available on the website of the European Community. Finally, regarding Gea.Vite Evaluation System is not public; the Ita.Ca's protocol is made upon the OIV-GHG (Organisation Internationale de la Vigne et du Vin—Greenhouse Gas) emission calculator that is publicly available (on the OIV website)

"Web Site": as shown in the table, all the programs have a specific website, a part from the Ita.Ca and Gea.Vite programs, for which information is contained in a specific part of the Agronomi Sata's website. Regarding

Organic Wine, it is possible to find information regarding Organic Farming on the website of the European Community [45]; more information regarding the specific product "wine", in Italy, is available on the website of Federbio [46].

"Label": communication to consumers using a sustainability label is not provided by all the programs. V.I.V.A. and Magis are going to release their sustainability label, which will enable consumers to recognize wines produced according to their specific protocols or rules. The ECO-Prowine projects will release a sustainability label by the end of 2014, as stated in the objectives of the program. "New Green Revolution" has created a collective brand. Its use is not mandatory for wineries; indeed, after the validation, they can ask to use the logo on bottles, advertising materials etc. (but they have to specify that the logo refers to the sustainability of the processes and not to the specific product). Finally, Vino Libero also provided a label that is applied to the neck of the bottles produced in compliance with its disciplinary.

2.5.7 VALIDATION AND THIRD PARTY INTERVENTION

Given the difference between validation and certification explained in Chapter 3, the analysis on sustainability programs has highlighted that, in general, all the programs considered expect a "check" on the performances of their members, to ensure they adhere to the program's guidelines and objectives. Nevertheless, there is a certain difference in the way these checks are provided. In some cases, only a check about the methodology's application is provided; other times, some threshold values are set, and the sustainability assessment outcomes have to be maintained within a certain limit (usually this is the case for programs having, as an output, a label or a certificate, that cannot be gained if the company is outside the set values/ required improvement has not been achieved).

Regarding the "Validation and third party intervention" parameters, the analysis has been conducted distinguishing between the following:

• In/Out: the validation/certification is guarantees adherence to the rules stated in the protocol/disciplinary. Maintaining the membership status is not a matter of thresholds, but rather related to do/not to do requirements.

- Gradual (scale): the check is made upon thresholds, and the achievement of levels of sustainability.

During this phase, the involvement of a third party for the validation or certification has been also considered. Results are shown in Table 9.

It is clear from the table that—given that a program can release a "certification" only when the guidelines have been established by a regulatory international Organisation (e.g., by the International Standard Organization)—only the Organic Wine can award a certification. For other programs, it is always better to use the word "validation".

TABLE 9: Results for the analysis regarding the validation, certification and third party involvement in the program.

Validation/Certification	Check on Values		Involvement of a 3 rd Party	
	In/Out	Gradual (scale)	Validation	Certification
Tergeo[1]		•	•	
Magis[2]		•	•	
SOStain		•		
V.I.V.A.		•	•	
ECO-Prowine[3]		•		
Ita.Ca+Gea.Vite[4]		•		•
Vino Libero	•			
New Green Revolution[5]		•	•	
Organic Wine	•			•
VinNatur	•			

Notes: The black spot indicates the type of verification or certification (if provided) of each program; the grey spot means the partial presence of the element [1]Tergeo: Validation on gradual values foreseen in the future, with the respect of a minimum threshold. Involvement of a 3rd party for validation foreseen in the future; [2]Magis: validation on gradual scale foreseen in the future to gain a different colored label; [3]ECO-Prowine: Sustainability label foreseen; [4]Ita.Ca/Gea.Vite: Certification can be provided but only for the Carbon Footprint (for which an ISO Standard exists); [5]New Green Revolution: a validation on gradual scale is required to gain the logo.

Companies joining the Ita.Ca program can ask for the ISO14064 certification, since the Carbon Calculator proposed by Ita.Ca is in compliance with the ISO rule.

Programs such as V.I.V.A. or New Green Revolution involve a third party verifier at a double level: for the validation of the protocol/disciplinary and for checks on companies joining the program.

Other programs—at least at the present moment—verify the compliance of the company with the program's requirements and/or the achievement of threshold values.

2.6 DISCUSSION

The results of this study indicate that, in general, no one sustainability program is "better" than the other, because any comparison always has to take into account the specific objective that each program aims to gain and the peculiarities of the companies they address (size, wine quality, etc.). However, considering the final objective of the study, which is to understand if it is possible to draw a common line between the sustainability initiatives, the analysis highlighted some important similarities and differences among the programs. Table 10 summarizes the main elements evaluated in the analysis.

2.6.1 COMMON ASPECTS

Integration of the social and economic pillars: compared with the past, when—talking about sustainability—the attention of farmers, producers, and stakeholders in general was mainly on the environmental or social impact. Nowadays (as shown by the analyses) the initiatives tend to aim to assess the companies' sustainability in a more complete and integrated way, considering all the three pillars. Not only environmental, but also social and economic sustainability are hence considered fundamental in order to say that a winery is sustainable.

Completeness in the boundaries definition. Almost all the analyzed programs provide a sustainability assessment both in the vineyard and the winery, even if the level of detail is very different among initiatives: for example, the Organic Wine protocol regulates mainly food safety aspects, whereas programs such as SOStain or V.I.V.A. aim to assess sustainability along all the chain, including aspects related to the winery and the wine production (e.g., water consumptions, waste management, etc.).

Management tools: questionnaires. The use of questionnaires and check-lists proposed to companies seems to be quite common, in order to collect data and/or understand and calculate the level, or the positioning of a winery in the path towards sustainability. Although in a different way and with a different level of detail, this management tool is used by Tergeo (matrix), Magis, V.I.V.A, SOStain, ECO-Prowine, Gea.Vite, and New Green Revolution.

Existing good practices and standards as a base. It is possible to state that the majority of sustainability initiatives are consistent from a scientific point of view since they are "inspired" by agricultural good practices, laws and standards in the wine sector etc. This is a good starting point: in any case, the "added value" of each program was analyzed according to its "originality" and innovation. In addition, the majority of initiatives are carried on in cooperation with Universities, Research Centers, etc. (although in a different grade).

2.6.2 DIFFERENCES

The main differences between programs can be found in:

Level of detail applied to analyze the different aspects of sustainability. Given that, as previously said, the three pillars of sustainability are considered by nearly all the programs, it is possible to detect big differences in the elements and indicators used by each program to assess sustainability within its pillars. Especially with regard to social and economic aspects, initiatives are characterized by many differences. Hence, programs such as SOStain (with its specific check-list for the evaluation of social and economic sustainability) and V.I.V.A. (specific indicators for social and economic sustainability have been created as a relevant part of the project) are strongly focused on all the pillars; other programs investigate these aspects in a superficial way, for example dedicating them just a few questions within an entire check-list. It is fundamental to take these differences into consideration, by virtue of the strong social impact of viticulture being based on the physical work and being strongly linked to the territory (landscape, tourism, etc.).

Grade of innovation. Being innovative, proposing something truly new in order to enhance sustainability in the sector: this should be the main parameter to be considered when analysing and comparing sustainability initiatives. As previously underlined, building a program on the base of existing laws and regulations is important (e.g., promoting safe conditions in the work place, fair salaries, avoid gender discrimination, complying with the minimum environmental requirement, and with the product safety laws): nevertheless, it is necessary that a program is able to go beyond, to be characterized by a minimum grade of innovation that is necessary proposing a mere "reorganization" of something that already exists and is defined by rules or standardized. This would only confuse consumers; why they should commit to a sustainability program, investing effort, money and time, to do something that is already mandatory by law.

In this sense, indicators, methodologies not previously used, and new technologies and products are examples of what we mean by "innovation".

V.I.V.A. is characterized by innovation in the use of new indicators—specifically created for the project (Vineyard and Territory)—and new communication tools (Quick Response (QR) Code and smart phone app). SOStain and New Green Revolution are characterized by the strong "territorial vocation" (Sicily and Umbria, respectively).

Ita.Ca was the first carbon calculator for the Italian wine sector, and Gea.Vite is characterized by the final objective: to extend the analysis at the organization's efficiency and the use of weighing factors in the sustainability's level assessment.

Tergeo is particularly innovative for its "double" nature: the Matrix of sustainability, just finished being defined by Tergeo's researchers, is an innovative tool that enables companies to assess their "positioning" in the sustainability path (but only for the vineyard area for the moment). Tergeo's "Applications" are innovative "by nature": indeed, they are approved by a Scientific Committee also on the base of their level of innovation. This platform, that aims to match researchers, producers' needs and innovative products and methodologies is, at the moment, unique in the Italian framework.

Transparency and communication. In this context, communication is not always considered a priority, being the focus of the initiatives posed on the improvement of sustainability in the product and/or organization. It is worth noting that communication is becoming increasingly important. Improving sustainability performance and being scientifically consistent are at the base of each program; but, at the same time, it is absolutely necessary to implement effective communication strategies. Considering the increasing interest of consumers towards green products and responsible consumption [47], it is necessary to allow consumers to know about the commitment of companies to sustainability improvement and to make them recognize the more sustainable products.

In the context of this analysis, transparency and communication have been evaluated considering the disclosure of the sustainability results (by means of a report), the possibility for the consumer to be informed about the analysis system (how the sustainability level of a company is evaluated by the program) and the presence of a label.

Regarding the *provision of a report*—that means that the program asks to its member to produce and publicly release a report about performances gained in the framework of the initiative—only SOStain satisfies this requirement. For example, Tasca D'Almerita, a big Sicilian winery participating in the program, has published a report on its web site presenting the results gained under the framework of the SOStain project [48].

The *availability of the evaluation system* is an important element to judge the transparency of a sustainability initiative. Usually, the simple fact of being part of a sustainability initiative—particularly when it is promoted by a well-known Organization, Research Center, or Institution—is a guarantee for consumers: but it is the program itself (its scientific-technical staff) that need to guarantee better sustainability performance to consumers. However, does the consumer know how the specific company was evaluated? Are the parameters considered in the analysis known? Unfortunately, it seems that the majority of initiatives do not disclose their evaluation system to the broad public. Sometimes, a Protocol or a Disciplinary is publicly available on the initiative's websites, but—aside from the general rules to be followed to be part of the program and behavior/substances admitted and forbidden in the vineyard and in the cellar—it is not possible to know how audits have been conducted and which evaluation parameters have been used. According to these statements, the only programs disclosing the evaluation system are V.I.V.A. (since its disciplinary—that is validated by a third party, in order to ensure they are built in a correct way—is available on the website) and ECO-Prowine (the "Manual" to use the online tool is publicly available on the project website; hence, consumers can read about all the elements considered in the analysis).

A *label of sustainability*, when it is the output of consistent scientific analysis (and it is not mere green washing), enables consumers to recognize the more sustainable companies and to reward them for the effort by purchasing their product. Although it is not possible to state that a sustainability label can totally influence the choice of a certain products, due to the strong influence of other parameters (particularly in the purchase of food products), and the difficulty to separate the effect of other factors from the effect of the label [49,50], it can be very useful to increase awareness regarding the topic of sustainability and sustainable consumption, and it can increase the influence of sustainability as a choice factor, as declared by consumers [51]. When the label is the expression of a membership to an initiative, or a well-defined sustainability program, and it is supported by an "entity" (e.g., a University, a Research Center, a Public Institution) in which consumers trust, the efficacy of the label can be increased.

From this perspective, the more relevant initiatives are:

- V.I.V.A: a complex label composed by a QR (Quick Response) code (a square code linked to the information about the program and the related product), the four indicators (pictures, not values), the logo and name of the initiative, the logo of the Italian Ministry of the Environment.
- Magis: the label at the moment is a simple logo representing the membership of the company to the program and the achievement of minimum requirement. In the future, the logo will be differently colored according to the sustainability level gained.
- ECO-Prowine: the label, foreseen within 2014, will be the sign that the company is committed to the sustainability program and a minimum set threshold has been gained, as well as an improvement. It will be composed of the logo of the project and a code to be inserted on the dedicated website to gain access to information about sustainability performance of the company.

- Vino Libero: the label, presenting the logo of the project, is the "proof" that a product has been made according to the "Vino Libero" rules.

2.7 CONCLUSIONS

Sustainability has become a key issue for the Italian wine industry. Starting from the establishment of a methodology for the comparison of sustainability initiatives in the Italian wine sector, this analysis has highlighted the differences and similarities between them. Indeed, a variety of systems, methodologies and tools are being implemented, for a variety of reasons with different objectives. For example, for the Organic certification, the focus is mainly on reducing environmental impact, whereas in V.I.V.A. sustainability is evaluated in its three dimensions (environmental, economic, and social). When evaluated against their own objectives, all the programs and systems can be judged as effective, but the results can be different if an analysis against the Three Bottom Line (TBL) principles is conducted, and if their completeness (referring to the identified elements presented in the "Glossary") is evaluated too.

The great spread of sustainability initiatives in the wine sector can be a great opportunity for the overall sector; but confusion and overlapping of initiatives, methodologies and results must be avoided. The key point is that a common notion of sustainability in the Italian wine sector should be promoted, together with a broader industry wide sustainability strategy, and in order to do this it is necessary to foster the cooperation of all the program representatives and researchers. Creating a common understanding of sustainability is crucial for producers as well as for the entire sector. This common understanding is necessary first of all for an effective and beneficial consumer communication [5,12] and to reduce the uncertainty linked to the presence of a wide range of certifications and sustainability labels on the market. Secondly, a common language and framework is needed, one that can be used by the largest possible number of producers and farmers in the sustainability path. The great number of sustainability initiatives and programs can be very confusing for companies, inducing the risk that the real characteristics of each program and the real benefits for the company and the business are not clearly understood by the company's management and ownership. Finally, a single, unique sustainability framework and brand could enhance the competitiveness of Italian wine on foreign markets, particularly on those promoting sustainable products. Sustainability could become a new "distinctive trait" of the Italian wine, the "flagship product" for a "made in Italy and sustainable" production.

Will it be possible to "merge" the different elements of the programs in order to create a single, uniform framework to spread sustainability in the Italian wine sector, at the same time promoting scientific consistency, clarity and transparency towards wineries and consumers? This is the question to be posed to the program promoters. A positive answer seems to be possible, as was also evident Intervention during the stakeholder consultation held during the International Conference SIMEI-Enovitis [52].

As was shown in similar studies [21], results of this kind of analysis show that each of these programs has its own strenghts, but it is not possible to say that one program is better than another: elements considered in one program are sometimes not analyzed in another that, may instead focus on other aspects. On the contrary, there is a general consensus on some aspects, such as the use of certain calculators or methodology to assess sustainability (for example, the calculation of the Carbon Footprint to assess the impact on the air quality or the use of check lists as a tool to evaluate the company's performances).

Starting from these assumptions and the results of the analysis, the authors believe that a unique framework could be created, but it wil be necessary to clearly define:

- The presence and the meaning of the characterizing elements of a real and complete sustainability program (protocol, indicators, the label, etc.) that could be suitable for the Italian wine sector;
- The main areas (e.g., Air, Water, Soil, etc.) to be considered;
- Within each area, shared "indicators" to be used to calculate and compare sustainability performances.

An integration between the programs and their elements, hence, is possible; however, it is necessary that, starting from inputs given by this study, further analysis and initiatives are promoted by stakeholders at an operating level, with the objective of understanding if synergies between programs are possible in practice. The work is surely not easy, due to the complexity of the sustainability concept in itself, that is extremely wide and somehow ambiguous. However, by creating an ongoing dialogue and exchange of experiences and opinions between the researchers and pro motors that are working for sustainability further analyses will bring concrete results.

REFERENCES

1. Gabzdylova, B.; Raffensperger, J.; Castka, P. Sustainability in the New Zealand wine industry: Drivers, stakeholders and practices. J. Clean. Prod. 2009, 17, 992–998.
2. Barber, N.; Taylor, C.; Strick, S. Wine consumers environmental knowledge and attitudes: Influence on willingness to purchase. Int. J. Wine Res. 2009, 1, 59–72.
3. Ohmart, C. Innovative outreach increases adoption of sustainable winegrowing practices in Lodi region. Calif. Agric. 2008, 62, 142–147, doi:10.3733/ca.v062n04p142.
4. Sinha, P.; Akoorie, M. Sustainable Environmental Practices in the New Zealand Wine Industry: An Analysis of Perceived Institutional Pressures and the Role of Exports. J. Asia-Pac. Bus. 2010, 11, 50–74, doi:10.1080/10599230903520186.
5. Szolnoki, G. A cross-cultural comparison of sustainability in the wine-industry. J. Clean. Prod. 2013, 53, 243–251, doi:10.1016/j.jclepro.2013.03.045.
6. Marshall, R.; Cordano, M.; Silverman, M. Exploring Individual and Institutional Drivers of Proactive Environmentalism in the US Wine Industry. Bus. Strat. Environ. 2005, 14, 92–109, doi:10.1002/bse.433.
7. Hoffman, M. Keeping the wineglass full: Exploring the intersection of sustainable viticulture and sustaining winegrower legacy in Lodi. Master Thesis, Iowa State University, Ames, IA, USA, 2009.
8. United Nations (UN). Resolution adopted by the General Assembly. 60/1. 2005 World Summit Outcome. Available online: http://unpan1.un.org/intradoc/groups/public/documents/un/unpan021752.pdf (accessed on 11 November 2013).
9. Castellucci, F. Resolution CST 1/2004, Development of sustainable vitiviniculture. Available online: http://www.oiv.int/oiv/files/3%20-%20Resolutions/EN/2004/CST%2001-2004.pdf (accessed on 10 October 2013).
10. Castellucci, F. Resolution CST 1/2008, OIV Guidelines for Sustainable Vitiviniculture: Production, Processing and Packaging of Products. Available online: http://www.oiv.int/oiv/files/3%20-%20Resolutions/EN/2008/CST%2001-2008.pdf (accessed on 10 October 2013).
11. Broome, J.; Warner, K. Agro-environmental partnership facilitate sustainable winegrape production and assessment. Calif. Agric. 2008, 62, 133–141, doi:10.3733/ca.v062n04p133.
12. Klohr, B.; Fleuchaus, R.; Theuvsen, L. Sustainability: Implementation programs and communication in the leading wine producing countries. Available online: http://www.newgreenrevolution.eu/documents/Sustainability%20Programs%20and%20communication_Bastian%20Klohr%201.pdf (accessed on 20 February 2014).

13. Istituto Nazionale di Economia Agraria (INEA). Sostenibilità ambientale, sociale ed economica della filiera vitivinicola. Available online: http://dspace.inea.it/handle/inea/679 (accessed on 11 November 2013). (In Italian).

14. Santini, C.; Cavicchi, A.; Casini, L. Sustainability in the wine industry: Key questions and research trends. Agric. Food Econ. 2013, 1, 1–14, doi:10.1186/2193-7532-1-1.

15. Bastianoni, S.; Marchettini, N.; Panzieri, M.; Tiezzi, E. Sustainability assessment of a farm in the Chianti area (Italy). J. Clean. Prod. 2001, 9, 365–373, doi:10.1016/S0959-6526(00)00079-2.

16. Pulselli, F.; Ciampalini, F.; Leipert, C.; Tiezzi, E. Integrating methods for the environmental sustainability: The SPIn-Eco Project in the Province of Siena. J. Environ. Manag. 2008, 86, 332–341, doi:10.1016/j.jenvman.2006.04.014.

17. Marchettini, N.; Panzieri, M.; Niccolucci, V.; Bastianoni, S.; Borsa, S. Sustainability indicators for environmental performance and sustainability assessment of the production of four Italian wines. Int. J. Sust. Dev. World 2003, 10, 275–282, doi:10.1080/13504500309469805.

18. Ardente, F.; Beccali, G.; Cellura, M.; Marvuglia, A. POEMS: A Case Study of an Italian Wine-Producing Firm. Environ. Manag. 2006, 38, 350–364.

19. Bosco, S.; di Bene, C.; Galli, M.; Remorini, D.; Massa, R.; Bonari, E. Greenhouse gas emissions in the agricultural phase of wine production in the Maremma rural district in Tuscany, Italy. Ital. J. Agron. 2011, 6, 93–100.

20. Vazquez-Rowe, I.; Rugani, B.; Benetto, E. Tapping carbon footprint variations in the European wine sector. J. Clean. Prod. 2013, 43, 146–155, doi:10.1016/j.jclepro.2012.12.036.

21. Hughey, K.; Tait, S.; O'Connell, M. Qualitative evaluation of three "environmental management systems" in the New Zealand wine industry. J. Clean. Prod. 2005, 13, 1175–1187, doi:10.1016/j.jclepro.2004.07.002.

22. Warner, K. The quality of sustainability: Agroecological partnerships and the geographic branding of California winegrapes. J. Rural Stud. 2007, 23, 142–155, doi:10.1016/j.jrurstud.2006.09.009.

23. Forbes, S.; Cohen, D.; Cullen, R.; Wratten, S.; Fountain, D. Consumer attitudes regarding environmentally sustainable wine: An exploratory study of the New Zealand marketplace. J. Clean. Prod. 2009, 17, 1195–1199, doi:10.1016/j.jclepro.2009.04.008.

24. Jang, Y.J.; Bonn, M.A. Consumer Purchase Intentions of Organic Wines. Available online: http://scholarworks.umass.edu/cgi/viewcontent.cgi?article=1272&context=gradconf_hospitality (accessed on 20 February 2014).

25. Fotopoulos, C.; Krystallis, A.; Ness, M. Wine produced by organic grapes in Greece: Using means-end chains analysis to reveal organic buyers' purchasing motives in comparison to non-buyers. Food Qual. Prefer. 2003, 14, 549–566, doi:10.1016/S0950-3293(02)00130-1.

26. Remaud, H.; Mueller, S.; Chvyl, P.; Lockshin, L. Do Australian Wine Consumers Value Organic Wine? Available online: http://academyofwinebusiness.com/wp-content/uploads/2010/04/Do-Australian-wine-consumers-value-organic-wine_paper.pdf (accessed on 12 February 2014).

27. Sirieix, L.; Remaud, H. Consumer perceptions of eco-friendly vs. conventional wines in Australia. Available online: http://academyofwinebusiness.com/wp-con-

tent/uploads/2010/04/SirieixRemaud-Consumer-perceptions-of-eco-friendly-wines. pdf (accessed on 20 February 2014).

28. Mueller, R.; Remaud, H. Are Australian wine consumers becoming more environmentally conscious? Robustness of latent preference segments over time. Available online: http://academyofwinebusiness.com/wp-content/uploads/2010/04/Mueller-Remaud-Are-Australian-wine-consumers-environmentally-conscious.pdf (accessed on 20 February 2014).

29. Schleenbecker, R.; Hamm, U. Consumers' perception of organic product characteristics. A review. Appetite 2013, 71, 420–429, doi:10.1016/j.appet.2013.08.020.

30. Associazione Italiana per l'Agricoltura Biologica (AIAB). Available online: http://www.aiab.it/index.php?option=com_content&view=category&layout=blog&id=28&Itemid=61 (accessed on 12 February 2014).

31. Specific files created for each program are available from the authors upon request.

32. It was not possible to find a clear definition of "Sustainability Protocol" in the literature. In this context, the meaning of the word "Protocol" has been taken from the Information & Communication Technology, communication and medical sectors. In these areas, indeed, this word is frequently used, although with different meanings. Although the different situations, generally speaking it is possible to notice that all the "protocols" have some common elements: (a) the existence of a precise goal (e.g., the communication between devices in the ICT, the understanding between individuals or entities in the communication field, the resolution of an emergency situation in the medical sector); (b) a set of rules to apply, code of procedures or parameters to own, in order to achieve that goal.

33. The meaning of all these aspects was carefully explained to the interviewed persons. All these elements were chosen according to the "areas" generally considered in the major international sustainability programs, particularly the California Code of Sustainable Winegrowing of the California Sustainable Winegrowing Alliance.

34. Unione Italiana Vini. Tergeo. Available online: http://www.uiv.it/progettotergeo/ (accessed on 11 November 2013).

35. Magis. Che cosa è Magis? Available online: http://magisvino.imagelinenetwork. com/default.cfm (accessed on 11 November 2013). (In Italian).

36. SOStain. II progetto. Available online: http://www.sostain.it/IT/SOStain.aspx (accessed on 11 November 2013).

37. Sustainability in the Italian viticulture. Available online: http://www.viticolturasostenibile.org/EN/Home.aspx (accessed on 11 November 2013).

38. ECO-Prowine. ECO-Prowine—Approach. Available online: http://www.ecoprowine.eu/approach/ (accessed on 11 November 2013).

39. Ita.Ca. Italian Wine Carbon Calculator. Available online: http://www.agronomisata. it/template.php?pag=18069 (accessed on 11 November 2013).

40. Gestione dell'Efficienza Aziendale, Available online: http://www.agronomisata.it/template.php?pag=18072 (accessed on 11 November 2013). (In Italian).

41. Associazione Vino Libero. Disciplinare tecnico di produzione vitivinicola integrata evoluta. Available online: http://www.vinolibero.it/wp-content/uploads/2013/04/Disciplinare-Tecnico-Vino-Libero.pdf (accessed on 13 November 2013). (In Italian).

42. New Green Revolution. Elaborazione, introduzione e applicazione di un nuovo protocollo di produzione. Available online: http://www.newgreenrevolution.eu/ (accessed on 11 November 2013). (In Italian).
43. European Commission. Commission Implementing Regulation (EU) No. 203/2012 amending Regulation (EC) No 889/2008 laying down detailed rules for the implementation of Council Regulation (EC) No 834/2007, as regards detailed rules on organic wine.
44. Associazione Viticoltori Naturali. Vinnatur. Available online: http://www.vinnatur.org/ (accessed on 11 November 2013). (In Italian).
45. European Commission. Agricultural and Rural Development. Organic Farming. Available online: http://ec.europa.eu/agriculture/organic/ (accessed on 12 February 2014).
46. FederBio. Il Vino Biologico. Available online: http://www.federbio.it/Vino_biologico.php. (accessed on 12 February 2014). (In Italian).
47. European Commission. Attitudes of European towards building the single market for green products. Available online: http://ec.europa.eu/public_opinion/flash/fl_367_en.pdf (accessed on 11 November 2013).
48. OPERA Research Center. SOStain Sustainable Development Report—Tasca d'Almerita 2012. Available online: http://tascadalmerita.it/pdf/report_2012.pdf (accessed on 10 November 2013).
49. Gallaraga Gallastegui, I. The use of eco-labels: A review of the literature. Eur. Environ. 2002, 12, 316–331, doi:10.1002/eet.304.
50. Vanclay, J.; Shortiss, J.; Aulsebrook, S.; Gillespie, A.; Howell, B.; Johanni, R.; Maher, M. Customer Response to Carbon Labelling of Groceries. J. Consum. Policy 2011, 34, 153–160, doi:10.1007/s10603-010-9140-7.
51. ECO-PROWINE. Life Cycle perspective for Low Impact Winemaking and Application in EU of Eco-innovative Technologies. Available online: http://www.ecoprowine.eu/media/uploads/d4.5_market_analysis_report.pdf (accessed on 10 November 2013).
52. Corbo, C. Dialogo e armonizzazione. Parole chiave per lo sviluppo sostenibile in viticoltura. Available online: http://www.uiv.it/wp-content/uploads/2013/12/p_03_05_simei-convegni.pdf (accessed on 12 February 2014). (In Italian).

Table 10 is not available in this version of the article. To view this additional information, please use the citation on the first page of this chapter.

CHAPTER 3

TRANSNATIONAL COMPARISON OF SUSTAINABILITY ASSESSMENT PROGRAMS FOR VITICULTURE AND A CASE-STUDY ON PROGRAMS' ENGAGEMENT PROCESSES

IRINA SANTIAGO-BROWN, ANDREW METCALFE, CATE JERRAM, AND CASSANDRA COLLINS

3.1 INTRODUCTION

This article aims to document and compare the most prominent sustainability assessment programs for individual organisations in viticulture worldwide and their certification processes. Sustainability concerns have become increasingly important since the publication of the "Our Common Future" report by the United Nations Commission on Environment and Development (WCED) in 1987 [1]. Sustainable development was defined as "economic growth that meets the needs of the present without compromising the ability of future generations to meet their own

Transnational Comparison of Sustainability Assessment Programs for Viticulture and a Case-Study on Programs' Engagement Processes. © *2014 by the authors; licensee MDPI, Basel, Switzerland.* Sustainablity, *6,4 (2014), doi:10.3390/su6042031. Licensed under Creative Commons Attribution 3.0 Unported License, http://creativecommons.org/licenses/by/3.0/.*

needs" [2,3]. Since then many countries have developed sustainability initiatives to promote sustainable development. Many of these initiatives have in turn generated regulations, especially on the environmental and social aspects of sustainability. Harmful consequences of chemical inputs from agriculture have been a common driver of many agricultural sustainability initiatives [4]. Because of the high value of wine grapes [5], wine grape growing regions have developed some of the most complex sustainability assessments and certifications for individual agricultural organisations. Most of these assessment programs incorporate a triple-bottom line approach, which evaluates entire production systems considering the interrelationship of economic, environmental and social factors [6].

To the best of our knowledge, similar comparisons to this study have not been previously published in other peer-reviewed journals. This article seeks to fill this research gap by describing the following sustainability programs for wine grape growing: Lodi Winegrowing Commission (LWC) Sustainable Workbook/Lodi Rules; Vineyard Team/ Sustainability in Practice (SIP); Low Input Viticulture and Enology (LIVE); California Sustainable Winegrowing Alliance (CSWA)/California Sustainable Winegrowing Program (SWP); VineBalance, New York State's Sustainable Viticulture Program/Long Island Sustainable Winegrowing (LISW); Sustainable Winegrowing (SWNZ) from New Zealand; Integrated Production of Wine (IPW) from South Africa; Sustainable Wine from Chile (SWC); and McLaren Vale Sustainable Winegrowing (MVSWGA) from Australia. Where data are available, these descriptions include names of key individuals whose personal enthusiasm and motivation are perceived as essential for the program's implementation.

Finally, this article presents an analysis of the engagement process of viticultural sustainability programs based on results derived from 14 focus groups with 83 top-level managers from wine grape production organizations. We discuss growers' expected benefits and motives to become part of sustainability assessment programs as well as the main inhibiting factors and desirable reporting system features that can potentially contribute to program funding and membership uptake.

3.2 METHODS OF MEASUREMENT AND ASSESSMENT IN SUSTAINABILITY PROGRAMS

Methods of measurement are likely to change over time because of the development of new technologies; however the fundamental principles of measurements and especially of validity are likely to remain the same. "Validity is an overall evaluative judgment, founded on empirical evidence and theoretical rationales" ([7] p. 33). The main validity concerns, within measurements, are "interpretability, relevance, and utility of scores, the import or value implication of scores as basis for action, and the functional worth of scores in terms of social consequences of their use" ([7] p. 33).

Sustainability programs have developed their own assessment methods, adopting and/or adapting other evaluation methods from other sustainability programs in viticulture, agriculture or from other fields such as education, accounting, and management. Minimum fundamental issues must be defined prior to the establishment of the assessment method, such as: a sustainability definition [8]; scope; context; objectives; and viewpoint of the assessment [9,10]. Having these issues defined, assessments must be constantly evaluated regarding their appropriateness, meaningfulness and usefulness [7]. Therefore, in the scope of this investigation, beneficial outcomes from inferences made from assessment results improve wine grapes growers' sustainability.

Hansen [8] concluded that sustainability is crucial to guide change in agriculture. In order to be able to positively impact on agricultural systems, rapidly respond to the need for change, ensure viability of agriculture over time and be a useful criterion to guide changes, sustainability not only needs to be defined but its characterization should be literal, system-oriented, quantitative, predictive, stochastic and diagnostic. In a previous phase of this project, following focus group meetings, a sustainable farm or vineyard was defined as "one that is able to economically provide for the farmer while maintaining its ability to consistently produce and improve quality over time". In a following phase, it was determined that assessment for sustainability must incorporate a triple bottom line including economic, environmental and social components.

Sustainability assessments roughly encompass four stages: (1) definition of assessment method; (2) definition of indicators; (3) attributed scores and weights or compliance and (4) certification (conformance). The term indicator is used broadly to indicate any direct, indirect, qualitative or quantitative defined measure of something to assess sustainability within a given system [11]. The assessment method determines which, and how, indicators are used. Scores and weights are subjective values attributed by the proponent of the assessment [12,13], usually based on scientific/expert knowledge and/or assessment goals and context. Compliance is related to fitness of the method and content and conformance is directly related to certification of compliance by an external authority [14].

We categorize assessment methods generally in four distinct types based on their overall focus (see Table 1): (1) process-based, (2) best practice-based, (3) indicator-based and (4) criterion-based. Each of these can be used individually or combined and each has weaknesses and strengths. Independently of chosen method, the establishment of benchmarks and performance measures is necessary [10,15] to assist wine growers to improve their sustainability by comparing to their peers and analysing their results and/or performance against program goals. Certification can be developed for any of these methods with a higher or lower degree of complexity; however, it is important to point out that the purpose of certification is marketing. It is to provide a seal of assurance [16] for society that the organisation conforms to a stated requirement [14].

Process-based assessments are usually based on the International Organisation for Standardization Standards (ISO) standards. In agricultural assessments, a typical example is the implementation of environmental management systems (EMS) through the ISO 14001 standard or through ISO-based locally developed guidelines. The greatest shortcoming of a process-based assessment is that it does not ensure performance outcomes [17,18]. The practical outcome of process-based methods is the production of written documentation (e.g., management plans). Furthermore, the ISO family of standards were developed mainly to provide a model for large enterprises to set and operate a management system. The ISO 14001 is a challenging [19] and costly task for small and medium size organisations [14].

TABLE 1: Methods of assessment of sustainability (examples from viticulture).

Assessment methods	Focus	Example
Process-based	process rather than the outputs of the activity	Is there is a plan to manage soil erosion?
Best practice-based	implementation of the task, therefore the output of the activity	Were cover crops planted to prevent erosion?
Indicator-based *	past input usage	Record of electricity usage
Criterion-based	compliance to a set of rules	Determines x% of the farm land dedicated to biodiversity

Note: * *"indicator" as used in this table, is not understood as a broad concept as described in the text, but purely as a quantitative value.*

The best practice-based assessment method's strongest point is the practical and immediate pathway to objectively deliver net sustainability gains [20]. Education is a core component of this method [21]. Among the described methods, it seems to be the easiest to engage farmers because of its focus on the sustainability output. Gibson [22] argues that best practice systems should be implemented gradually but the process is not without risk(s). The greatest challenge of this method is to ensure that factors that are not priorities in conventional decision-making process are not left behind. In an example of environmental sustainability assessment, Gibson points out that the effects on the community might not be prioritised. To overcome the problem, definition, scope and trade-off rules must be clearly established prior to the development of the assessment.

Indicator-based assessments rely on reporting of numerical values related to past input use. The weakness of the system is related to the meaningless value of indicator collection when not linked to reference levels. When not related to reference levels, indicators become just a set of collected data [10]. Carbon/greenhouse gas accounting and water footprint methods are typical examples of indicator-based assessments. These examples of the usage of indicator-based assessments oversimplify life cycle assessment methods and are insufficient to understand the dynamics of the interrelationships of system outcomes and resource (inputs) use [23,24]. On the other hand, the strength of this method relates to the small time required for data recording and ability to readily compare data [24].

Criterion-based assessments are assessment methods focused on compliance with legislation or sets of rules from the sustainability assessment program employed [25]. The strength of the method seems to also be its weakness: the method clearly excludes non-compliant (with rules) participants and establishes a clear message of group exclusivity. However, the exclusion can undermine possible participation by growers who are in most need of help to improve their sustainability.

3.3 RESEARCH METHOD

3.3.1 DESCRIBING/DOCUMENTING SUSTAINABILITY PROGRAMS

The most relevant sustainability programs for viticulture worldwide are documented in this article. The description of the programs is based on interviews, observations, and secondary sources. Between December 2011 to January 2014, 10 in-person semi-structured interviews and two semi-structured email interviews were conducted with people either currently or formerly in charge of the sustainability program described in this article. The names and characteristics of interviewees and participants are withheld to honour confidentiality commitments, except where specific permission has been given to reference them. All such references are cited throughout this text as "personal communication".

Each interview followed a semi-structured schedule which commenced with demographic questions, and then progressed to key questions concerning the creation of the sustainability program they were involved with as well as its assessment methodology, original motivations, certification, engagement processes to maintain the program and strategies to engage new members. Thereafter, each interview focused on the specific responses of the participant. Interviewees also provided observations and opinions of the current situation of the programs. All interviewees were re-contacted in January 2014 to update program statistics and validate texts from their specific programs.

3.3.2 EXPECTED BENEFITS, ENGAGEMENT STRATEGIES, INHIBITING FACTORS AND REPORTING SYSTEMS OF SUSTAINABILITY PROGRAMS

The results from this study are part of a larger 3-stage study in which stage (1) aimed to define sustainability through an Assisted Focus Group Method of Enquiry (AFGME) [9], stage (2) produce a list of indicators for sustainability assessment through an Adapted Nominal Group Technique (ANGT) [10], and stage (3)—reported in this paper—aimed at discussing the engagement process of viticultural sustainability programs through a traditional focus group approach and document and compare the most prominent sustainability assessment programs for individual organisations in viticulture worldwide. Table 2 outlines the questions used in the stage 3 focus group discussions.

TABLE 2: Focus group question: stage 3 used for this article.

(1) What potential benefits would induce you to participate in sustainability program? What benefits would you expect to receive for your business from participating in a sustainability program for your vineyard?
(2) What reasons would cause you to not participate in a sustainability program?
(3) If you were responsible for implementing a sustainability program, what would you do to engage growers to participate?
(4) Assuming that results oriented foundation is considering sponsoring your program. How would you convince them to fund the project?
(5) Assuming you are the external sponsor. What would you want to be measured?

The stage 3 group discussions were conducted from December 2011 to November 2012 with top-level managers (e.g., CEOs, Chief Winemakers and Chief Viticulturists) of grape growing organisations from five countries: Australia, Chile, New Zealand, South Africa and United States [9].

The countries were selected because of the existence of sustainability assessment programs for viticulture "at the farm-gate" (for individual organizations). This group of countries is known as "New World" wine countries. To the best of our knowledge, at the time of this investigation, there were no similar programs in "Old World" wine countries.

Multiple checks and balances were built into the research design to ensure validity and transferability of the study results [26]. Triangulation was achieved through use of multiple data sources including interviews, focus groups, participant observation, documentation and archival analysis. Cross-comparison analyses were conducted, and findings and results were presented to and discussed with industry and academic panels on multiple occasions.

3.4 DATA ANALYSIS

The program descriptions below were informed by the interview transcriptions and personal observations. Program websites and official program documents were also used as secondary sources to develop the information from the data gathered through the interviews. All data was then analysed in a three-phase process in which first-order analysis combined descriptive- and pattern-coding, second-order analysis for data-reduction using thematic-coding, then third-order analysis using tag cloud analysis, which is described in detail below. The coding process was aided by use of NVivo 10 [27], a qualitative data management software package.

Tag cloud analysis is usually used for indexing and searching websites [28]. This was adapted for this study, using tag clouds analytically to aid researcher evaluations of data emphasis and participants' prioritization of benefits, hidden factors and critical values relating to the stage 3 questions.

Question content and moderators' utterances were discharged from the coding which exclusively analysed participant responses. Four tag-clouds were created from the seventy most frequently occurring exact words. The tag clouds display the most frequently used words in larger fonts in a circular layout, randomly organized to optimize display space. Tag clouds are pictorial heuristic representations of text with the aim to present a problem in a simplified but sufficient way. Its interpretation is straightforward and requires little explanation [29]. It helps to find solutions to problems through the display of patterns of a given problem [30]. The tag clouds generated and used for analysis in this study are displayed in the Results and Discussion section of this paper.

3.5 RESULTS AND DISCUSSIONS

3.5.1 SUSTAINABILITY ASSESSMENT PROGRAMS FOR INDIVIDUAL ORGANISATIONS IN VITICULTURE

The first results presented here are the descriptions of the sustainability programs investigated for the study. They are presented in the order in which the initiatives for the programs commenced. There is a deliberate emphasis, where possible, on recognising the names of persons who were initiators and drivers of the programs. A finding arising from this research was that the individual persons who initiate and drive programs are critical to the process and its success. In other words, without the initiating and driving people, and the early adopters, it is unlikely these programs would exist as they are.

3.5.1.1 LODI WINEGROWING COMMISSION SUSTAINABLE WORKBOOK AND LODI RULES

Lodi Winegrape Commission (LWC) was created in 1991 with the core objective of promoting the Lodi wine region in California, United States and its wines [31]. At that time, Integrated Pest Management (IPM) was identified as one of the most important issues for wine grape growers. The Grassroots IPM program was launched in 1992 [3,32] and a consultant with a PhD in IPM conducted the group until 1995 when the Commission's budget was reorganized. An IPM organisation was contracted replacing the consultant and Dr Cliff Ohmart became part of the project with the objective to expand it. Ohmart, with a group of innovative growers, was the driving force behind the sustainability initiatives in Lodi. In 1995 the program had its objectives expanded by implementing and tracking the results of a series of sustainable winegrowing projects [33]. The first assessment methodology was based on the Farm*A*Syst model, developed by the United States Department of Agriculture (USDA) agencies, regional organisations, universities and local governments. The content of the model was tailored to fit Lodi's purposes [33].

By the end of its first year, 40 growers were directly involved with the project and monitored weekly. The group represented about 1000 hectares and 70 distinct vineyard sites. Many new topics beyond pest and disease management such as ecosystem management and human resources, among others, were included as well as incorporating the knowledge and data gathered from the demonstration vineyards' data [34]. The proposed content was reviewed by a committee of growers, vineyard consultants, University of California Farm Advisors and Scientists, vintners and a wildlife biologist [35]. The first edition of the assessment workbook was launched in 2000. The workbook was about growing quality wine grapes efficiently: the strategy was quality and sustainability was the means [34].

After the launch of the assessment methodology, 40 workshops were organised and scores from 265 growers were compiled in a system that became the basis for the database of the sustainability program. A second printed version of the workbook was published in 2008 and a third in 2013. The self-assessment workbook has the educational purpose of optimising wine grape quality and costs. The Lodi Rules, a third-party certification scheme, was launched in 2005 to respond to the growers' demand for a marketing application for the self-assessment workbook. The certification process encompasses two components: the Lodi Rules (practice standards), and a Pesticide Environmental Assessment System (PEAS), a risk assessment tool that measure the total impact of all organic and synthetic pesticides used during the year by each individual participant grower [36].

In the workbook, each self-assessment topic has four options (plus non-applicable). Self-assessment topics range from questions about soil and water management to ecosystem and human resources. The self-assessment options which range from 1 (least sustainable) to 4 (most sustainable) should be interpreted exclusively within each assessment topic. The Lodi Rules (certification) are designed to lead to measurable improvements in the health of the surrounding ecosystem, society-at-large, and wine quality. To achieve certification, growers must achieve 70% of the total possible (maximum) score plus at least 50% of the total score of each chapter. Protected Harvest, a third party non-profit organization, independently audits Lodi Rules. Vineyards must be audited annually through a rigorous process of in-site inspection prior to harvest and pesticide and nutrient usage post-harvest [35].

Lodi growers were surveyed by the LWC in 1998 and 2003. The surveys and data gathered from the growers' assessments helped the identification of winegrowers' regional strengths and weaknesses and, therefore, educational needs. Ohmart is a believer of self-assessments for growers: "if you self-assess, you invariably learn something by doing it" [34].

3.5.1.2 SUSTAINABLE WINEGROWING NEW ZEALAND—SWNZ

Sustainable Winegrowing New Zealand (SWNZ) has its origins in a pilot project started by a group of seven growers from the Hawke's Bay Winegrowers Association in 1995 [37]. The original motivation was to assess vineyard chemical usage, inspired by international demands and constraints learned from fresh food exporters [37]. At that time, these growers had access to, and adopted, a system called the Wäidenswil Integrated Production Scorecard developed in Switzerland [38]. The idea spread rapidly among other wine grape regions and a working group was formed with the objective to develop a sustainability program. In 1997 the group had approximately 120 vineyards self-assessing their operations. Certifications by third party started in 2000. A wineries standard was introduced in 2002 [39]. In 2004, Ms. Sally van der Zijpp was employed as the National Coordinator for the program, a position that she still holds. The scorecard model was fully reviewed and changed in 2007 to embrace the reality of New Zealanders' growers: new assessment areas were added and the scoring methodology was completely changed. The online system was launched in 2007.

The program defines sustainability as "delivering excellent wine to consumers in a way that enables the natural environment, the businesses and the communities involved, to thrive" [39]. The New Zealand Winegrowers' Sustainability policy states that wine must be made from 100% certified grapes in fully certified winemaking facilities and certification must be through an independently audited third program (SWNZ or one of the recognized organic or biodynamic certifications). The program aims to provide a "best practice" model, and it is also a quality assurance scheme that addresses consumer concerns and aims to protect the market for wines from New Zealand [40].

To become part of the program members must self-assess their operations online annually and provide supporting documentation for their responses. There is also a data collection of indicators such as water and input use (electricity, fuel records and spray diary). As a premise, the program avoids collection of data that can be gathered through other sources such as government. It also avoids collecting data that will not be analysed or help growers to improve their sustainability in a practical manner. The SWNZ flexibly incorporates practices across a range of business sizes and regions and meets the International Organisation of Vine and Wine (OIV) and International Federation of Wine and Spirits (FIVS) guidelines, in spite of not being accredited by them. The decision not to be accredited is an on-going debate amongst growers because they are unsure about the direct benefits of such accreditations that would increase the program's costs.

The program is based on three pillars: monitor, measure, and manage. Currently, the measures that SWNZ focuses on are water, energy and agrochemical use. Members are required to supply their spray diaries. Reports produced from growers' data are analysed and reported back to growers and also used by New Zealand Wines to represent the industry needs, for research purposes and discussions with government.

The self-assessment consists of three sets of questions: major, minor and best practices: Majors are mandatory, minors are generally relevant practices and best practices are the next step up. Questions can be answered in the following ways: yes or compliant; no or non-compliant and non-applicable (NA). Questions are followed by a list (to be ticked) of applied strategies, which growers select.

Compliance with all major questions and 80% of minor questions are required to achieve certification. If 100% of major questions are not achieved, corrective actions are required to pass. A second on-site inspection may or may not be requested, depending on each situation. Questions are supported by a guidance text and pertinent excerpts from the Standard. The program also sets maximum chemical sprays per target and demands comprehensive justification in case of extra spray needs [37]. Questions, results and benchmarks are not readily available to the external public but widely used in the NZSW's website texts including growers' individual profiles. The program is focused in engaging all members of the wine growing community.

Program auditors, from independent third-party organisations, work closely with the program managers. Vineyards need to meet the program requirements at an initial inspection in order to be certified, then are inspected again every three years. Auditors are encouraged to provide advice to growers to help them to meet the Standard's requirements. This is similar to the educational approach taken in the Integrated Production of Wine (IPW) program in South Africa (described below), but to a smaller extent. Certificates are issued by SWNZ based on inspection results. Certification is still voluntary [41,42], however since 2010, the New Zealand Winegrowers, the body responsible for promoting the brand New Zealand Wines, made vineyard and wine accreditation to the SWNZ (or one of the recognized organic or biodynamic certifications) a pre-requisite to participation in promotional events. As a result, 90% of the wines produced in New Zealand became part of the SWNZ.

3.5.1.3 VINEYARD TEAM (SUSTAINABILITY IN PRACTICE—SIP)

In 1994, a group of growers, wineries and service providers in California, in the United States volunteered to create the Central Coast Vineyard Team (CCVT). Two years later, the Positive Points System (PPS), a self-assessment on sustainable vineyard practices was launched and 20 vineyard assessments were performed. In 1999, their database had 200 growers. In 2000, the membership program began. The group grew steadily with a strategy of engaging the community and vineyard neighbours in informal meetings and viticultural educational initiatives [43].

In 2004, the CCVT began the development of a third party certification program called Sustainability in Practice (SIP) Certification. SIP Certification was designed to be a distinguishing program with requirements for certification, which authenticate vineyard practices and distinguish their wines in the market. According to the SIP Certification Manager, SIP it is not a certification that every vineyard can achieve. All of the questions are practice-, as opposed to process-, based and are auditable [44]. The PPS was used as foundation to the development of the certification. The pilot project for the certification program was launched in 2008 and 14 vineyards, representing 1200 hectares, became certified in the region. Certification is now extended to the whole state of California.

The program's standards are annually updated and peer-reviewed by a vast group of Universities, Government departments and industry associations. Currently, the program only assesses vineyards but wineries are able to certify their wines, which allows them to use a SIP Certified seal on a wine bottle provided that a chain of custody audit shows that the final product is made with at least 85% SIP Certified fruit. A certification for sustainable winery production is being developed. Certification must be renewed annually in a three-year cycle: on-site inspection in the first year and evaluation of demanded records, a combination of paper audits, interviews, and on-site inspections in years 2 and 3 [43,45].

3.5.1.4 LOW INPUT VITICULTURE AND ENOLOGY (LIVE)

The creation of the Low Input Viticulture and Enology (LIVE) in Oregon, USA, has its roots in a mid -1990s presentation arranged by Dr Carmo Vasconcelos, a Portuguese researcher at the Oregon State University, and conducted by Dr Ernst F. Boller, a founding member of the International Organization for Biological and Integrated Control (IOBC). After Boller's presentation, a few individuals from the audience realised that they were already practising and sharing similar principles to those presented by Boller [46]. The creation of an official assessment program based on IOBC guidelines was initially proposed by a winegrower, Mr Al MacDonald, who was also involved with the University. Low Input Viticulture and Enology program (LIVE), a voluntary organization, was then established in 1997 by a group of Oregon winegrowers led by Mr Ted Casteel [47].

The pilot project started with about 20 vineyards and the group was voluntarily inspected through a partnership developed with the Oregon State University. The objective was to understand their level of compliance with the guidelines. In 1999 LIVE was incorporated and certified by

IOBC to certify individual farmers. In the same year the inspections were conducted by independent third party contractors with IPM—integrated pest management expertise. In 2006, the program was expanded to include growers from Washington State. In 2007, LIVE hired Mr Chris Serra as a paid Program Manager who was promoted to Executive Director in 2011 (LIVE, 2013). All Board members and technical committee members were and still are volunteers.

The assessment system is freely available on-line on the LIVE website [47]. Transparency is the key part of their strategy to engage growers and consumers: the first to join the program and the latter to trust what the LIVE brand stands for [48]. Growers need to join the program to have access to a username and password to access all functionalities of the on-line system and to have their data saved and considered for inspection by the program management. All educational resources and administrative documents are available as well. The program has the objective of promoting viticulture in conjunction with environment preservation and conservation of the vineyard and surrounding areas, a farm's economic viability and support to its social, cultural and recreational aspects. Also, to sustain healthy and high quality grapes with great emphasis on minimizing pesticide residues, by encouraging biological diversity and use of natural regulating mechanics and unwanted side effects from agro-chemical handling.

The program assessment is comprised of mandatory record keeping (pesticide, fertilizer, and irrigation), 5% of farm area set aside as a biodiversity and ecological compensation zone and a checklist of 13 chapters, each one with a series of topics, called "control points" [48]. The approved pesticides lists are specific to two vineyard locations based on climate: Region I refers to cool-weather maritime climate and region II refers to warm-weather continental viticultural climate. The check list follows a colour scheme rationale where Red control points are 100% required, which means that full compliance is mandatory to become part of the program. LIVE requires 90% of the Yellow control point and 50% of the Green control points. The system was developed to avoid members concentrating too heavily on any one given area of assessment [47].

Certification is only achieved after completion of two years of farming under LIVE standards. Farmers have to be inspected in the first two years of the program. After passing the second year inspection, they can be certified by an independent third party, if program requirements are met. Certification must be renewed every three years but any member is subjected to random inspections at any time. Additionally, members certified or not, must submit their records every year.

3.5.1.5 INTEGRATED PRODUCTION OF WINE (IPW)

The Integrated Production of Wine (IPW) scheme was promulgated by a South African governmental Act in November 1998. IPW is one of the three schemes managed by the Wine and Spirit Board of South Africa (WSB). WSB is also responsible for the Wine of Origin (WO) claims (origin, cultivar and vintage assurance) and the Estate Brandy Scheme [49]. The first IPW certifications started two years after the program's promulgation in 2000. The work developed by of the Agricultural Research Council (ARC) was used as the background of the program content. As the main agricultural research organisation in the country, the ARC has been conducting and funding research as well as disseminating information and education since its establishment in 1990 [50].

The Wine of Origin (WO) scheme was mandatory, highly regulated and regimented at the time IPW started. According to Ms Sue Birch, former head of Wines of South Africa (WOSA), WO was not itself a marketing message for the wines produced in South Africa. Likewise, IPW did not have a strong marketing direction or intent. The WOSA, in collaboration with IPW, was behind the introduction and design of a new seal "integrity and sustainability guaranteed" in 2010, incorporating the WO seal's attributes. The new seal ensures not only origin but also 100% certification under the IPW program. This seal is voluntary. The WO seal still exists, for wine being currently bottled from vintages previous to 2010 or for wines that failed to meet the IPW requirements or blended with uncertified grapes [51]. Although the original driver for the program was not retailers' demands the seal [49,50] added integrity and a clear message about the wines produced for the retailers [52].

The IPW program is based on two main documents: the guidelines and the manual. The guidelines present recommendations of what should be done, as well as minimum standards and the manual is a practical document showing the pathways for the implementation of the guidelines and completion of the self-assessment for further third party auditing and WSB certification. The wine labels drive the certification process. Wineries must be compliant with the guidelines, as well as 100% of the grapes used to produce the wine. Each bottle has a seal, which is uniquely numbered [53] which ensures integrity and traceability of the process at the consumers' level through the South African Wine Industry Information and Systems (SAWIS) website [53].

The farm/vineyard component of the IPW program consists of a set of guidelines focused on critical aspects for good agricultural practices related to grape production [54] and minimum compliance with the South African legislation (environmental related issues, food safety, labelling and social aspects). Farms with vineyards are verified annually through the completion of the self-assessment and require farm and production records. The guidelines and manuals are reviewed and updated bi-annually. The program assessment is compliant with FIVS and OIV.

Growers must reach 60% of the total points of the program to comply and become IPW certified. The self-assessment is undertaken on an annual basis and independently audited on a spot check basis [54]. Growers are allowed to score zero, two, three and five, in a scale of zero to five, for each criteria. Only auditors are allowed to score one or four. The self-assessment is then sent back through the online system with the pertinent documentation. One third of the wine producers are inspected annually, therefore all members are inspected in a three-year cycle [51]. The number of vineyard inspections will be driven by the origin of the grapes.

In South Africa, there are about 70 Producer Cellars, which are wineries that receive and process grapes on behalf of a group of wine grape growers [55]. Each Producer Cellar has an IPW coordinator, responsible for liaising with IPW and meeting the program requirements. The chief viticulturist usually fills this role. The Producer Cellars produce about 90% of the total wine in South Africa [50].

Environscientific, the auditing body for the WSB, conducts the audits and advises the WSB on who can/may be certified if/when found compliant. They are an independent group formed by scientists (with at least a Master's degree) with demonstrated field experience (at least 5 years). The auditors are not allowed to be involved in any agricultural products sales. Auditors are paid by IPW, unless the grower fails and needs to be re-inspected. In this case, the grower has to pay for the re-inspection. Additional supporting documentation is accepted by the auditing body after the initial audit, within a specific time frame, if this was the reason for failure. If inspected growers do not reach the pass mark by about 5%, a shorter re-audit can also be arranged. Unlike ISO14001 audits where auditors are not allowed to provide any advice, the IPW auditors point out pathways to reach the pass mark, provide information about minimum requirements of South African legislation used by the program assessment, share scientific knowledge and suggest training when the need is perceived. The core objective is to help growers to meet the requirements while ensuring credibility of the program [51]. The consultative audits, conducted as part of the South African IPW program, are one of the most complex and strict auditing processes of its kind.

In South Africa, two other schemes, created about 10 years ago, are also directly related to wine grape sustainability: Biodiversity and Wine Initiative (BWI) and the Wine and Agricultural Industry Ethical Trade Association (WIETA) Code. The first one is related to the conservation of the Cape Floral Kingdom (CFK), the richest and also the smallest plant kingdom on the planet [56] and WIETA is related to fair labour practices [57]. BWI requires IPW accreditation, as a condition to become part of the group. In the South African wine industry, "the ultimate goal is to have one seal, issued by the Wine and Spirit Board, that certifies the Wine of Origin information (vintage, date, variety), the environmental sustainability (IPW) and the ethical treatment of workers (WIETA)" [58].

3.5.1.6 CALIFORNIA SUSTAINABLE WINEGROWING ALLIANCE—CSWA/CALIFORNIA SUSTAINABLE WINEGROWING PROGRAM (SWP)

In the late 1990s a group of the wine industry executives determined that sustainability was one of the important issues that needed to be addressed by the wine industry in California [59]. Several sustainability initiatives were already in place in wine regions such as Lodi and the Central Coast in California. The California Sustainable Winegrowing Program (SWP) was originated in 2001 through a partnership of the Wine Institute and the California Association of Winegrape Growers. The California Sustainable Winegrowing Alliance (CSWA) was formed in 2003, a year after the first edition of the SWP workbook was published, with the objective to implement the SWP [60]. Part of the Lodi and Central Coast sustainability programs and other related regional and statewide efforts, were adapted and adopted by CSWA for use in its Code of Sustainable Winegrowing Program Self-Assessment Workbook [32]. The Certified California Sustainable Winegrowing program was launched in 2010 as a third-party certification to verify adoption of sustainable practices and continuous improvement. The SWP was developed as a statewide sustainability assessment program. It was felt that a state-wide program would create a common base for sustainability goals in the state and also promote sustainability of California vineyards and wineries as a group [59]. The statewide initiative also aimed to become an important educational channel for the wine industry providing an objective pathway to continuously improve organisations' sustainability through better operational and management practices.

The vision of the SWP is "the long-term sustainability of the California wine community". To place the concept of sustainability into the context of winegrowing, the program defines sustainable winegrowing as "growing and winemaking practices that are sensitive to the environment (Environmentally Sound), responsive to the needs and interests of society-at-large (Socially Equitable), and are economically feasible to implement and maintain (Economically Feasible)" [61]. The program's development is guided by values such as to: increase and optimise grape quality; protect and conserve the environment; maintain the long-term viability of agricultural lands and community; ensure economic and social wellbeing of farmers and employees; and support research and education among others.

Currently the program encompasses two sets of assessments: indicator collection and the self-assessment workbook. The workbook assessment data is publically reported in statewide sustainability reports presenting counts of responses as percent distribution of responses. These reports are available on line and are an indication of the Californian self-assessment results for the workbook topics. The indicators, called "Performance Metrics", are water use, energy use, greenhouse gas emissions and nitrogen use. At the time of publication reports on the performance metrics were still not available. However, benchmarks will be generated per acre and per ton of fruit production for vineyards, and per case for wineries, and will be available when they have sufficient data to produce statewide benchmarks. Most information regarding the assessment and administrative documentation is freely available on-line.

The current workbook version (3rd Edition) was released in January 2013. The online assessment is only available to California participants through a user name and password. The assessment topics are presented in increasing scenarios (options/categories) from 1 (least sustainable but within regulatory compliance, if regulations exist) to 4 (most sustainable). From the results of the self-assessment, growers are encouraged to produce an action plan to set their own sustainability goals for improvement. The workbook is available for sale through the website for non-participants. The key component of the engagement process for participants is education. More than four hundred seminars and workshops have been organized about vineyard and wineries issues throughout California to provide education on sustainability [59].

The program is described as having participants rather than members. Participants self-assess their operations and most of them report the results back to the CSWA to produce the program's sustainability reports emphasising strengths and weaknesses of the state. Most growers joined the SWP because of its educational benefits [59].

Many participants became certified to meet customer demand for sustainable certification including retailers, distributors, restaurants, and consumers. Independent third party auditors are accredited by the CSWA to conduct audits. CSWA has a mix of practice-based and process-based certification centred on the continuous Plan-Do-Control-Act (PDCA) developed from Deming's Plan-Do-Study-Act (PDSA or Shewhart) improvement cycle [62]. To retain certification status, growers must pass the initial certification audit, update the online self-assessment and action plan annually (with targets and times), and complete an annual audit. In most cases the on-site audits are on a three-year cycle so, in intermediate years, auditors review annual SWP assessment and action plan during an off-site audit. The certification program also includes a minimum of 50 vineyard and 32 winery prerequisites that must be achieved. There are minimum scores and rules for each one of these prerequisites [61].

3.5.1.7 VINEBALANCE, NEW YORK STATE'S SUSTAINABLE VITICULTURE PROGRAM AND LONG ISLAND SUSTAINABLE WINEGROWING

The VineBalance program, launched in 2004 was a result of a series of initiatives, initially driven by water quality concerns in both the Finger Lakes and Long Island [63]. In 1997, Dr Tim Martinson, from Cornell University became the local extension educator with the Finger Lakes Grape Program. He was tasked with writing an "Agricultural Environmental Management (AEM)" worksheet for grape growing. The AEM for grapes was inspired by the work developed by dairy farms in reservoir watersheds in state of New York, particularly around Keuka Lake.

Meanwhile, in 1992, Long Island grape growers where developing management guidelines to emphasize good stewardship practices for the region [64]. The assessment developed by Martinson as well as the LIVE and Lodi programs inspired the development of the guidelines. In 2004, the initiatives developed by Martinson and Long Island growers merged.

At that time, Martinson was approached by the National Grape Cooperative to develop a sustainable practices workbook for grapes. The National Grape Cooperative is a subsidiary of the Welch Food Inc., that represents about 1300 members; producers of grape juice and table grapes [65]. The National Grape Cooperative has adopted VineBalance as their production standard [66]. The project was funded by a larger grant from the New York Farm Viability Institute and resulted in the VineBalance program, a joint initiative of the Finger Lakes Grape Program, Lake Erie Regional Grape Program and Long Island's Grape Extension Program.

VineBalance was developed to answer industry groups' demand to develop "an outreach and educational program to promote the adoption of sustainable viticultural practices in New York State's vineyards". The grower self-assessment workbook sections were developed using materials from two previous programs in New York: NYS Agricultural Environmental Management (AEM) worksheets and the Long Island Sustainable Practices Workbook. During the winter of 2005–2006, a steering committee composed of extension, research, industry and growers' representatives from National Grape Cooperative, Centerra Wine Co., as well as

Finger Lakes and Long Island vineyards reviewed the original topics and added new content to the program to address the diversity of the distinct wine grape growing regions (Long Island, Lake Erie and Finger Lakes) in New York State. In 2006, 15 growers volunteered to become part of the pilot assessment using the new workbook. Feedback from growers was used to improve the content. In 2007, the VineBalance's New York Guide to Sustainable Viticulture Practices Grower Self-assessment Workbook was publically launched [67].

The workbook is the program's foundation. It has eight chapters containing 134 topics. Each topic has four options from 1 (most desired, most sustainable) to 4 (least desired, least sustainable) plus NA (non-applicable). Most questions have an explanatory section about the rationale used to develop the promoted practice plus additional resources for further education. It assesses a combination of specific production practices used to manage soil, vines, water, pest and disease and promotes education about sustainable options for improving growers' sustainability. The self-assessment is followed up by the production of individual action plans (with templates provided by the program) for the growers [67]. Martinson emphasises that the action plan, developed from self-assessment, is the key component to promote positive sustainability outcomes: "What should change? What can you afford to change?" [68].

Martinson defines himself as a "university extension person" and from their standpoint at the Cornell University; they felt that it was up to the industry groups to decide how they wanted to use the workbook developed by them to communicate with their consumers. For him, the success of VineBalance can be measured by the adoption of their assessment methodology by the industry. VineBalance does not have a certification scheme, as its main objectives are to educate and promote the adoption of sustainable practices. However two different groups, Welch's with the juice grape growers and Long Island Sustainable Winegrowing with high-end wine vineyards have adopted VineBalance using two different approaches. Since 2012, the Long Island Sustainable Winegrowing (LISW), a non-profit organisation started a certification process based on the VineBalance workbook [69]. The process was started by a group of four wineries, which worked with Alice Wise's team from the Cornell Cooperative Extension to write a specific code for certification [63]. Eleven growers were certified in the first year through independent third party audits [70].

The certification started in 2011. Initially, it only covered the Green chapter. By the end of 2012, the Red and Orange chapters were added to the certification process. Independent third party certifiers that are accredited by the program conduct the audits. A minimum three-month implementation period, prior to the first certification cycle is required. Certification must be renewed every two years. The process certifies the company's sustainable management from the winery's viewpoint, giving them the right to use the "Certified Sustainable Wine of Chile" seal. The certification has different rules depending on ownership, based upon the minimum percentage of total surface area included in the certification process, as the program distinguishes vineyards owned or long-termed leased (type A) by wineries and external vineyards (type B). The program stipulates a progressive increase in the proportion of total vineyard area under the certification process, required to reach certification [71].

3.5.1.8 WINES OF CHILE—SUSTAINABILITY PROGRAM

In 2009, the Wines of Chile, a non-profit organisation representing 95% of the bottled wine exported from Chile, released the Wines of Chile Strategic Plan 2020. The Plan points out sustainability as one of its key principles and empowered the Consorcio Tecnológico (Technological Consortium)—the technical arm of the industry—to develop a sustainability program. Retailers' demands for sustainability initiatives and certification were the main drivers for the creation of the program [72].

A joint project between industry representatives and the University of Talca started the development of the Sustainability Code. The Code became the foundation of the Wines of Chile Sustainability Program, which encompasses a series of initiatives, with the objective of establishing a sustainable wine industry in the country. The Code covers three areas: Vineyard (Green Area), Winery and Bottling plant (Red Area), and Social (Orange Area), and provides a checklist of control points and a compliance standard which establishes the requirements in the three areas. The green area focuses on natural resources, pest and disease, agrochemicals and job safety and has 18 critical points of assessment [71,73]. The red area contains chapters about energy, water management, contamination prevention and waste. Finally, the orange chapter considers all social issues and includes relationships with the workers, community, environment and clients.

The program accepts two levels of participation: Level 1 (training and education) and Level 2 (certification). The certification started in 2011. Initially, it only covered the Green chapter. By the end of 2012, the Red and Orange chapters were added to the certification process. Independent third party certifiers that are accredited by the program conduct the audits. A minimum three-month implementation period, prior to the first certification cycle is required. Certification must be renewed every two years. The process certifies the company's sustainable management from the winery's viewpoint, giving them the right to use the "Certified Sustainable Wine of Chile" seal. The certification has different rules depending on ownership, based upon the minimum percentage of total surface area included in the certification process, as the program distinguishes vineyards owned or long-termed leased (type A) by wineries and external vineyards (type B). The program stipulates a progressive increase in the proportion of total vineyard area under the certification process, required to reach certification [71].

3.5.1.9 MCLAREN VALE SUSTAINABLE WINEGROWING AUSTRALIA

The McLaren Vale Sustainable Winegrowing Australia (MVSWGA) program has its origins in the early 2000s. Since that time, the McLaren Vale Grape Wine and Tourism Association (MVGWTA) developed a series of viticultural initiatives with the objective to improve viticultural practices, fruit quality and financial viability in the region. These initiatives included seminars and workshops; a growers' bulletin (CropWatch) providing information from nine weather monitoring stations and pest and disease alerts for the region, research trials and information days. The Association also released a Financial Benchmark for McLaren Vale growers in 2005, and a Pest and Disease Code of Conduct in 2006, which was voluntarily endorsed by the growers in 2007. In this same year the Soil Management, Water Management and Preservation of Biodiversity Codes were also released.

The program creation was influenced by a visit from Ohmart (who developed Lodi Rules) to McLaren Vale in the mid-2000s. Ohmart's visit was hosted by Mr James Hook, who was then employed by the MVGWTA. In 2008, the then-Chair of the Association, Mr Dudley Brown, formalized the project with the argument that: "while this (all viticultural activities promoted by the MVGWTA) yielded great on farm results, we were unable to measure and discuss the outputs of our investment with ourselves or the outside world because we were not measuring the results". From this realisation, the Generational Farming program was born with the purpose of monitoring and measuring results and promoting best viticultural practices based on sound science. Mr Jock Harvey, local grower and a former Chair of the MVGWTA was the project leader with the goal of developing Hook's outline into a regional sustainability program, including a certification scheme.

Ms Jodie Pain took on the Viticultural Officer role at the Association in 2008 and continued to develop the project. Pain developed the assessment book with the voluntary assistance of a group of growers put together by Harvey. In 2009, Generational Farming was officially launched and an assessment book (workbook) was made available for the growers in the region. Hook continued to contribute to the program, authoring two of its six chapters. At the time, about 50 growers decided to self-assess their operations. By the end of 2010, the MVGWTA decided they needed an employee dedicated to the Generational Farming program. Viticulturist Ms Irina Santiago was hired as a part-time employee for this role. The data from 41 growers (representing 56 vineyard sites) were collected in 2011 and Santiago reviewed and revised the assessment methodology and developed a reporting system. The workbook was re-written by local growers and the program was re-named to McLaren Vale Sustainable Winegrowing Australia (MVSWGA) to be more easily found by others searching for their program. Volunteer workbook authors were growers with either extensive experience and/or formal education in viticulture.

The new method of assessment is similar to that of the Lodi and CSWA workbooks, in that, it replaces yes/no questions with scenario questions ranging from zero (explicitly unsustainable) to four (most sustainable) as well as non-applicable (NA). This methodology differs from the Lodi and CSWA workbook methods with the addition of having a "zero" scoring option. Growers from McLaren Vale followed the methodology to develop program content based on assessment topics that most impacted their sustainability. The content of the assessment is updated annually and peer-reviewed by independent experts, mostly from universities and governmental departments recognised globally for excellence in the relevant fields.

The MVSWGA method of assessment has three main principles: (1) assessment over time; (2) grower sustainability levels identified on a continuum and not on a pass/fail basis; (3) the assessment and reporting system must be useful for the grower to understand their sustainability status and be able to improve it. In contrast to the other certifications which have a single category of compliance, the MVSWGA places growers into four certification categories: category 1—red, needs attention; category 2—yellow, good; category 3—green, very good; and category 4—blue, excellent. The sustainability level is determined by attributing a weight to each topic, section and chapter from the assessment method. It is expected that very few growers can reach the blue level in the program. The program's content also changes annually to incorporate any relevant and commercially feasible scientific findings to the assessment. To continue in a certain category growers must update and improve their operations to align with the current content of the assessment of a certain category. This way, the workbook does not only show the pathway to improve sustainability in every assessment topic but also promotes continuing improvement through content update. The program assessment is compliant with FIVS and OIV [74].

Ten percent of program members are randomly selected annually and audited by a third-party. These audits are paid for by the MVGWTA, including the on-site inspections. Audits are in place to ensure credibility of the growers' sustainability levels based on their responses. There are specific rules and penalties that, in extreme cases, can lead to a member's exclusion in case of discrepancies between inspections and the self-assessment answers and data reporting. Audits are also available to members who wish to become certified. Certification audits are carried out every three years, whereas self-assessment, random inspection process and data reporting through the on-line system are annual. The online system uses

GPS coordinates to identify each vineyard block and relate it to the spray diaries. The reporting system cross-tabulates regional disease pressure on vineyards (i.e., spray targets) and chemical usage, as data is entered.

The MVSWGA program uses a systemic assessment that combines relevant indicators and best-practices and processes to indicate a clear pathway for growers to improve their sustainability at their own pace, using a triple bottom line approach (economic, environment and social).

3.5.2 COMPARISON OF PROGRAMS

The main characteristics of the most relevant sustainability programs for viticulture are displayed in Table 3. The programs from Chile, South Africa and New Zealand have a national scope while the others are regional. McLaren Vale Sustainable Winegrowing Australia is the only program that has its scope limited to a single wine region. All the other regional programs have at least a statewide scope. Lodi is the pioneer of sustainability initiatives among all programs and LIVE conducts the oldest certification scheme. VineBalance is the only program that does not hold a certification scheme but other initiatives that do lead to certification (such as newly started Long Island Sustainable Program) were derived from it.

Program website addresses, number of members, certified vineyards and area they represent is also shown in Table 3. All programs are voluntary. However, New Zealand Wine has a quasi-compulsory situation [42], as to be included in international marketing, promotional and awards events, wines from vintage 2010 onwards must be certified. This creates and helps explain the strong adoption rate (90%) in New Zealand.

Of the programs reported in this paper, South Africa's program is the only one that is regulated. South Africa's WO scheme was already in place and mandatory when the IPW started. When the IPW and WO scheme merged, IPW embraced the traceability and integrity features of WO and added legislation compliance and sustainability topics to the scheme/ assessment. Wine grape growers in South Africa seem to be extremely conscious about the importance of preserving and conserving the natural resources of the country as well being able to ensure to (predominantly) international consumers that they are "doing the right thing". Because of the level of organisation of the wine industry, IPW also became one of the tools to enforce South African legislation for farms and wineries. In spite of not being mandatory, the big cooperatives, largely, only buy grapes that are certified. Also, it is currently difficult to sell wines from South Africa without the IPW seal. All of these factors helped the broad adoption (92%) of IPW by winegrowers.

The assessment types used by the sustainability programs for viticulture are not completely comparable against each other. However, even when it is not explicitly stated in a program's literature for "assessment topics and content" (e.g., IPW), it was apparent to researcher observation and through discussion regarding tacit embedded assumptions in articulated goals, that all programs embrace, to a certain extent, the triple bottom line approach (economic, environmental and social). For instance, in South Africa (IPW), an embedded protection system for chemical operators that is not explicated in the assessment literature is assumed knowledge and is fully assessed during the audits. Similarly, in South Africa, there is a stated intent for future purposes (see Section 5.1.5) to integrate the separate programs that will make these tacit assumptions more explicit.

Table 4 summarizes program content and the number of assessment topics of programs and/or certification. The economic component is directly evaluated by Lodi, SIP and SWP in their business management chapters. All the best practice-based programs assess economic sustainability through the analysis of the adoption rate of best practices and the potential to reduce costs by optimizing resource use and fruit quality. Economic sustainability is understood through a diverse correlation of data from best-practices benchmarks.

Many programs analyse economic sustainability using value indicators to understand measurable outcomes (e.g., yield/area and inputs/area). MVSWGA reports this type of data and SWP is currently developing such correlations. Economic sustainability is also evaluated through regional socio-economic indicators such as average grape price per ton per variety, land price, wine bottle prices from the region, new planting areas, planted area, longevity of current vineyards, and similar measures. The governments of Australia, New Zealand, United States and South Africa collect and publish this sort of data which is used by the programs.

Programs located in countries with no specific chemical usage legislation (allowed, restricted and prohibited inputs as well as withholding periods) for wine grape growing, usually developed their own chemical list. These lists, in turn, became a requirement to meet their sustainability standards. Australia seems to have one of the strictest chemical usage legislation specific to vineyards and based on a sum of all export markets requirements. "Agrochemicals Registered for Use in Australian Viticulture" [75] is published annually by the Australian Wine Research Institute (AWRI) and distributed for free for all members of the wine industry and also as an insert in a trade magazine. This resource is also publically available online [75].

All programs use best-practice assessments, however, they use it in different combinations, as demonstrated in Table 1. While all programs have sustainability focus, motivations and outcomes vary. Some are driven primarily by education, others by certification. Most have a combination of these two goals in some form. For instance, VineBalance is a program with no certification, but Long Island adopted the VineBalance program and standards and used them to develop a certification scheme. IPW, LIVE and SWNZ are certification schemes in nature but enable participation, education and support without certification for wine growers who are unable to meet (or are in process of meeting) certification standards. Some programs may or may not lead to certification by the individual wine grower's choice—CSWG, Lodi, SWC and Vineyard Team. These programs have a certification scheme in place but certification is independent of sustainability program participation. Similarly, MVSWGA also has optional certification. However, all members (certified or not) are subject to random third-party audit to validate responses from self-assessment. MVSWGA is the only program that embraces four different levels of sustainability certification.

Overall, the idea of creating each one of these programs came from a group of progressive/innovative growers who were aware of the need for operational improvement of their activities. Certifications were developed to ensure external credibility (marketing) of what was happening in their vineyards. For those programs that developed sustainability assessment for wineries, the main driver was to communicate (or, market) the sustainability message in a more systemic way. Increasing environmental concerns, as expressed through large retailers' demands, were also taken into consideration. It is important to raise the issue that certification only attests to compliance with the standard of a nominated program, and makes no claims to individual standards of vineyards that have chosen not to participate or certify.

The sustainability programs that are currently in place, with great membership uptake, have a strongly motivated and technical manager with a powerful interpersonal network. The importance of the interpersonal networks in innovation adoption (sustainable practices in the context of this investigation) is exhaustively discussed by Rogers [76]. This is the case in New Zealand, Lodi, McLaren Vale and Oregon for instance. These program managers seem to be strongly supported by the wine grower community. It was observed that this situation seemed to be driven by growers' perception of their program managers' strong technical skills in viticulture and ability to manage all aspects of the program. The important role of these program managers can also be perceived as their greatest weakness, as it is uncertain the direction such programs will take when these manager are not in their roles anymore. Succession planning was never discussed during this study, so retirement or withdrawal of entrepreneurial and driving leaders is likely to have a severe negative effect on program performance, similar to the negative effects of a lack of succession planning in private businesses [77,78].

3.5.3 CREATION OF SUSTAINABILITY ASSESSMENT PROGRAMS IN VITICULTURE: ENGAGEMENT PROCESSES; ENABLING AND INHIBITING FACTORS

Third order analysis of the focus group discussions used a qualitatively analysed and content-analysis driven visualisation that resulted in four tag clouds that were subsequently analysed in three segments that are described in the following sections: benefits; inhibiting factors; and engagement process.

3.5.3.1 BENEFITS (QUESTION 1 FROM FOCUS GROUP)

Eighty-three top-level managers from the wine industry from five countries were asked in 14 focus group sessions about the potential benefits to induce their participation in a sustainability program. They were also asked to list specific benefits they would expect to receive from a chosen program. The qualitative analysis of the transcripts shows that the educational aspect is the most important benefit gained by participants in sustainability programs and one of the core reasons for participation. Education was expressed as an objective opportunity to self-improve (Figure 1a). According to participants, education is the main consequence of the sustainability self-assessment and benchmarks derived from the collection of their peers' results as well as interaction with peers and training promoted by the program's management. This result endorses the viewpoint of Ohmart who emphasises the improvement opportunity growers receive by just being part of sustainability programs [34]. All programs listed in this article had origins directly related to the need to promote operational improvement in their vineyards.

There was a limitation in the analysis software that originally made it difficult to align quantitative results with the qualitative results. This was significant in the results on the topic "education" as seen in Figure 1a, which displays the tag cloud created in Nvivo10, from Stage 3, question 1 of the focus group discussions. The term "education" only appears as the 32nd most recurrent word in the analysis of all transcripts and is displayed in very small font on the bottom left of the tag-cloud—a result inconsistent with the clearly established importance of education in all other findings, which is why the qualitative results are critical in understanding the focus group results. Content analysis methods and the role of the investigator in making analytical choices to produce meaningful results have been widely studied in the academic literature [79,80,81]. If a qualitative analysis of the transcript was not conducted, and the results relied solely on the tag-cloud interpretation, it could lead to a misinterpretation of the results.

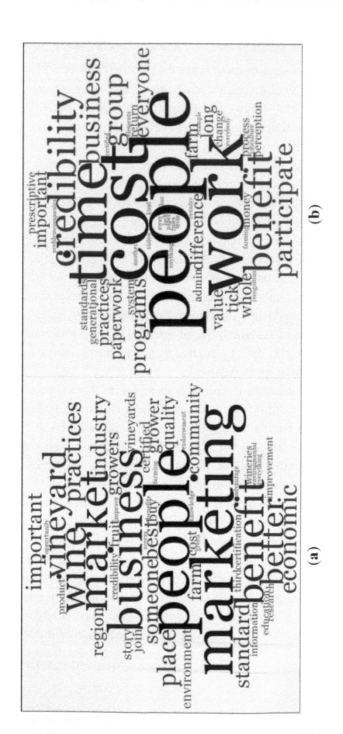

FIGURE 1: (a) Benefits and (b) inhibiting factors for growers' participation in wine growing sustainability programs.

The qualitative analysis of the transcripts showed that many displayed words were directly expressing educational aspects, such as "better", "practices", "standards", "vineyard", "improve", "improvement", "information", "knowledge", "technical", "benchmarks", among others—giving us a deeper and richer insight into the importance of education to the participants of this study. Each content analysis must be seen as a unique situation, and tag-clouds are still powerful displays of contents [28,29,82], so are used here to display these results.

The term "people" is the most recurrent term followed by "marketing". In the context of the interviews, "people" represents program managers, program peers and community members where vineyards are located as well as consumers. This result seems quite understandable as programs are created for people (in the context of this investigation, wine growers) by people to promote sustainability in vineyards. Sustainability encompasses the economic (business, quality, practices, product, fruit, buy, wineries, amongst others), environmental and social (community, region, people) components. Marketing helps to create the bridge between the winegrowing/making processes with the external world, to assure consumers. The qualitative analysis also pointed out marketing as the second most expected benefit. The terms "marketing", "market", "wine", "certified", "certification", "endorsement", "buy", "credibility" and "story" are intrinsically related to public external validation, and therefore the accountability of sustainability assessments. All the terms displayed in the tag-cloud are intrinsically related, intersecting many of the benefits listed by participants. All these tag-cloud results align consistently with, and highlight the key factors brought out in the qualitative analysis—that educational focus, people, and marketing are core drivers and benefits for participation in sustainability programs.

3.5.3.2 INHIBITING FACTORS (QUESTION 2 FROM FOCUS GROUP)

The inhibiting factors for participation are presented in Figure 1b. "People" is displayed as the most import factor to drive wine growers away from a sustainability program. In this context, "people" represented two very specific situations directly related to the credibility of the program: (1) the program managers, if they are seen as someone that lacks appropriate background or experience to run the program, or be able to interact or provide any technical benefit for the wine growing community; and (2) program peers, when they are not really sustainable but try to use the (good) perceived image of other members to increase their own value, compromising the credibility of the group as a whole. Participant 56 says: "I could be practicing a very high level of sustainability and because I'm practicing in a very high level and someone is just saying they're sustainable…they're benefiting from the kind of practices that I'm utilising but they are not really doing anything…" Participant 40, from another group says: "Accreditation (to a program) is often a risk…you have people who abuse the system and if there is a scandal involved with accreditation then you're all painted with the same brush…in our business, we try as much as possible to put maximum effort in between all accreditations so that our customers will entrust (sic) us rather than our accreditations" [83].

In summary, the other inhibiting factors are cost; time consuming paperwork; lack of appropriateness (and a low bar) of the assessment; lack of useful information provided back to growers; absence of business improvement or marketing benefit, programs that are too prescriptive; and confusion between sustainability and farming system choices (e.g., organic). There are many pathways to achieve sustainability in winegrowing, which is a context-dependent situation. Not all innovations (in this context, sustainable practices) are desirable for all situations. In agriculture, for instance, the needs and reality of small-sized organisations differ greatly to the ones from large commercial farms [76]. We suggest that the role of sustainability programs should not be telling growers how to grow grapes but contribute to their education to help them to optimize quality and costs, comply with legislations, minimize impacts on environment and ensure a healthy working environment for employees.

The prescriptive factor that might inhibit growers to become part of the sustainability program was discussed by Andrew Jefford in the Decanter Magazine, when writing about a visit to a traditional French vineyard where the owner, Jean Orliac, expresses deep dissatisfaction about certifications in general: "For us, agriculture is an art modeste, needing lots of experience and reflection… The role of the winegrower is to be in some sense a free man." Jefford states: "a small-scale, independent winegrower in almost any country on earth is an unusually free individual—meaning that, once debt is repaid, they are economically beholden only to themselves, and that their work involves making decisions (key to existential notions of freedom) rather than conforming to a pattern of behaviour acquired from or imposed by others" [83].

3.5.3.3 ENGAGEMENT PROCESS (QUESTION 3 FOCUS GROUP)

Just as the major strengths in programs and leaders tend to be the greatest weaknesses, in the same way, the factors that promote sustainability programs and participation are also the same factors that inhibit wine growers participation. The balance between cost and time vs. benefits and credibility will drive wine growers' participation in sustainability assessments. Figure 2 shows the tag-cloud created from the third question of this section of the discussion. For example, when growers were asked about the strategies they would use if in charge of engaging other growers to become part of a sustainability program, the majority of the growers referred to the benefits they had listed before. Additionally, they would emphasise success stories from members involved with sustainability programs.

Many participants mentioned that current members should be the main focus of the program management. Participants pointed out that their stories and the changes promoted by adoption of sustainable practices would drive the engagement of new members. Furthermore, it was emphasised that it was important to demonstrate that the group is stronger than individual growers. For instance, among the economic benefits, accreditation of the program with wineries and retailers and a consequent payment of a bonus price for the grapes would contribute to membership uptake.

3.5.3.4 REPORTING AND SPONSORSHIPS (QUESTIONS 4 AND 5)

When asked about how the program should report results to obtain funding from external sponsors, including government(s), "benefits to people" was the most important outcome that should be demonstrated through measurements (Figure 3). Not only benefits to the growers themselves, thorough the perpetuity of their businesses, but especially social benefits promoted by the program to their employees and community. The direct and indirect social-economic impact of the grape-growing activity as well as the benefits to the environment should be used as the main reasons to attract sponsorship. According to participants, the benefits to the environment could be demonstrated through preservation and conservation actions, including water as well as chemical reduction from vineyard management practices improvement. Program popularity among growers, measured through membership in relation to the total number of growers, should also be taken into consideration when approaching sponsors for the program. When programs are voluntary, membership and acreage numbers seem to be the most direct measurement of program relevance for growers.

3.6 CONCLUSIONS

Most research on sustainability emphasises the environmental impacts of productive processes. However, environmental issues were not the main drivers for the conception of sustainability assessment programs for viticulture. The environmental aspect is incontestably important and all programs have embraced environmental sustainability as part of their assessments. Nevertheless, successful programs like those described in this study have been created to increase growers' overall sustainability, mainly through the direct and indirect education they promote and the overall economic benefit to their business caused by overall improvement of their operations. The universities involved played an essential role in the development of these programs aiming to improve grape growers' sustainability. Ultimately, viticultural research should be driven by the need to keep the wine industry alive, over time, or in other words sustainable.

This study lays the foundation for multiple avenues of future research. A deliberately imposed limitation of this study included the exclusion of "old world" wine growers as there was no known relevant sustainability assessment program for viticulture at the individual organisational level. Therefore, there is clearly a need for investigation into issues of "old world" wine growing sustainability.

FIGURE 2: Engagement process for growers' participation in wine growing sustainability programs.

FIGURE 3: Suggested results reported by wine grape growers to obtain funding for wine growing sustainability programs.

Opportunities to further develop the research reported here exist in two dimensions: linearly within the wine industry chain or laterally within other agricultural pursuits or other fields. The linear dimension could be developed to further look mainly at: (1) the impact and usefulness of sustainability programs on improving growers' sustainability; and (2) the impact at the farm level of the increasing demand for sustainable schemes by retailers as a requirement for wine purchase and exports. The lateral dimension could be similarly developed as the linear dimension, but from the development of a sustainability definition and elaboration of appropriate and meaningful indicators in other agricultural crops or other fields where sustainability programs are being implemented or reviewed. These two dimensions would contribute to the sustainability of world agriculture.

Agriculture is the cultivation and harvesting of crops [84] and the primary purpose of agriculture is to meet the demand for agricultural products, mainly food, but also raw materials for fibre production [85] to maintain and enrich life. In the context of this study, the purpose of wine grape growers is to produce grapes to produce wines and to do so sustainably; i.e., "be able to economically provide for the farmer while maintaining its ability to consistently produce and improve quality over time" [9]. The main finding from this study is threefold: that the success of each of these programs is largely due to the people driving the programs (program managers, innovative growers and/or early adopters); the way these people communicate and engage with their stakeholders and peers; and the usefulness of the developed program to improve sustainability. This is consistent with the findings from a study conducted in 2009 by Gabzdylova et al. [86] in New Zealand, which suggested that "people" is one of the main drivers of sustainability initiatives in the wine industry.

Sustainability assessment programs in viticulture only make sense if they are useful to help growers to improve their sustainability in the context of the community and environment in which they are located.

REFERENCES

1. Shen, L.; Kyllo, J.M.; Guo, X. An integrated model based on a hierarchical indices system for monitoring and evaluating urban sustainability. Sustainability 2013, 5, 524–559, doi:10.3390/su5020524.

2. World Comission on Environment and Development. Report "Our Common Future"; United Nations: Tokyo, Japan, 1987.
3. Szolnoki, G. A cross-national comparison of sustainability in the wine industry. J. Clean. Prod. 2013, 53, 243–251, doi:10.1016/j.jclepro.2013.03.045.
4. Forbes, S.L.; Cohen, D.A.; Cullen, R.; Wratten, S.D.; Fountain, J. Consumer attitudes regarding environmentally sustainable wine: An exploratory study of the new zealand marketplace. J. Clean. Prod. 2009, 17, 1195–1199, doi:10.1016/j.jclepro.2009.04.008.
5. Jackson, R.S. Wine Science: Principles and Applications, 3rd ed. ed.; Academic Press: Waltham, MA, USA, 2008; p. 751.
6. Elkington, J. Partnerships from cannibals with forks: The triple bottom line of 21st-century business. Environ. Qual. Manag. 1998, 8, 37–51, doi:10.1002/tqem.3310080106.
7. Messick, S. The Once and Future Issues of Validity: Assessing the Meaning and Consequences of Measurement; Educational Testing Service: Lawrence Township, NJ, USA, 1988; Volume 33, p. 45.
8. Hansen, J.W. Is agricultural sustainability a useful concept? Agric. Syst. 1996, 50, 117–143, doi:10.1016/0308-521X(95)00011-S.
9. Santiago-Brown, I.; Jerram, C.; Metcalfe, A.; Collins, C. What does sustainability mean? Knowledge gleaned from applying mixed methods research to wine grape growing. J. Mixed Methods Res. 2014. in press.
10. Santiago, I.; Metcalfe, A.; Jerram, C.; Collins, C. Economic, environmental and social indicators to assess sustainability of individual agricultural systems: A wine grape growing case studyUnpublished work.
11. Bragança, L.; Mateus, R.; Koukkari, H. Building sustainability assessment. Sustainability 2010, 2, 2010–2023, doi:10.3390/su2072010.
12. Reig-Martínez, E.; Gómez-Limón, J.A.; Picazo-Tadeo, A.J. Ranking farms with a composite indicator of sustainability. Agric. Econ. 2011, 42, 561–575, doi:10.1111/j.1574-0862.2011.00536.x.
13. Rametsteiner, E.; Pülzl, H.; Alkan-Olsson, J.; Frederiksen, P. Sustainability indicator development—Science or political negotiation? Ecol. Indic. 2011, 11, 61–70, doi:10.1016/j.ecolind.2009.06.009.
14. Font, X. Environmental certification in tourism and hospitality: Progress, process and prospects. Tourism Manag. 2002, 23, 197–205, doi:10.1016/S0261-5177(01)00084-X.
15. Sainsbury, K.; Sumaila, U.R. Incorporating ecosystem objectives into management of sustainable marine fisheries, including"best practice"reference points and use of marine protected areas. In Responsible Fisheries in the Marine Ecosystem; Sinclair, M., Valdimarsson, G., Eds.; Food and Agriculture Organization (FAO): Rome, Italy, 2003; p. 343.
16. Haynes, J.; Cubbage, F.; Mercer, E.; Sills, E. The search for value and meaning in the cocoa supply chain in costa rica. Sustainability 2012, 4, 1466–1487, doi:10.3390/su4071466.
17. Cary, J.; Roberts, A. The limitations of environmental management systems in australian agriculture. J. Environ. Manag. 2011, 92, 878–885, doi:10.1016/j.jenvman.2010.10.055.

18. Gunningham, N. Incentives to improve farm management: Ems, supply-chains and civil society. J. Environ. Manag. 2007, 82, 302–310, doi:10.1016/j.jenvman.2005.06.016.

19. International Organization for Standardization ISO 14000—Environmental management. Available online: http://www.iso.org/iso/iso14000 (accessed on 26 February 2014).

20. Bond, A.; Morrison-Saunders, A.; Pope, J. Sustainability assessment: The state of the art. Impact Assess. Proj. Apprais. 2012, 30, 53–62, doi:10.1080/14615517.2012.661974.

21. Landers, A. Sustainable plant protection within sustainable fruit growing-an american experience. Agric. Eng. Int. CIGR J. 2008, 8, 1–12.

22. Gibson, R.B. Sustainability assessment: Basic components of a practical approach. Impact Assess. Proj. Apprais. 2006, 24, 170–182, doi:10.3152/147154606781765147.

23. Halog, A.; Manik, Y. Advancing integrated systems modelling framework for life cycle sustainability assessment. Sustainability 2011, 3, 469–499, doi:10.3390/su3020469.

24. Van der Werf, H.M.G.; Petit, J. Evaluation of the environmental impact of agriculture at the farm level: A comparison and analysis of 12 indicator-based methods. Agric. Ecosyst. Environ. 2002, 93, 131–145, doi:10.1016/S0167-8809(01)00354-1.

25. Kajikawa, Y.; Inoue, T.; Goh, T.N. Analysis of building environment assessment frameworks and their implications for sustainability indicators. Sustain. Sci. 2011, 6, 233–246, doi:10.1007/s11625-011-0131-7.

26. Miles, M.B.; Huberman, A.M. Qualitative Data Analysis: An Expanded Sourcebook, 2nd ed. ed.; Sage Publications: Thousand Oks, CA, USA, 1994.

27. QSR International. Nvivo Qualitative Data Analyis Software, 10; QSR International Pty Ltd: Doncaster, Vic, Australia, 2012. Available online: http://www.qsrinternational.com (accessed on 26 February 2014).

28. Sinclair, J.; Cardew-Hall, M. The folksonomy tag cloud: When is it useful? J. Inf. Sci. 2008, 34, 15–29, doi:10.1177/0165551506078083.

29. Bateman, S.; Gutwin, C.; Nacenta, M. Seeing things in the clouds: The effect of visual features on tag cloud selections. In Proceedings of the Nineteenth ACM Conference on Hypertext and Hypermedia, Pittsburgh, PA, USA, 19–21 June 2008; pp. 193–202.

30. Shah, A.K.; Oppenheimer, D.M. Heuristics made easy: An effort-reduction framework. Psychol. Bull. 2008, 134, 207–222, doi:10.1037/0033-2909.134.2.207.

31. Hoffman, M. Keeping the Wineglass Full: Exploring the Intersection of Sustainable Viticulture and Sustaining Winegrower Legacy in Lodi, California. Master Thesis, University of Iowa, Iowa City, IA, USA, 2009.

32. Zucca, G.; Smith, D.E.; Mitry, D.J. Sustainable viticulture and winery practices in California: What is it, and do customers care? Int. J. Wine Res. 2009, 2, 189–194.

33. Ohmart, C. Innovative outreach increases adoption of sustainable winegrowing practices in Lodi region. Calif. Agric. 2008, 62, 142–147, doi:10.3733/ca.v062n04p142.

34. Ohmart, C.; SureHarvest, Davis, CA, USA.. In-person interview by I. Santiago-Brown 2012.

35. Lodi Winegrape Comission Lodi rules for sustainable winegrowing. Available online: http://www.lodiwine.com/certified-green (accessed on 26 February 2014).

36. Ohmart, C. Lodi rules certified wines enter the marketplace. Pract. Winery Vineyard 2008, 29, 32–42.

37. Van der Zijpp, S.; SWNZ, Blenheim, New Zealand.. In-person interview by I. Santiago-Brown 2012.

38. Hughey, K.F.D.; Tait, S.V.; O'Connell, M.J. Qualitative evaluation of three 'environmental management systems' in the new zealand wine industry. J. Clean. Prod. 2004, 13, 1175–1187.

39. New Zealand Wines Sustainability. Available online: http://www.nzwine.com/sustainability/ (accessed on 26 February 2014).

40. Renton, T.; Manktelow, D.; Kingston, C. Sustainable winegrowing: New Zealand's place in the world. In Proceedings of Romeo Bragato Conference, Christchurch Convention Centre, Christchurch, New Zealand, 12–14 September 2002; p. 7.

41. Sinha, P.; Akoorie, M.E.M. Sustainable environmental practices in the New Zealand wine industry: An analysis of perceived institutional pressures and the role of exports. J. Asia Pac. Bus. 2010, 11, 50–74, doi:10.1080/10599230903520186.

42. Flint, D.J.; Golicic, S.L. Searching for competitive advantage through sustainability: A qualitative study in the new zealand wine industry. Int. J. Phys. Distrib. Logist. Manag. 2009, 39, 841–860, doi:10.1108/09600030911011441.

43. CCVT Central coast vineyard team. Available online: http://www.vineyardteam.org (accessed on 26 February 2014).

44. Vukmanic-Lopez, B.; SIP, Atascadero, CA, USA.. Email interview by I. Santiago-Brown 2014.

45. SIP Sustainability in practice (SIP) certification program. Available online: http://www.sipcertified.org (accessed on 26 February 2014).

46. Macdonald, A.; Northwest Viticulture Center, Salem, MA, USA.. In-person interview by I. Santiago-Brown 2012.

47. LIVE Low input viticulture and enology. Available online: http://liveinc.org (accessed on 26 February 2014).

48. Serra, C.; LIVE, Portland, OR, USA.. In-person interview by I. Santiago-Brown 2012.

49. Rossouw, J.; Distell, Cape Town, South Africa.. In-person interview by I. Santiago-Brown 2012.

50. Schietekat, D.; IPW, Stellenbosch, South Africa.. In-person interview by I. Santiago-Brown 2012.

51. Van Schoor, L.; Environscientific, Stellenbosch, South Africa.. In-person interview by I. Santiago-Brown 2012.

52. Birch, S.; WOSA, Stellenbosch, South Africa.. In-person interview by I. Santiago-Brown 2012.

53. SAWIS. South african wine industry information & systems. Available online: http://www.sawis.co.za (accessed on 26 February 2014).

54. IPW Integrated production of wine (IPW). Available online: http://www.ipw.co.za/ (accessed on 26 February 2014).

55. SAWIS. Sa wine industry statistics NR 37. Available online: http://www.sawis.co.za/info/download/Book_2013_eng_web.pdf (accessed on 26 February 2014).

56. WWF South Africa Biodiverstiy and wine. Available online: http://www.wwf.org. za/what_we_do/sustainable_agriculture_/biodiversity___wine_initiative/ (accessed on 26 February 2014).

57. WIETA. Wieta certified fair labour practice wines. Available online: http://www. wieta.org.za/wieta_wine_brands.php (accessed on 26 February 2014).

58. WOSA. Wines of south africa. Available online: http://www.wosa.co.za/ (accessed on 26 February 2014).

59. Jordan, A.; Francioni, L.; CSWA, San Francisco, CA, USA.. In-person interview by I. Santiago-Brown 2012.

60. California Sustainable Winegrowing Alliance California wine community sustainability report executive summary. Available online: http://www.sustainablewinegrowing.org/docs/cswa_2004_report_executive_summary.pdf (accessed on 26 February 2014).

61. CSWA. Sustainable winegrowing program. Available online: http://www.sustainablewinegrowing.org/sustainable_winegrowing_program.php (accessed on 26 February 2014).

62. Deming, W.E. Out of Crisis; MIT Press: Cambridge, MA, USA, 2000.

63. Cattell, H. Long island sustainable winegrowing debuts. Available online: http:// www.winesandvines.com/template.cfm?section=news&content=100062 (accessed on 26 February 2014).

64. Moreno-Lacalle, J. Long island sustainable winegrowing: The road to certification. Available online: http://blogwine.riversrunby.net/long-island-sustainable-winegrowing-the-road-to-certification (accessed on 26 February 2014).

65. Amanor-Boadu, V.; Boland, M.; Barton, D.; Anderson, B.; Henehan, B. Welch Foods, Inc. And National Grape Cooperative Association. Agribusiness Case Studies; Arthur Capper Cooperative Center: Manhattan, KS, USA, 2003. Available online: http://www.agmanager.info/agribus/research/case_studies/Welch%20Foods. pdf (accessed on 26 February 2014).

66. Martinson, T.; Cornell University, Ithaca, NY, USA.. Email interview by I. Santiago-Brown 2014.

67. VineBalance New York guide to sustainable viticultural practices. Available online: http://www.vinebalance.com (accessed on 26 February 2014).

68. Pirro, J.F. The vinebalance program: Working toward sustainability viticulture in the northeast. Available online: http://www.growingmagazine.com/article-9366.aspx (accessed on 26 February 2014).

69. Goldberg, H.G. Certifying good growers. Available online: http://www.nytimes. com/2012/05/13/nyregion/certifying-the-east-ends-good-wine-growers.html?_ r=4& (accessed on 26 February 2014).

70. Nigro, D. Long island wineries set out to prove their green cred. Available online: http:// www.winespectator.com/webfeature/show/id/46836 (accessed on 26 February 2014).

71. I+D Vinos de Chile Consortium Sustainability code of the chilean wine industry. Available online: http://www.sustentavid.org/en (accessed on 26 February 2014).

72. Carbonell, C.; SWC, Santiago, Chile.. In-person interview by I. Santiago-Brown 2013.

73. WOC. Wines of chile sustainability program. Available online: http://www.winesof-chile.org (accessed on 26 February 2014).
74. Hayes, P. Foreword. In McLaren Vale Sustainable Winegrowing Australia; McLaren Vale Grape Wine and Tourism Association: Mclaren Vale, Austrilia, 2012; p. vii.
75. Australian Wine Research Institute (AWRI). Agrochemicals Registered for Use in Australian Viticulture ("Dog Book"); AWRI: Glen Osmond, Australia, 2014; Volume 2013–2014.
76. Rogers, E.M. Diffusion of Innovations, 5th ed. ed.; Free Press: New York, NY, USA, 2003.
77. Trow, D.B. Executive succession in small companies. Admin. Sci. Q. 1961, 6, 228–239, doi:10.2307/2390756.
78. Gilding, M.; Gregory, S.; Cosson, B. Motives and outcomes in family business succession planning. Entrepren. Theor. Pract. 2013, doi:10.1111/etap.12040.
79. Luyt, R. A framework for mixing methods in quantitative measurement development, validation, and revision: A case study. J. Mixed Methods Res. 2012, 6, 294–316, doi:10.1177/1558689811427912.
80. Onwuegbuzie, A.J.; Dickinson, W.B.; Leech, N.L.; Zoran, A.G. Toward more rigor in focus group research: A new framework for collecting and analyzing focus group data. Int. J. Qual. Meth. 2009, 8, 1–21.
81. Krippendorff, K. Reliability in content analysis. Hum. Commun. Res. 2004, 30, 411–433.
82. Kaser, O.; Lemire, D. Tag-cloud drawing: Algorithms for cloud visualization. Available online: http://arxiv.org/pdf/cs/0703109v2.pdf (accessed on 26 February 2014).
83. Jefford, A. The existential winegrower. Available online: http://www.decanter.com/news/blogs/expert/584657/jefford-on-monday-the-existential-winegrower-Sl-4DOwhl8ys6MuQB.99 (accessed on 26 February 2014).
84. International Labour Organization. Safety and Health in Agriculture; International Labour Organization: Geneva, Switzerland, 1999. Report VI (1).
85. European Environamental Agency. 22. Agriculture. Available online: http://www.eea.europa.eu/publications/92-826-5409-5/page022new.html (accessed on 26 February 2014).
86. Gabzdylova, B.; Raffensperger, J.F.; Castka, P. Sustainability in the new zealand wine industry: Drivers, stakeholders and practices. J. Clean. Prod. 2009, 17, 992–998, doi:10.1016/j.jclepro.2009.02.015.

Tables 3 and 4 are not available in this version of the article. To view this additional information, please use the citation on the first page of this chapter.

PART II

ELEMENTS OF SUSTAINABLE VITICULTURE: FROM LAND AND WATER USE TO DISEASE MANAGEMENT

CHAPTER 4

ADOPTION OF ENVIRONMENTAL INNOVATIONS: ANALYSIS FROM THE WAIPARA WINE INDUSTRY

SHARON L. FORBES, ROSS CULLEN, AND RACHEL GROUT

4.1 INTRODUCTION

Feder and Umali (1993) noted that as environmental issues in agricultural businesses have gained attention, increasing focus is being applied to examining the adoption of environmental innovations. Clearly the need for the adoption of environmental practices and innovations is greater than ever, especially in agriculture where the level and severity of environmental problems continue to rise. Conventional wine production practices result in similar environmental issues to those incurred in other agricultural businesses, including groundwater depletion, water pollution, effluent run-off, toxicity of pesticides, fungicide and herbicide use, habitat destruction, and loss of natural biodiversity. This study adds to the current knowledge regarding the adoption of environmental innovations, specifically in the wine industry.

Waipara is a rapidly growing wine region located north of Christchurch on New Zealand's South Island. The Greening Waipara Project began in

Adoption of Environmental Innovations: Analysis from the Waipara Wine Industry. Forbes SL, Cullen R, Grout R. Wine Economics and Policy **2**,1 *(2013), DOI: 10.1016/j.wep.2013.02.001. Reprinted with permission from the authors.*

2005 and around 32 of the Valley's vineyards and wineries are now participating. The Project stemmed from initiatives by Lincoln University's Bio-Protection Research Centre, the Waipara Valley Winegrowers Association, the Hurunui District Council and Landcare Research to make use of 'nature's free services'. In addition, the Greening Waipara Project was initiated because the Waipara wine region is less well known than other high profile wine regions within New Zealand and one aim was to give Waipara wines a clear point of difference. The Project has developed and introduced seven environmental innovations that could be implemented by wine companies in the Waipara region. These practices are based on utilising nature's services in areas including pollination, biological control of pests, weed suppression, improved soil quality, filtering of wastes and conservation of native species. The Project has issued brochures which claim that the adoption of the practices will reduce agrichemical and labour costs, support eco-tourism, and help with the marketing of Waipara wines.

This paper examines how many of the vineyards and wineries have adopted the innovations and the implications of the implemented practices in terms of business costs and benefits. The remainder of this paper is structured as follows. Firstly, details are provided of the seven environmental innovations developed by the Greening Waipara Project. A review of the environmental innovation literature includes both a focus on agriculture in general and a specific focus on the wine industry. Details of the research method adopted in this study then follow. Presentation of the results is followed by the discussion and conclusions.

4.2 GREENING WAIPARA ENVIRONMENTAL INNOVATIONS

The Greening Waipara Project developed and introduced a total of seven environmental innovations for Waipara vineyards and wineries to adopt. One of these innovations was based on the use of biological control practices to control leafrollers (*Planotortrix* and *Ctenopseustis genera*) in vineyards. The wine industry in New Zealand has identified leafrollers as an important insect pest as they cause leaf, flower and fruit damage, and open berries to infection by the fungus *Botrytis cinerea* (Berndt et al.,2006).

Crop losses attributed to leafroller damage in the New Zealand wine industry have been estimated to cost up to NZ$360/ha in a dry year and signifi-
cantly more in wetter seasons (Lo and Murrell 2000). The usual practice to
control leafrollers in vineyards is the application of a broad-spectrum insec-
ticide. The Greening Waipara Project innovation used inter-row plantings
of flowering plants (e.g. buckwheat) to attract parasitoid wasps, a natural
enemy of leafrollers, into the vineyards. Research at trial sites revealed that
adding annual flowering plants, such as buckwheat, into a vineyard ecosystem increased the impact of parasitoids on leafrollers (Berndt et al., 2006).

Another innovation introduced by the Greening Waipara Project involved the plantings of native groundcovers to control under vine weeds
and thus reduce the need for herbicide applications. Other benefits that
were expected to arise from these plantings included increasing the diversity and abundance of beneficial insects, reduced runoff and improved soil
structure. A third innovation focused on the restoration of natural habitats
in and around vineyards and wineries. The aim of this innovation was for
the native plant species to assist with the conservation of native fauna and
flora, as well as soil retention, weed suppression and eco-tourism. The
Project has planted more than 20,000 native plants into the Waipara Valley. Other innovations developed by the Greening Waipara Project included the use of mulches (i.e. pea straw, linseed straw and grass clippings)
under vines to manage Botrytis, improvements in the filtering of winery
waste water, and the development of windbreaks through hedging.

The seventh innovation involved the introduction of Biodiversity
Trails on selected winery properties. These Trails were established to provide winery customers with a unique and informing experience at Waipara
wineries. Each Trail was developed close to a tasting room or restaurant
and led the visitor through areas of vines and native plants, and included
information boards where they could learn more about biodiversity and
Greening Waipara.

4.3 ADOPTION OF ENVIRONMENTAL INNOVATIONS

Mosher (1978) formally defined adoption as the process through which a
person is exposed to, considers, and finally rejects or accepts and practic-

es an innovation. More recently, Rogers (2003) defined adoption as the implementation of transferred knowledge about a technological innovation. Adoption can thus be thought of as the final stage of the technology transfer process. Adoption occurs when a person has decided to make full use of a new technological innovation as the best way to address a need (Rogers, 2003). This would suggest that wine producers with the greatest need to resolve or control a problem would be most likely to adopt a related innovation. Feder and Umali (1993, p. 216) defined an innovation as "a technological factor that changes the production function and regarding which there exists some uncertainty, whether perceived or objective (or both)".

The characteristics of an individual innovation influence the rate of its adoption. These characteristics are the levels of relative advantage, compatibility, complexity, trialability and observability (Rogers, 2003). Relative advantage can be measured economically, but can also include advantages in terms of prestige, convenience or satisfaction. Compatibility is achieved when an innovation is consistent with existing values, past experiences and the needs of the potential adopters. Complexity is the degree to which an innovation is difficult to understand, implement and maintain. Trialability relates to whether the innovation can be experimented with on a limited basis, whilst observability is the degree to which the results of the innovation are visible to others. Prior research suggests there are numerous factors which influence whether an agricultural innovation is adopted or not, and many of these can be seen to relate to the innovation characteristics developed by Rogers (2003).

Vanclay and Lawrence (1994) suggested that there are fundamental differences between commercial innovations and environmental innovations which affect adoption by agriculturists. Sassenrath et al. (2008) also noted that some innovations are driven by a desire to improve yields, whilst others are concerned for the environment. Environmental innovations are those which focus on improvements to land management. Although environmental innovations may result in some direct economic benefits, the costs associated with the adoption of these innovations are often high and are typically borne by the individual farmer. In contrast, commercial innovations are focused on increased productivity of agricultural activities and the benefits arising from adoption typically outweigh the costs. The authors argue that as the costs of adopting environmental innovations may outweigh the commercial benefits for an individual farmer then adoption will not be in the farmer's economic interest and the result will be large-scale non-adoption (Vanclay and Lawrence, 1994). This argument would appear to hold true, as Buttel et al. (1990) reported that the environmental innovations in agriculture that have been most widely adopted are those which are commercially oriented, such as minimum tillage. Clearly, the profit motive is one of a range of factors which can influence the adoption of new technologies and practices in agriculture.

A variety of factors have been found to influence the adoption of agricultural innovations. Sassenrath et al. (2008) noted that the adoption of innovations by agriculturists is an interaction between a range of external and internal factors, such as political and social pressures and monetary constraints. Whilst there is little doubt that agriculturists seek increased profitability through innovations, they also tend to be quite risk-averse. Agricultural innovations which reduce risk and are simple to establish are thus likely to be those which are most readily adopted by farmers (Sassenrath et al., 2008). A review of agricultural literature revealed that adoption of innovations by farmers is generally related to the process of learning about the innovation, the relative advantage of the innovation over existing practices and the ease of innovation trialability (Cullen et al., 2008). Vosti et al. (1998) stated that socioeconomic aspects of a technological innovation would influence its adoption. These studies again highlight the importance of commercial or economic factors. Similarly, Feder and Umali (1993) noted that the factors which constrain the adoption of agricultural innovations included lack of credit, limited access to information and inputs, and inadequate infrastructure. Australian researchers have developed a tool for predicting an agricultural innovation's likely peak extent of adoption and the likely timeframe for reaching that peak (Kuehne et al., 2011). The authors suggest that multiple variables in the tool sit in four quadrants: (1) population-specific influences on the ability to learn about the innovation; (2) relative advantage for the population; (3) learnability characteristics of the innovation; and (4) relative advantage of the innovation. Variables that reside within these quadrants include group involvement, skills and knowledge, awareness, trialability, innovation complexity, observability, profit orientation, environmental orientation, risk orientation, upfront cost, profit benefit, ease and convenience, and environmental costs and benefits (Kuehne et al., 2011).

Several factors have also been found to influence the adoption of environmental innovations by agriculturists. Studies that have examined the factors which influence the adoption of biological pest control practices have suggested that adoption is moderated by the perception of risk (Griffiths et al., 2008 and Shadbolt, 2005). In general, the adoption of biological pest control practices by agriculturists has been found to be quite limited (Falconer and Hodge, 2000 and Pietola and Lansink, 2001). Other factors which were found to influence the adoption of biological pest control innovations include the efficacy of the innovation, the possibility of price premiums in the marketplace, and reduced expenditure on agrichemicals and labour (Griffiths et al., 2008 and Shadbolt, 2005). Other studies have reported that costs are the dominant reason why agriculturists do not adopt environmentally sustainable practices (Curtis and Robertson, 2003 and Rhodes et al., 2002). A study of New Zealand dairy farmers and their propensity to adopt sustainable management practices provides a summation of the factors frequently mentioned in the adoption literature. Besswell and Kaine (2005) reported that farmers recognised the environment was important, but they were not convinced that some of the practices being promoted as environmentally friendly were actually practical. Adoption was found to depend primarily upon the farmer's perception of the benefits that would arise, and these related to the commercial and practical realities of the innovation to the farmer.

The previous sections summarise literature relating to the adoption of agricultural innovations in general and environmental agricultural innovations in particular. Other studies have examined the factors which drive the adoption of sustainable, ecological or environmental practices within the wine industry. The identified drivers for adoption of environmental innovations include the attitudes and norms of the manager (Marshall et al., 2010), increased profits (Hughey et al., 2004), and improved environmental performance (Delmas et al., 2006). Marketing reasons, such as gaining a competitive advantage, creating product differentiation, and improved or maintained market access, have also been found to drive the adoption of environmental practices in the wine industry (Adrian and Dupre, 1994,

Bhaskaran et al., 2006, Marshall et al., 2005, Molla-Bauza et al., 2005 and Nowak and Washburn, 2002).

The literature review above highlights the varying reasons why agriculturists might adopt innovations. In particular, previous research suggests that farmers will be less likely to adopt environmental innovations if they will not result in economic benefits. The Greening Waipara Project claimed that adoption of the environmental innovations will reduce costs and assist with marketing Waipara wines. The literature would support the idea that adoption of the innovations amongst Waipara wine companies would be high as the Project has stated that economic benefits would be gained. The first research question examined by this study is thus: What has been the level of adoption of the seven environmental innovations by Waipara vineyards and wineries?

4.4 BUSINESS IMPACT OF ENVIRONMENTAL INNOVATIONS

The economic impact arising from the implementation of environmental innovations is a key factor for agriculturists to consider, and one which has not been extensively explored in the literature. In their seminal paper, Constanza et al. (1997) suggested that ecosystem services are not fully captured or adequately quantified in traditional economic analysis; they estimated that the value of biological control of pests globally was US$417 billion per year. Pimentel et al. (1997) estimated that services arising from biodiversity in the United States contributed $319 billion each year, whilst globally the benefits amounted to $2929 billion annually. In addition, Pimentel et al. (1997) reported that the growing ecotourism industry contributed between US$0.5 and US$1 trillion per year to the global economy. Dyllick and Hockerts (2002) noted that although the value of ecosystem services is quite considerable, this value is not necessarily well understood.

Similarly, Cullen et al. (2008) stated that economic assessments of biological pest control programmes are rarely conducted and therefore poorly understood. Pannell et al. (2006, p. 1409) stated that "... the benefits and costs of some conservation practices are not clearly observable" and hence decision making regarding adoption of these innovations by farmers may be impeded. A few previous studies have provided some support for small economic savings being gained through the adoption of biological pest control programmes (Kellermann, 2007 and Thomas et al., 1991), whilst others have reported that these programmes are not cost effective for farmers (Schmidt et al., 2007).

Research from the wine industry also reports mixed results in terms of the costs and benefits relating to the adoption of environmental, ecological or sustainable practices. Delmas et al. (2006) reported that increased costs of 10–15% can be expected in the first four years of adopting sustainable vineyard practices. The study also reported an increase in labour costs of 30% due to planning, preparation and maintenance. Conversely, Marshall et al. (2010) reported that the adoption of environmental practices in the wine industry would result in economic benefits through reduced consumption of raw materials, increased productivity, decreased energy consumption and waste reductions. Hughey et al. (2004) suggested that environmental strategies are becoming an important marketing tool in international markets where consumers are more environmentally aware. It has also been suggested that wine businesses can gain price premiums through the adoption of environmental practices (Adrian and Dupre, 1994 and Fairweather et al., 1999). The California wine industry has invested time and money to develop sustainable production techniques; this industry seeks to increase the market value and perceived quality of their wines through branding it as sustainable (Warner, 2007).

The literature reports mixed results in terms of the business impact arising from the adoption of environmental innovations in agriculture. The second research question examined in this study is thus: What impact has the implementation of the seven environmental innovations had on economic, marketing and operational factors at the Waipara vineyards and wineries?

4.5 METHOD

This study gathered data via a self-completed, structured questionnaire mailed to all of the vineyards and wineries in Waipara in early December 2009. Follow-up postcards were sent to the vineyards and wineries in early 2010 in order to increase the response rate. A total of 14 companies responded to the questionnaire, resulting in an acceptable response rate of 44%. Five of the 14 companies in the sample were vineyards without attached wineries; the largest vineyard was approximately 450 ha in size and the next largest was just 55 ha. Of the nine wineries, eight had annual wine sales of less than 200,000 l. The sampled wineries thus reflect the nature of the New Zealand wine industry as a whole, which is comprised predominantly of small producers. Most of the companies noted that they had been involved with the Greening Waipara Project for two or three years and had joined for a number of reasons, including the planting of native species and the provision of shelters for birds. One respondent noted that they had no particular reason for joining. Although the sample is small in number, there is no reason to believe that it is not representative of the Waipara wine companies in terms of adoption of the environmental innovations. The authors received anecdotal evidence from an Analyst employed by the Greening Waipara Project about low rates of adoption, and this corresponds with our results.

The questionnaire began with general questions that were used to categorise the winery or vineyard operation. Section B examined whether the company had implemented a Biodiversity Trail or had any desire to do so. Respondents indicated which of the innovations they had implemented in Section C of the questionnaire and subjectively rated the effectiveness of each implemented innovation using a 4-point likert scale (i.e."ineffective", "somewhat effective", "very effective" and "unsure"). The final section asked each respondent to indicate what impact (i.e. increase, decrease or no effect) the adoption of each innovation had on their business in terms of various listed factors (e.g. labour costs, domestic sales, and water use).

4.6 RESULTS AND DISCUSSION

The first research question examined the level of adoption of the seven environmental innovations by Waipara vineyards and wineries. Table 1 indicates the number of companies that have adopted the seven innovations introduced by the Greening Waipara Project and each respondent's rating of the effectiveness of each implemented innovation.

TABLE 1: Adoption of the environmental innovations.

Innovation	Adopted				Not Adopted
	Ineffective	Somewhat effective	Very effective	Unsure	
Inter-row plantings to prevent leaf rollers	1	1			12
Under vine weed control through native ground-covers		1			13
Windbreaks through hedging		2			12
Winery waste water filtering		1			13
Conservation of native fauna and flora (native plantings)	2	6	2	1	3
Botrytis management through mulching			1		13
Biodiversity Trail		1			13

The results in Table 1 indicate there was a very low level of adoption for all the innovations except for the conservation of native fauna and flora. One respondent commented that they had been a member of the Greening Waipara programme for two years and had not implemented any of the innovations. Only one of the 14 respondents had implemented a Biodiversity Trail, although five respondents expressed an interest in building one in the future. The innovations to manage under vine weeds through native groundcovers, winery waste water filtering, and *Botrytis* management through mulching had also only been implemented by a single respondent.

The results in Table 1 suggest that the Greening Waipara innovations may not have delivered clear economic benefits to adopters, in line with previous literature that has reported the low adoption rates by farmers of environmental innovations that lack economic benefits (Buttel et al., 1990 and Sassenrath et al., 2008). Rogers (2003) stated that adoption will occur when there is a need to address. The low level of adoption across six of the Greening Waipara innovations would thus also suggest that these innovations were not addressing needs that were of serious importance to decision makers. The low adoption rates also infer that the innovations did not have the desired levels of relative advantage, compatibility, complexity, trialability and observability (Rogers, 2003).

The innovation that was most widely adopted was that of conservation of native fauna and flora through native plantings. It should be noted that this innovation did not require the vineyards and wineries to make any contribution in terms of financial or labour inputs; the Greening Waipara Project paid for the thousands of native plants that were planted around the participating properties and supplied the labour to plant these. It could be argued that the uptake of this innovation has been greatest because the companies had not been required to make any financial investment or contribution. As previous studies have reported that costs are a major reason for the non-adoption of environmental practices (Curtis and Robertson, 2003 and Rhodes et al., 2002), the high adoption rate of the native plantings innovation is likely to relate to the low costs involved. In line with Rogers (2003) study, it could be argued that adoption of this innovation has provided a relative advantage, compatibility and observability with no level of risk to the decision maker. Risk-averse agriculturists will more readily adopt low risk innovations (Griffiths et al., 2008, Sassenrath et al., 2008 and Shadbolt, 2005); thus the low risk native planting innovation was the one which was most likely to be widely adopted.

Whilst adoption has been generally low, Table 1 also indicated that where the innovations had been implemented their overall level of effectiveness has been quite poor. Further research would be necessary to fully understand the reasons behind the effectiveness ratings that the respondents have provided, although Table 2 may provide some of the answers. Table 2 illustrates the impact (i.e. Increase, No effect or Decrease) that the adopted innovations have had on various economic, marketing and operational factors at each company.

The majority of the adopted innovations have had little or no effect on the companies in terms of economic, operational or marketing factors (see Table 2). Indeed, the windbreaks innovation has had no effect at all on the two businesses that had adopted the innovation, and nor did the waste water filtering innovation have any effect on the single business that had adopted it.

From an economic perspective, several respondents noted that some of the innovations resulted in increased costs for companies. For instance, the adoption of some of the innovations has led to increased labour and vineyard floor management costs, and has introduced additional costs in terms of maintaining the implemented innovations. On a positive note, some of these increased costs may be offset by the reduced agrichemical costs which some respondents noted they gained as a result of adopting some of the innovations. One respondent noted that they did not have the necessary funds to purchase the equipment they would need in order to implement the inter-row plantings innovation. Another respondent commented that they have to water the new native plants and this incurs them an extra cost in terms of labour and water. They also noted that they have not adopted any of the other innovations as their understanding was that costs would increase by too much for their business. The results of this study provide support for previous research which has suggested that the adoption of environmental practices will increase costs (Delmas et al., 2006 and Schmidt et al., 2007), as well as partial support for literature which has reported that adoption can lead to economic benefits (Kellermann, 2007, Marshall et al., 2010 and Thomas et al., 1991).

In terms of marketing, none of the adopted innovations has had an effect on important aspects such as the wine price, consumer demand, cellar door sales, domestic sales, or international sales. Some respondents noted that adoption of innovations such as inter-row plantings, under vine weed control, and a Biodiversity Trail had increased their access into new domestic and international markets. This result supports the previous study of Hughey et al. (2004) who had suggested that environmental practices are an important marketing tool for New Zealand wineries in international markets. Whilst some companies have included a comment about the innovations on the back label of their bottles, it should be noted that there is no standardised Greening Waipara symbol or logo that companies can include on their front labels. The lack of marketing benefits to arise from the innovations is thus likely attributable to poor consumer awareness and recognition.

From an operational perspective, the innovations have had no effect on yield per hectare, but in some instances they have resulted in an increased level of water use. There were mixed results reported for both wine quality and the level of vineyard soil erosion. Overall, adoption of the innovations has generally had little effect on operational factors. Again, the low adoption rate may reflect the lack of operational benefits; this suggestion is supported by previous research in the New Zealand dairy industry which suggests that farmers will primarily consider commercial and practical realities when deciding whether to adopt environmental practices (Beswell and Kaine, 2005).

Preliminary findings suggest that the level of economic cost associated with many of the Greening Waipara Project innovations may outweigh the economic benefits. However, in many instances, the innovations have been implemented by winery and vineyard owners who are personally committed to preservation of the environment and are prepared to pay an economic cost in order to support these beliefs. Whilst the literature notes that the costs of environmental innovations are often borne by the individual farmer, Vanclay and Lawrence (1994) also noted that large-scale non-adoption will occur if the costs of the environmental innovation do not outweigh the commercial benefits for the farmer; this appears to be what the results of this study indicate has happened with non-adoption of the Greening Waipara practices. Personal beliefs alone may not be enough to ensure that the implemented innovations will continue to exist given the present difficult financial times faced by wine companies. There is no doubt that the innovations introduced by the Greening Waipara Project can have a positive effect on the sustainability of the environment; however, if they are not also sustainable at a business level they are unlikely to be implemented or maintained.

4.7 CONCLUSION

This research has studied the adoption of environmental innovations in the Waipara wine growing region and found that of the seven innovations, only one has been widely adopted. The adoption of that innovation, conservation of native flora and fauna, has been heavily subsidised by the Greening Waipara Project and its adoption has been almost costless and risk free for wine growers. In addition, the success of this innovation was relatively easy for wine companies to 'measure'; the initial plantings and subsequent growth of native habitat areas is a particularly visual innovation for both staff and other stakeholders to see and enjoy. The performance of some of the other innovations was not generally so easy for companies to measure.

The adoption of innovations in agriculture has been widely studied. It is obvious that agricultural businesses need to focus on economic viability in order to survive. Environmental innovations have been developed for many types of agriculture including the wine industry, and they will be considered for adoption only if they bring an environmental and economic advantage. It is clear that the Greening Waipara Project promoted the seven innovations they developed based on environmental or ecological improvements. The low level of adoption of the other six environmental innovations, together with comments provided by industry respondents, indicates that the innovations do not provide a sufficient economic advantage to businesses. Their non-adoption is consistent with the results from other New Zealand and international environmental innovation adoption research. The results of this study suggest that economic, marketing or operational factors were not considered by the Project and the lack of resulting benefits or increasing costs in these areas are instrumental in the low levels of adoption reported herein. This study advocates that economic, marketing and operational factors are considered during the development and promotion of future environmental innovations.

This leads to an interesting proposition for further research. It would be useful to examine whether environmental innovations which have been developed with economic, marketing and operational factors in mind achieve a higher adoption rate than those innovations which focus on environmental improvement alone.

REFERENCES

1.   M. Adrian, K. Dupre. The environmental movement: a status report and implications for pricing. SAM Advanced Management Journal, 59 (2) (1994), pp. 35–40

2.   L.A. Berndt, S.D. Wratten, S.L. Scarratt. The influence of floral resource subsidies on parasitism rates of leafrollers (Lepidoptera: Tortricidae) in New Zealand vineyards. Biological Control, 37 (2006), pp. 50–55

3.   Beswell, D., Kaine, G., 2005. Adoption of environmental best practice amongst dairy farmers. In: Proceedings of the 11th Annual Conference of the New Zealand Agricultural and Resource Economics Society Inc., 26–27 August, Nelson.

4.   S. Bhaskaran, M. Polonsky, J. Cary, S. Fernandez. Environmentally sustainable food production and marketing. British Food Journal, 108 (8) (2006), pp. 677–690

5.   F.H. Buttel, O.F. Larson, G.W. Gillespie. The Sociology of Agriculture. Greenwood Press, New York (1990)

6.   R. Constanza, R. d'Arge, R. de Groot, S. Farber, M. Grasso, B. Hannon, K. Limburg, S. Naeem, R.V. O'Neill, J. Paruelo, R.G. Raskin, P. Sutton, M van den Belt. The value of the world's ecosystem services and natural capital. Nature, 287 (1997), pp. 253–260

7.   R. Cullen, K.D. Warner, M. Jonsson, S.D. Wratten. Economics and adoption of conservation biological control. Biological Control, 45 (2008), pp. 272–280

8.   A. Curtis, A. Robertson. Understanding landholder management of river frontages; the Goulburn Broken. Ecological Management and Restoration, 4 (1) (2003), pp. 45–54

9.   Delmas, M.A., Doctori-Blass, V., Shuster, K., 2006. Caego vinegarden: how green is your wine? UC Santa Barbara: Donald Bren School of Environmental Science and Management Program on Governance for Sustainable Development. Available from: <http://www.escholarship.org/uc/item/5k657745>.

10.   T. Dyllick, K. Hockerts. Beyond the business case for corporate sustainability. Business Strategy and the Environment, 11 (2002), pp. 130–141

11.   J.R. Fairweather, H.R. Campbell, J. Manhire. The "Greening" of the New Zealand Wine Industry: Movement Towards the Use of Sustainable Management Practices. Department of Anthropology, University of Otago, Dunedin (1999)

12.   K. Falconer, I. Hodge. Using economic incentives for pesticide usage reductions: responsiveness to input taxation and agricultural systems. Agricultural Systems, 63 (2000), pp. 175–194

13.   G. Feder, D.L. Umali. The adoption of agricultural innovations: a review. Technological Forecasting and Social Change, 43 (1993), pp. 215–239

14.   G.J.K. Griffiths, J.M. Holland, A. Bailey, M.B. Thomas. Efficacy and economics of shelter habitats for conservation biological control. Biological Control, 45 (2008), pp. 200–209

15.   K.F.D. Hughey, S.V. Tait, M.J. O'Connell. Qualitative evaluation of three environmental management systems in the New Zealand wine industry. Journal of Cleaner Production, 13 (2004), pp. 1175–1187

16.   Kellermann, J.L., 2007. Ecological and Economic Services Provided by Birds on Jamaican Blue Mountain Coffee Farms. MSc Thesis Humboldt State University, California, USA.

17.   Kuehne, G., Llewellyn, R., Pannell, D., Wilkinson, R., Dolling, P., Ewing, M., 2011. ADOPT: a tool for predicting adoption of agricultural innovations. In: Proceedings of the 55th Annual National Conference of the Australia Agricultural & Resources Economics Society, 8–11 February, Melbourne.

18. P.L. Lo, V.C. Murrell. Time of leafroller infestation and effect on yield in grapes. New Zealand Plant Protection, 53 (2000), pp. 173–178

19. R.S. Marshall, M.E.M. Akoorie, R. Hamann, K. Sinha. Environmental practices in the wine industry: an empirical application of the theory of reasoned action and stakeholder theory in the United States and New Zealand. Journal of World Business, 45 (4) (2010), pp. 405–414

20. R.S. Marshall, M. Cordano, M. Silverman. Exploring individual and institutional drivers of proactive environmentalism in the US wine industry. Business Strategy and the Environment, 14 (2) (2005), pp. 92–109

21. B.B. Molla-Bauza, L.M. Martinez, A.M. Poveda, M.R. Perez. Determination of the surplus that consumers are willing to pay for an organic wine. Spanish Journal of Agricultural Research, 3 (1) (2005), pp. 43–51

22. A.T. Mosher. An Introduction to Agricultural ExtensionSingapore University Press, Singapore (1978)

23. L.I. Nowak, J.H. Washburn. Building brand equity: consumer reactions to proactive environmental policies by the winery. International Journal of Wine Marketing, 14 (3) (2002), pp. 5–19

24. D.J. Pannell, G.R. Marshall, N. Barr, A. Curtis, F. Vanclay, R. Wilkinson. Understanding and promoting adoption of conservation practices by rural landholders. Australian Journal of Experimental Agriculture, 46 (2006), pp. 1407–1424

25. K.S. Pietola, A.O. Lansink. Farmer response to policies promoting organic farming techniques in Finland. European Review of Agricultural Economics, 28 (2001), pp. 1–15

26. D. Pimentel, C. Wilson, C. McCullum, R. Huang, P. Dwen, J. Flack, Q. Tran, T. Saltman, B. Cliff. Economic and environmental benefits of biodiversity. Bioscience, 47 (11) (1997), pp. 747–757

27. H.M. Rhodes, L.S. Leland, B.E. Niven. Farmers, streams, information and money: does informing farmers about riparian management have any effect? Environmental Management, 30 (5) (2002), pp. 665–677

28. E.M. Rogers. Diffusion of InnovationsThe Free Press, New York (2003)

29. G.F. Sassenrath, P. Heilman, E. Luschei, G.L. Bennett, G. Fitzgerald, P. Klesius, W. Tracy, J.R. Williford, P.V. Zimba. Technology, complexity and change in agricultural production systems. Renewable Agriculture and Food Systems, 23 (4) (2008), pp. 285–295

30. N.P. Schmidt, M.E. O'Neal, J.W. Singer. Alfalfa living mulch advances biological control of soybean aphid. Environmental Entomology, 36 (2007), pp. 416–424

31. Shadbolt, A.A.K., 2005. 'Greening Waipara': Viticulturist's Attitudes and Practices Associated with a Region-Wide Ecological Restoration Scheme. BSc (Honours) Dissertation Lincoln University, Christchurch, New Zealand.

32. M.B. Thomas, S.D. Wratten, N.W. Sotherton. Creation of 'island' habitats in farmland to manipulate populations of beneficial arthropods: predator densities and emigration. Journal of Applied Ecology, 28 (1991), pp. 906–917

33. F. Vanclay, G. Lawrence. Farmer rationality and the adoption of environmentally sound practices; a critique of the assumptions of traditional agricultural extension. European Journal for Agricultural Education and Extension, 1 (1) (1994), pp. 59–90

34. S.A. Vosti, J. Witcover, S. Oliveira, M. Faminow. Policy issues in agroforestry: technology adoption and regional integration in the western Brazilian Amazon. Agroforestry Systems, 38 (1/3) (1998), pp. 195–222

35. K.D. Warner. The quality of sustainability: agroecological partnerships and the geographic branding of California winegrapes. Journal of Rural Studies, 23 (2007), pp. 142–155

Table 2 is not available in this version of the article. To view this additional information, please use the citation on the first page of this chapter.

CHAPTER 5

IMPROVING WATER USE EFFICIENCY IN GRAPEVINES: POTENTIAL PHYSIOLOGICAL TARGETS FOR BIOTECHNOLOGICAL IMPROVEMENT

J. FLEXAS, J. GALMÉS, A. GALLŬ, J. GULHAS, A. POU, M. RIBAS-CARBO, M. TOMAS, AND H. MEDRANO

5.1 INTRODUCTION

The wine and viticulture industry are currently enduring a period of profound changes worldwide. The viticulture expansion worldwide and the increasing environmental impacts of crop practices are of particular interest for the sustainability of the cropping areas. Within this frame, Bisson et al. (2002) have highlighted that, in addition to a product enjoyable in all sensorial aspects, consumers expect wines to be healthy and produced in an environmentally sustainable manner. They suggested that, as consumers become more aware of the vulnerability of our global environment, the demand for sound agricultural production practices will increase. Hence, in the near future, the wine industry will not only depend on wine quality, but also on managing environmentally friendly vineyards.

Printed with permission from Flexas J, Galmés J, Gallé A, Gulías J, Pou A, Ribas-Carbo M, Tomàs M, and Medrano H. Improving Water Use Efficiency in Grapevines: Potential Physiological Targets for Biotechnological Improvement. Australian Journal of Grape and Wine Research, *16,S1 (2010), DOI: 10.1111/j.1755-0238.2009.00057.x.*

Among environmental problems related to viticulture, water scarcity is one of the major worldwide limitations to agriculture in general (Araus 2004, Morison et al. 2008), and to grape production in current viticulture areas in particular (Chaves et al. 2007). About two-thirds of the major viticulture areas of the world have annual precipitation below 700 L/m^2 (Figure 1). Moreover, a large proportion of vineyards in these areas are located in regions with a seasonal drought that coincides with the grapevine growing season (e.g. Mediterranean climate-type areas). In these areas, progressive soil water deficits and high leaf-to-air vapour pressure gradients, together with high irradiance and temperatures, exert large constraints on yield and quality. Climate change predictions suggest that the viticulture areas subject to water deficit will increase further in the near future. For instance, for the whole viticulture area in Europe, Schultz (2000) predicted that doubling atmospheric CO_2 will result in decreases in soil moisture content from 20% or more in Central Europe to 70% in the Iberian Peninsula and the Balearic Islands. Indeed, the number of dry days per year has increased in southern Europe (Chaves et al. 2007). Increasing water scarcity will have an impact on viticulture, from changing the optimum ranges for different grape varieties (Schultz 2000) to forcing viticulturists to rely on irrigation more (Chaves et al. 2007). In turn, irrigation management will have impacts on the wine industry, because the effects of excess irrigation on grape quality are at least controversial, with several reports suggesting that excess water leads to reduced quality through decreases in colour and sugar content, and imbalanced acidity (Matthews et al. 1990, Medrano et al. 2003, Salón et al. 2005), as well as interfering with the normal timing of flavonoid development (Castellarin et al. 2007).

5.2 IMPROVING VINEYARD WATER USE EFFICIENCY BY CROP MANAGEMENT

In this scenario, reducing water use for irrigation and increasing water use efficiency (WUE)—i.e. improving the yield to water consumption ratio—becomes a major priority in agriculture (Costa et al. 2007, Morison et al. 2008). These include improvements in agronomical management practices (Gregory 2004), developing regulated deficit irrigation (RDI) systems

(Costa et al. 2007), which are often supported by physiologically based monitoring tools, such as infrared thermometry (Grant et al. 2007), trunk diameter sensors (Conejero et al. 2007) or sap flow meters (Fernández et al. 2008), or introducing physiologically based agronomic techniques such as partial root drying (PRD, Chaves et al. 2007).

These agronomic techniques have also been tested and developed in grapevines, and some are already being used in field trials and commercial vineyards. For instance, not only canopy management is an important agronomic technique being widely used in viticulture to regulate the microenvironment around the clusters, and hence, fruit sanitary conditions, yield and quality, but also light absorption by the canopy, and hence, canopy photosynthesis and water loss, i.e. WUE (Smart 1974, Carbonneau 1980, Williams and Ayars 2005). Other techniques related to vineyard management have started to be tested but are still far from common agronomic practice, such as the use of mulching (Buckerfield and Webster 2001, Hatfield et al. 2001, Gregory 2004) or growing herbaceous species in vineyard inter-row cropping, which may help regulate soil evaporation and runoff, as well as root development and nutrient availability (Monteiro and Lopes 2007). To date, the effects of this later technique on yield, grape quality and WUE under Mediterranean conditions, where competition for water could lead to severe water stress, have been infrequently reported (Monteiro and Lopes 2007, Gulías et al. 2008). The results suggest that reduction of vegetative vigour by water and nutrient competition between grapevine and the cover crop could be a way to reduce grapevine water consumption late in the season when this resource is scarce.

The use of PRD, a dripping system that irrigates each side of the grapevine root, system alternatively is becoming more common, as it has been shown to induce root abscisic acid (ABA) synthesis in the dried zone as well as in the berries (Dry and Loveys 1998, Antolín et al. 2006, 2008). PRD induces partial stomatal closure without reducing leaf water status and results in increased WUE (Stoll et al. 2000, de Souza et al. 2005, Chaves et al. 2007). Trials in Australia using this type of irrigation have shown a significant reduction in vegetative growth while maintaining fruit yield (Dry et al. 2001) and increasing quality parameters such as colour (Bindon et al. 2008), therefore saving on irrigation while increasing quality.

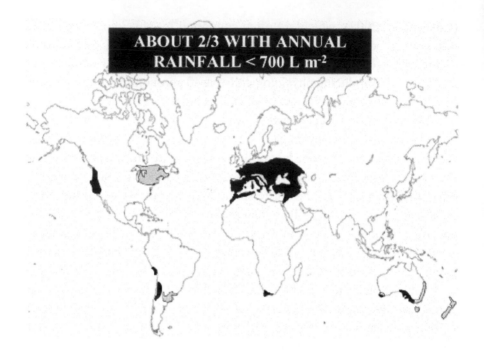

FIGURE 1: Global distribution of the major viticultural areas (grey and black) with those suffering from moderate-to-severe drought (i.e. those with a current annual rainfall of less than 700 L/m², black).

The applications of RDI in grapevines have recently been reviewed elsewhere (Costa et al. 2007), and several different physiological indicators of potential use for RDI scheduling in grapevines have also been reviewed (Cifre et al. 2005). Among them, sap flow meters were among the first to be proposed and tested in irrigated (Yunusa et al. 2000, Tarara and Ferguson 2006) and water-stressed (Escalona et al. 2002) grapevines. Ginestar et al. (1998) applied sap flow measurements to establish different irrigation coefficients to Shiraz grapevines, obtaining a gradient of grape yield and qualities. The usefulness of trunk diameter sensors to assess water stress and water consumption in grapevines was first demonstrated by Escalona et al. (2002) in potted plants, while only recently, studies have begun under field conditions (Patakas et al. 2005, Intrigliolo and Castel

2008). Remote sensing indicators have also been proposed and are presently being evaluated in field trials, such as passive chlorophyll fluorescence (Flexas et al. 2002a, Dobrowsky et al. 2005), infrared thermometry (Grant et al. 2007) and hyperspectral reflectance indices (Evain et al. 2004, Rodríguez-Pérez et al. 2007). Also, stem water potential has been proposed as an indicator for irrigation scheduling (Choné et al. 2001, Patakas et al. 2005), being a widely used indicator despite the practical problems of their use for commercial vineyards.

Besides these management techniques, in crops other than grapevines, plant breeding and genetic engineering are also seen as pivotal in improving WUE (Condon et al. 2004). However, as pointed out by Vivier and Pretorius (2002), grapevine improvement has been untouched by classical breeding programmes in the sense that relatively few new cultivars have become commercially successful, especially in the wine industry where commercial production relies in a few selected and ancient cultivars. Given the difficulties of acceptance of newly developed cultivars by traditional breeding, it is envisaged that it will be even more difficult to introduce genetically transformed genotypes. Nevertheless, there is little doubt that feeding the increasing human world population will require, in the near future, large increases in crop yields, while doing so in an environmentally sustainable manner would require, among other issues, a parallel increase in WUE, which ultimately will depend on biotechnologically improved crops, including genetically modified plants (Morison et al. 2008, Murchie et al. 2009). Viticulture cannot elude these requirements, and indeed, drought resistance and increased WUE have already been identified as the most important priority targets for grapevine biotechnology, together with pest and disease control, and improved grape quality (Vivier and Pretorius 2002). Nevertheless, as pointed out by Blum (2005), drought resistance and WUE are not synonymous, and it is important not to confound each other, although this is often the case in the literature (Morison et al. 2008). Indeed, both characteristics are often conferred by plant traits that are mutually exclusive (Blum 2005). For instance, *Arabidopsis* mutants with an increased plant nuclear factor, Y(NF-Y), present a higher drought tolerance than wild-type plants, but with identical WUE (Nelson et al. 2007). Contrarily, *ERECTA* mutants, which present an enhanced WUE, do not necessarily present an improved drought tolerance (Masle et al. 2005). In other cases, however, increased drought tolerance can be associated with higher WUE, like in P_{SAG12}-IPT transgenic tobacco (Rivero et al. 2007) or transgenic *Arabidopsis* expressing the bZIP transcription factor ABP9 (Zhang et al. 2008). The present review focuses on traits that can potentially confer increased WUE, not drought resistance.

Although until now, no increase in WUE has been achieved in grapevines by genetic modification, and despite the fact that grapevine was initially proven to be recalcitrant to such manipulations, the success in grapevine transformation in recent years (Vivier and Pretorius 2002) and the current availability of the completely sequenced grapevine genome (Jaillon et al. 2007) suggest that opportunities are open to achieve this goal in the near future (Troggio et al. 2008). In addition, initial studies have emerged describing water stress effects on gene expression and protein profiling (Cramer et al. 2007, Vincent et al. 2007), which may be viewed as a first step to identifying potential gene targets for genetic engineering of WUE.

Within this context, the aims of the present review are (i) to define the concept of WUE, linking plant to leaf-level WUE and describing its variability in grapevines; and (ii) to identify and discuss, from a physiological perspective, the potential target processes whose genetic manipulation may lead to improved WUE.

5.3 PLANT WUE: CONCEPT AND COMPONENTS

As stated above, making agriculture sustainable requires a major reduction in water use in many regions. WUE is a key parameter to determine how efficiently the agricultural sector is using water. WUE (i.e. the amount of carbon gained per unit water used) is currently a priority for the United Nations policy, in what is called the 'Blue Revolution' and summarised as 'more crop per drop'.

Figure 2 shows a theoretical diagram of the components of WUE and their possible interdependencies. On one hand, crop WUE (WUE_C) depends on total water consumed during the growing season. This is the sum of the amount of water lost without being used by the plant, plus the transpired water. The former occurs through soil evaporation, runoff, etc. and can be avoided or reduced by agronomic methods previously described, such as drip irrigation, mulching or vineyard intercropping (Gregory 2004). Whole-plant WUE (WUE_{WP}) considers only water used by the plant, i.e. transpired water (i.e. $WUE_{WP} = WUE_C$ − soil evaporation and runoff). Transpired water depends on aspects related to canopy growth and structure, such as leaf angle, leaf area index or shoot positioning, which ultimately determine light interception or the energy load for transpiration, and on leaf transpiration (E). E depends on the atmospheric evaporative demand, represented by the leaf-to-air vapour pressure deficit, and on leaf conductance, i.e. cuticular (gc) and, especially, stomatal (gs)

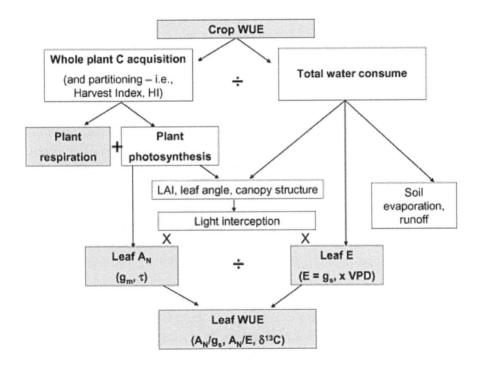

FIGURE 2: Theoretical diagram showing the dependency of crop water use efficiency (WUE_C) on different processes and its links with leaf-level WUE (some important parameters related to different processes are indicated in parenthesis). Grey boxes indicated processes extensively reviewed in the present review. See text for details. LAI, leaf area index; VPD, vapour pressure deficit.

conductances. Canopy structure and light interception are commonly regulated by management techniques, including the selection of a proper training system and pruning (Smart 1974, Carbonneau 1980). Despite their interest for improving WUE (Baeza et al. 2005, Williams and Ayars 2005), canopy structure depends on complex whole-canopy components. This makes it difficult to identify specific targets for genetic manipulation and, hence, is out of the scope of this review. Leaf conductance and its diurnal and seasonal regulation, in contrast, depend on well-identified physiological traits that may be targeted for transformation and WUE_C/WUE_{WP} improvement.

The second component of WUE_{WP} is the whole-plant carbon and biomass acquisition, as well as its partition to yield components (i.e. fruits) or harvest index. Considering the carbon balance of fruits only, a yield WUE (WUE_Y) is defined, which is the ultimate goal for improving WUE. Grapes are an extremely efficient carbon sink. Despite constituting only 20–30% of the total plant dry mass, they import about 80–90% of the total assimilates obtained in photosynthesis, and this strong sink capacity is maintained under water stress conditions (Bota et al. 2004). Therefore, partitioning seems to already be optimised in grapevines so, although it may slightly vary among genotypes, it is unlikely to be an optimum target for genetically improving WUE_Y. Instead, plant net carbon acquisition depends on two processes—photosynthesis and respiration—that, unlike carbon partitioning, strongly vary with grape genotypes (Flexas et al. 1999a, Bota et al. 2001) and are differentially affected by water stress (Flexas et al. 2006a). That makes them a more likely candidate to be good targets for genetic improvement of WUE_{WP} and WUE_Y.

As previously noted (Figure 2), a clear link emerges between leaf-level WUE (WUE_{leaf}) and WUE_{WP}. The former is often approached using the 'instantaneous' WUE_{leaf}, i.e. the ratio of net assimilation (A_N) to leaf transpiration (E), or the 'intrinsic' WUE_{leaf}, i.e. the ratio of A_N to stomatal conductance (g_s), which allows comparison to photosynthetic properties at a common evaporative demand. More recently, the carbon isotopic composition ($\delta^{13}C$) in leaf dry matter was used as a long-term indicator of WUE_{leaf} (Condon et al. 2004). The definition of the intrinsic WUE_{leaf} summarises the potential targets for genetic improvement of WUE at the leaf level: net carbon assimilation (i.e. the difference between carbon gain in photosynthesis and carbon losses in respiration) and stomatal conductance. It has to be pointed out that improving WUE_{leaf} may not necessarily result in improving WUE_{WP} and WUE_y, because of the interference of canopy and ambient processes (Figure 2). Nevertheless, comparing different grapevine cultivars grown in a glasshouse, good correlations have been found between WUE_y and WUE_{leaf} as approached by $\delta^{13}C$ (Gibberd et al. 2001). Also, in a preliminary experiment from our group with five contrasted cultivars, we have found a good correlation between WUE_{WP} (including roots) and intrinsic WUE_{leaf} (Magdalena Tomàs, unpublished data, 2009). Although the relationship was positive and significant, it varied between irrigated and water-stressed plants (Figure 3), indicating the complexity of the leaf to whole-plant WUE relationship.

5.4 GENETIC VARIABILITY OF WUE IN GRAPEVINES

Before considering theoretical aspects for genetically improving WUE_{leaf}, it is important to evaluate whether WUE_{leaf} presents some natural variation in grapevines and whether this has a genetic basis. In species other than grapevines, differences between genotypes in A_N/g_s and other estimates of WUE have been reported to have a genetic basis (Martin et al. 1989, Masle et al. 2005), and breeding for high WUE has been and continues to be a main objective for many crops (Condon et al. 2004). In grapevines, despite the fact that they have not been the subject of intense breeding for high WUE, substantial genotypic variations in intrinsic WUE_{leaf} have been described. For instance, A_N/g_s in irrigated plants was shown to range between 38 and 64 mmol CO_2/mol H_2O when comparing 20 Mediterranean cultivars (Bota et al. 2001), but higher values have been described in other cultivars such as Grenache and Syrah (Schultz 2003a), the species *Vitis riparia* (Flexas et al. 1999a) or the rootstock R-110 (Pou et al. 2008). The highest reported values in irrigated grapevines, ca. 100 mmol CO_2/mol H_2O, have been reported for the cultivar Rosaki (Rodrigues et al. 1993), while the lowest values, ca. 25 mmol CO_2/mol H_2O, have been shown for cultivar Kékfrankos (Zsófi et al. 2009). A_N/g_s typically increases under water stress conditions, commonly to 100–200 mmol CO_2/mol H_2O (Bota et al. 2001, Flexas et al. 2002b, Pou et al. 2008, Zsófi et al. 2009).

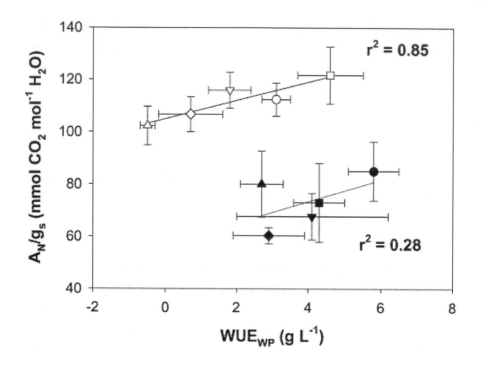

FIGURE 3: The relationship between intrinsic WUE_{leaf} (AN/gs) and whole-plant WUE_{WP} (based on vegetative growth including roots, first-year plants not producing grapes) in five different cultivars of *Vitis vinifera* growing outdoors in pots during summer in Mallorca (Balearic Islands, Spain). Closed symbols represent irrigated treatment at field capacity and open symbols to non-irrigated treatment defined by the leaf maximum daily gs (about 0.05 mol $H_2O/m^2/s$). Different cultivars symbols are: Grenache (circles), Callet (squares), Tempranillo (triangles up), Malvasia of Banyalbufar (triangles down) and Cabernet Sauvignon (diamonds). Values represent means ± standard error of six (AN/gs) or four replicates (whole-plant WUE). Regression lines are displayed with their r^2. WUE, water use efficiency.

Also, $\delta^{13}C$ shows large variation among cultivars. Gaudillère et al. (2002) showed differences of up to 3% in fruit $\delta^{13}C$ in a comparison of 32 cultivars, which were used to classify those cultivars in terms of their WUE. Slightly lower variations (2%) have been shown for leaf $\delta^{13}C$ in different comparisons of 19 (Gibberd et al. 2001) and 5 (Magdalena Tomàs, unpublished data, 2009) different cultivars under irrigation. Similar variations (2–3%) are observed for any single cultivar when subjected to water stress (de Souza et al. 2003, Chaves et al. 2007, Pou et al. 2008), when δ13C becomes less negative, so that it correlates positively with A_N/g_s (Chaves et al. 2007).

Much less data are available in grapevines for WUE_{WP} and WUE_Y, but variation is still present, ranging from 2.5 to 3.4 g of dry matter/Kg H_2O for WUE_Y (Gibberd et al. 2001) and from 2.5 to 6 g of dry matter/Kg H_2O for vegetative growth-based WUE_{WP} (Magdalena Tomàs, unpublished data, 2009; see Figure 3). However, contrary to A_N/g_s and $δ^{13}C$, WUE_{WP} decreases rather than increases under water stress (Figure 3), suggesting that variations in WUE_{leaf} resulting from reduced gs do not result in increased WUE_Y.

In summary, there is substantial evidence for genetic variability of WUE_{leaf} in grapevine cultivars (Bota et al. 2001, Gibberd et al. 2001, Gaudillère et al. 2002) and also depending on grapevine rootstocks (Satisha et al. 2006), which supports the possibility of improving WUE_{leaf} by genetic engineering.

5.5 POTENTIAL PHYSIOLOGICAL TARGETS FOR BIOTECHNOLOGICAL IMPROVEMENT OF WUE

As stated earlier, intrinsic WUE_{leaf} (A_N/g_s) is a good basis to summarise the potential targets for genetic improvement of WUE_{leaf}. Because A_N shows a direct but curvilinear dependency of g_s (Figure 4), any reduction in g_s results in an increased WUE_{leaf}. Therefore, genetically manipulating g_s would result in improved WUE_{leaf}, although one associated with reduced photosynthesis and potential yield. Although it has been pointed out that genotypic variation in WUE_{leaf} often derives from variations in g_s and not A_N (Gibberd et al. 2001, Blum 2005), the results shown in Figure 4 demonstrate that higher A_N is also a factor for increased WUE in grapevines at constant g_s. The data scattering in Figure 4 represents the genotypic and environmental variations in intrinsic WUE described in the previous section. This figure also illustrates how WUE could be increased while maintaining, or even increasing, yield, which would require an increase of AN at a given value of g_s, i.e. genotypic modifications in $A_N–g_s$ relationship (Parry et al. 2005). Because A_N is the net carbon assimilation, i.e. the difference between carbon gain in photosynthesis and carbon losses in respiration, both respiration and photosynthesis are potential targets for increasing A_N at a given value of g_s.

Because respiration occurs continuously in all living cells, its reduction may result in substantial increases in whole-plant carbon gain. In turn, increased photosynthesis at any given gs can be potentially achieved by either increasing total leaf area or increasing leaf photosynthetic capacity (Sharma-Natu and Ghildiyal 2005, Long et al. 2006). However, in grapevines, there is a limit for the capacity of improving photosynthesis by increasing total leaf area, and this may result in a penalty in terms of WUE_{WP}. Indeed, Escalona et al. (1999a, 2003) showed that leaves occupy-

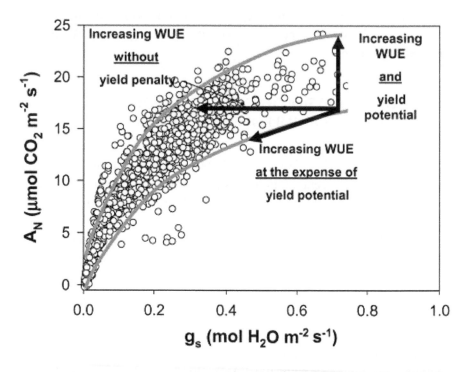

FIGURE 4: The relationship between net photosynthesis (A_N) and stomatal conductance (g_s) in grapevines. Open circles are real data for many different grapevine cultivars and conditions (modified from Cifre et al. 2005), and grey lines are the idealised upper and lower limits for this relationship. Black arrows indicate potential ways to increase WUE (see text for details). WUE, water use efficiency.

ing the inner part of the canopy constitute up to 35–50% of total leaf area, but contribute less than 5% of total net canopy carbon gain. Theoretical calculations suggest that selective pruning of these leaves would result in a 4–5% increase in WUE_{WP}.

It may be more interesting to increase leaf photosynthetic capacity at any given g_s. Potentially, this could be achieved by diverse methods (reviewed by Parry et al. 2005, Sharma-Natu and Ghildiyal 2005, Long et al. 2006, Peterhansel et al. 2008, Murchie et al. 2009). These include improving carboxylation efficiency and/or improving CO_2 diffusion in the mesophyll through either inducing C_4-like photosynthetic metabolism in C_3 plants (Long et al. 2006) or increasing the mesophyll diffusion conductance to CO_2 from sub-stomatal cavities to chloroplasts (g_m; Flexas et al. 2008). In turn, carboxylation efficiency can be achieved through increased Ribulose-1,5-bisphosphate carboxylase/oxygenase (Rubisco) catalytic rate (Parry et al. 2007) and/or specificity for CO_2 (Galmés et al. 2005, Parry et al. 2007), reducing or bypassing photorespiration (Long et al. 2006, Kebeish et al. 2007), increasing the capacity for RuBP regeneration (Feng et al. 2007, Peterhansel et al. 2008) or reducing the photoprotective state upon high-to-low light transitions (Long et al. 2006). However, considering grapevine physiology and its response to water stress, only a few of these appear to be really attainable targets for grapevine improvement.

Grapevine transpiration, photosynthesis, photorespiration, photoprotection and respiration responses to water stress have previously been described (Flexas et al. 2002b, Medrano et al. 2002, 2003). Some important features are shown in Figure 5, which average real data obtained for two different Mediterranean cultivars under commercial vineyard conditions. In the next sections, these realistic data will be taken as the basis for discussion and simulation of the potential effects of genetically engineering different physiological mechanisms in grapevines. First, water stress leads to progressive decreases in g_s (Figure 5, right to left), which results in leaf temperature increases from 3 to 8°C (Flexas et al. 1999b). The diffusion of CO_2 through the leaf internal structure, or mesophyll conductance to CO_2 (g_m), decreases concomitantly with g_s (Flexas et al. 2002b), but the relationship between both conductances is curvilinear, so that the higher the intensity of water stress, the higher gm is compared with g_s (Figure 5). Decreases of g_s and g_m as water stress intensifies result in less CO_2 available in the chloroplasts (C_c), which can be typically reduced to half or less of the values under irrigation (Figure 5). Together, decreased C_c, increased leaf temperature and sustained high irradiance lead to increased photorespiration and photosystem heat dissipation associated with xanthophylls de-epoxidation, both of which are thought to exert photoprotection of the photosynthetic apparatus (Flexas et al. 1999a,b, 2002a,b, Medrano et al. 2002). In contrast, leaf respiration is barely affected by water stress in grapevines (Escalona et al. 1999b), although there are no studies analysing the effects of water stress in the respiration of other organs that could be affected, particularly roots.

Coming back to physiological targets to increase photosynthesis, only a few appear to be reliable for grapevines subjected to water stress in view of the responses described above. For instance, because grapes often grow under high light conditions and only illuminated leaves contribute significantly to carbon gain (Escalona et al. 2003), both photorespiration and the photoprotective state associated with xanthophyll de-epoxidation may be necessary to avoid further damage to photosynthesis, particularly under water stress conditions (Medrano et al. 2002). Similarly, although increasing sedoheptulose-1,7-bisphosphatase (i.e. increasing the capacity for RuBP regeneration) in leaves has been shown to enhance photosynthesis in some species under irrigation and salt stress (Feng et al. 2007), grapevine photosynthesis operates in the Rubisco-limited region (i.e. photosynthesis limited by CO_2 availability in the chloroplasts) and not the RuBP-limited region, particularly under water stress (Flexas et al. 2002b, 2006a). For the same reason, increasing Rubisco specificity for CO_2 may be preferable to increasing its abundance, activation state or catalytic rate (Galmés et al. 2005, Parry et al. 2007). Instead, increasing mesophyll conductance to CO_2 (g_m) may result in increased C_c without increasing the water expenses that would be associated with increased g_s, and C_4 metabolism would be favourable because of increased leaf temperature, which further increases photorespiration under water stress conditions (Figure 5). Figure 6 summarises the known effects of combined water stress and increased leaf temperature on photosynthesis, showing the logics for transforming C_3 into C_4 plants, increasing g_m and improving Rubisco specificity for CO_2 as the most interesting choices for genetically improving photosynthesis and WUE_{leaf} under water stress in grapevines.

In summary, we will focus on discussing four potential targets for genetically engineering improved WUE_{leaf} and WUE_{WP} in grapevines: stomatal physiology, plant respiration, mesophyll conductance to CO_2 and Rubisco specificity for CO_2. Although C_4 metabolism could be an interesting target as well, particularly in view of the high leaf temperatures achieved by grapevines under water stress, the scarce success achieved in other species to such transformation suggests that this would be an unachievable goal in grapevines.

5.6 REGULATION OF TRANSPIRATION

5.6.1 CUTICULAR AND NIGHT CONDUCTANCE

The amount of water transpired by leaves depends on cuticular conductance and stomatal conductance. Water transpired through the cuticle contributes to lowering WUE_{leaf} because this is almost totally impermeable to CO_2, so that water is lost without any gain in photosynthesis (Boyer et al. 1997). Nevertheless, cuticular conductance is generally quite low (0.005 mol H2O/m²/s or less) in Vitis (Boyer et al. 1997, Flexas et al. 2009), so the gain in WUE_{leaf} by reducing cuticular conductance will be very small or insignificant. In contrast, stomatal conductance is high and strongly regulated, and intrinsically linked to photosynthesis, therefore constituting a potential target for biotechnological improvement of WUE_{leaf}.

It is important to highlight that stomatal conductance, contrary to previous thoughts, can be maintained at night, when no photosynthesis occurs, constituting a significant source of water loss and reduced WUE_{leaf} (Caird et al. 2007). Therefore, night conductance values (g_{night}) are often higher than what would be expected if the entire conductance under these conditions was cuticular (g_c). Values for g_{night} have been shown to range from 0.03 to 0.06 mol H_2O/m²/s for *Vitis berlandieri* and *Vitis rupestris*, and to values as high as 0.205 mol H_2O/m²/s in *V. riparia* (Caird et al. 2007). Considering that the maximum values of daytime stomatal conductance in grapevines are often lower than 0.4 mol H_2O/m²/s, these values of g_{night} are quite high. To the best of our knowledge, no data for g_{night} in *Vitis vinifera* have been previously reported. Here, we show (Table 1) data of

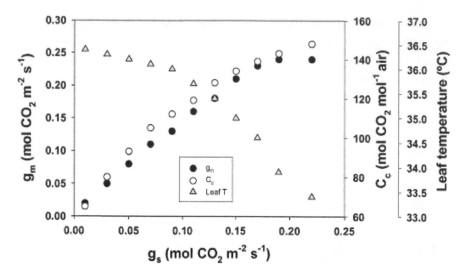

FIGURE 5: Common responses of several leaf parameters to progressive water stress in field-grown grapevines in Mallorca (Balearic Islands, Spain). The progression in intensity of water stress is represented as the decline of stomatal conductance (gs, from right to left in the x-axis). The represented parameters include leaf temperature (grey triangles), mesophyll conductance to CO_2 (g_m, solid circles) and the mean CO_2 concentration in the chloroplasts (C_c, open circles). Data are averaged for discrete intervals of g_s and for two cultivars (Tempranillo and Manto Negro), showing similar response to water stress. Original data for each of the two cultivars are published elsewhere (Flexas et al. 2002b).

g_{night} and leaf transpiration in the night (E_{night}) for two different cultivars of *V. vinifera*, Tempranillo and Manto Negro. They were subjected to various irrigation regimes, as recorded in a vineyard in Mallorca in August 2000 (Dr Jaume Flexas et al., unpublished data, 2000). The values range from 0.009 mol $H_2O/m^2/s$, i.e. close to cuticular conductance values, to 0.107 mol $H_2O/m^2/s$, i.e. similar to typical daytime values for mild-to-moderate water-stressed plants (Flexas et al. 2002b). Value for E_{night} ranged from 0.3 to 1.2 mmol $H_2O/m^2/s$. This represents a substantial loss of water. For instance, during a 9-h night, a typical individual grapevine plant with 6/m^2 of leaves would lose from 1 to 4 L of water.

Remarkably, differences in g_{night} and E_{night} were apparent between cultivars and treatments. While the values were generally lower in Manto Negro, a cultivar regarded as drought resistant, than in Tempranillo, they were

Water stress effects on photosynthesis

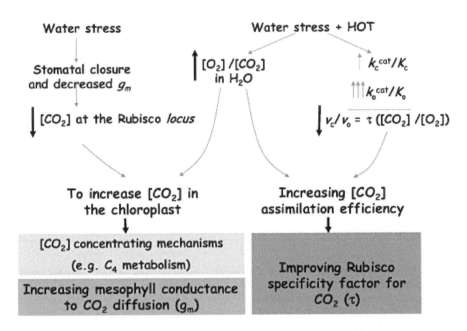

FIGURE 6: Diagram showing the physiological bases for decreased photosynthesis under combined water stress and high temperature (as often found in many semi-arid viticultural areas), and highlighting (bold) the theoretical targets to improve photosynthesis under these conditions.

more strongly responsive to water stress in the latter, so that under severe water stress, Tempranillo consumed less water by night than Manto Negro. These results suggest that g_{night} is a regulated character with a genetic basis, and, therefore, a potential target for biotechnological improvement. Unfortunately, although it is known that g_{night} responds to similar factors as daytime g_s, including CO_2, ABA and water stress (Caird et al. 2007), the physiological basis of g_{night} are poorly understood, so that possible molecular targets for genetic engineering are unknown. Moreover, there are some criticisms concerning the methodology used for determining g_{night}, so that a part of the observed differences could be because of unavoidable effects on these measurements (e.g. breaking down the boundary layer, alteration of chamber humidity). Clearly, more efforts are needed to characterise the importance of gnight in grapevines and to understand the molecular mechanisms underlying its regulation.

TABLE 1: Values of night leaf conductance (g_{night}) and transpiration (E_{night}) determined by 0500 h (local time) in a commercial vineyard in Mallorca, in August 2000.

	gnight (mol $H_2O/m^2/s$)	Enight (mmol $H_2O/m^2/s$)
Tempranillo 100% ETo	0.107 ± 0.009	1.19 ± 0.07
Tempranillo 30% ETo	0.036 ± 0.007	0.52 ± 0.09
Tempranillo 0% ETo	0.009 ± 0.001	0.16 ± 0.02
Manto Negro 100% ETo	0.033 ± 0.003	0.45 ± 0.03
Manto Negro 30% ETo	0.029 ± 0.006	0.41 ± 0.08
Manto Negro 0% ETo	0.019 ± 0.003	0.29 ± 0.04

Data were collected using a Li-6400 (Li-Cor Inc., Lincoln, NE, USA), in full darkness, at an atmospheric CO_2 concentration of 365 µmol CO_2/mol air, air temperature of 23°C and leaf-to-air vapour pressure deficit of 1.5 KPa. Two cultivars (Tempranillo and Manto Negro) and three different irrigation treatments (consisting in applying drip irrigation twice a week, respectively at 100%, 30% or 0% of the potential evapotranspiration (ETo) of the previous week) were analysed.

5.6.2 DAYTIME STOMATAL CONDUCTANCE

Much more is known about the regulation of daytime g_s. The well-known declining response of g_s to water stress and the concomitant increase in intrinsic WUE_{leaf} have allowed the development of management practices such as RDI and PRD, and of g_s-based or g_s-dependent physiological indicators for irrigation scheduling like sap flow or infrared thermometry (Cifre et al. 2005). But in addition to this empirical approach, the substantial knowledge about the molecular basis for g_s regulation opens the possibility of genetic engineering of this leaf trait. Indeed, in species other than grapevines, several mutants and transgenic plants have been described with altered g_s and/or altered g_s responses to the environment (Schroeder et al. 2001, Nilson and Assmann 2007). In grapevines, this goal has yet to be achieved, but efforts have just been initiated (Matus et al. 2008).

In grapevines, g_s is well known to be regulated by the concentration of ABA (Loveys and Kriedemann 1974, Rodrigues et al. 2008), either synthesised by roots in response to water stress and transported in the xylem to leaves (Correia et al. 1995, Lovisolo et al. 2002, Pou et al. 2008) or locally synthesised in buds and leaves (Soar et al. 2004, 2006). The capacity for ABA biosynthesis and ABA-mediated stomatal closure depends on the cultivar, being higher in isohydric cultivars like Grenache and lower in anysohydric cultivars like Shiraz (Soar et al. 2006). Many different genes and proteins involved in ABA synthesis, ABA catabolism and ABA-mediated stomatal closure are known, which have been used in other species to genetically modify the control of transpiration (Saez et al. 2006, Verslues and Bray 2006), and that can potentially be used in grapevines as well. Genes involved in ABA synthesis include 9-cis-epoxycarotenoid dioxygenase (especially *NCED3*), while genes involved in ABA catabolism include four cytochrome P450 monooxygenases, of which *CYP707A3* is specifically induced under water stress (Nilson and Assmann 2007). Among genes involved in ABA signalling and ABA-mediated stomatal closure, protein kinases (like *OST1*) act as positive regulators while type 2C protein phosphatases (*PP2C*) and farnesyltransferases (like *ERA1*) act as negative regulators (Schroeder et al. 2001). In addition to ABA, drought-induced variations in xylem pH have also been related to changes in g_s (Rodrigues et al. 2008). Although the molecular mechanism for these changes is unknown, Wilkinson and Davies (2008) have shown that xylem pH can be manipulated using buffered foliar sprays, resulting in modified gs, and have therefore proposed this method for managing WUE_{leaf} in a horticultural context.

In addition to chemical signalling, losses of hydraulic conductivity of xylem vessels have been suggested to induce down-regulation of g_s in grapevines under water stress (Schultz 2003a, Rodrigues et al. 2008), but especially during recovery after water stress (Lovisolo et al. 2008a, Pou et al. 2008). In particular, root hydraulic conductivity (Vandeleur et al. 2009) and the conductivity of leaf petioles (Lovisolo et al. 2008a) have been related to g_s. Differences in root hydraulic conductivity and vulnerability among cultivars of *V. vinifera* have been described, which can be either that of the cultivar itself (Vandeleur et al. 2009) or the rootstock on which the cultivar is grafted (Alsina et al. 2007, Lovisolo et al. 2008a,b). Although part of hydraulic losses may be because of embolism-induced cavitation of xylem vessels, an important regulatory role of aquaporins has been demonstrated (Galmés et al. 2007, Lovisolo et al. 2008b, Vandeleur et al. 2009). Indeed, during water stress, drought-adapted *V. berlandieri × V. rupestris* rootstocks present a higher proportion of aquaporin-dependent, rapidly reversible down-regulation of hydraulic conductivity and a lower proportion of embolism-dependent loss of conductivity than drought-sensitive *V. berlandieri × V. riparia* rootstocks (Lovisolo et al. 2008b).

The genes for several plasma membrane intrinsic protein aquaporins (*PIP1* and *PIP2*) as well as tonoplast intrinsic proteins are well characterised. In leaves, all these aquaporin genes are down-regulated during water stress and up-regulated after re-watering (Galmés et al. 2007). In roots, instead, some are unresponsive and some are up-regulated during water stress and down-regulated after re-watering (Galmés et al. 2007). This differential behaviour illustrates the different function of aquaporins in leaves and roots, and the complex nature of their regulation (Kaldenhoff et al. 2008). A recent study (Vandeleur et al. 2009) has demonstrated that the low stomatal control associated with anisohydric behaviour in Chardonnay depends on the maintenance, during water stress, of constant amplitude in the diurnal variations of root hydraulic conductance, which is instead reduced in the isohydric cultivar Grenache. These differences are associated with different expression patterns of aquaporins. In addition, aquaporins have also been suggested to be directly involved in guard cell movements, although this has yet to be demonstrated (Kaldenhoff et al. 2008), and are certainly involved in mesophyll conductance to CO_2 and photosynthesis (see next section). Clearly, aquaporins have diverse and important roles in processes associated with WUE at different levels, and are therefore a key target for genetically improving WUE.

In summary, genetic engineering for regulated gs and improved WUE_{leaf} can be achieved based on genes involved in ABA synthesis and signalling, as well as modified aquaporin genes. Nevertheless, as discussed in previous sections, improving WUE_{leaf} by means of reducing g_s may result in decreased photosynthesis and yield and, sometimes, in decreased WUE_{WP}. Therefore, increasing WUE_{WP} by means of decreasing plant respiration or increasing leaf photosynthesis at any given level of transpiration is likely to be more successful.

5.7 DECREASING PLANT RESPIRATION

Although plant respiration is a fundamental process for plant life, sustaining plant growth (growth respiration) and maintenance, which includes ion transport and compartmentation, protein turnover and tissue acclimation to environmental change (Amthor 2000, Sharma-Natu and Ghildiyal 2005), it is also associated with important carbon losses, which decreased net carbon gain and, hence, WUEWP. In absolute values, the rates of respiration are much lower than photosynthesis rates. However, because photosynthesis occurs in leaves and during the day only, while respiration occurs in all plant organs and through the entire plant life, the overall carbon losses by respiration are substantial. Also, the percentage of carbon gained in photosynthesis that is lost by whole-plant respiration typically ranges from 30 to 90% over the whole season, depending on the plant types and environmental conditions (Amthor 2000). Such estimates have not been accurately performed in grapevines, because very few studies have addressed respiration rates in this species, especially in roots (Comas et al. 2000, Huang et al. 2005). For leaves, it has been shown that respiration rates do not differ between sun leaves with high photosynthesis rates and shade leaves (Escalona et al. 1999a, 2003, Zufferey et al. 2000). The latter are often reported to present low photosynthesis, contributing to low WUE_{leaf} in the latter leaves and in the whole plant, although this may depend on their use of sunflecks, which may depend on many factors that are largely unknown (Kriedemann et al. 1973, Intrieri et al. 1995). Moreover, contrary to photosynthesis, leaf respiration is not impaired under water stress (Escalona et al. 1999b, 2003), and indeed, it is strongly enhanced by increased leaf temperature (Zufferey et al. 2000), a condition occurring under water stress (Figure 5).

For all these reasons, and because maintenance respiration uses a substantial proportion of the total carbon assimilated, it has been suggested that crop production and WUE_{WP} could be increased by reducing this component of respiration in favour of growth respiration. In particular, ca. 30–

40% of total maintenance respiration is associated with cyanide-resistant alternative oxidase (AOX; Florez-Sarasa et al. 2007), and temperature does not affect mitochondrial electron partitioning in non-stressed leaves (Macfarlane et al. 2009). Because this pathway does not contribute to ATP synthesis and growth, it has been suggested that reducing AOX could result in significant increases of plant carbon balance and WUE_{WP} (Loomis and Amthor 1999, Sharma-Natu and Ghildiyal 2005). However, the proportion of respiration occurring via the AOX increases under drought in some species (Ribas-Carbo et al. 2005), and it has been suggested that AOX has a potential role in more fundamental processes including photosynthesis (Juszczuk et al. 2007) and photoprotection (Bartoli et al. 2005). Owing to these suggestions, and to the fact that measurements of AOX activity in grapevines are lacking, we suggest that further studies are needed to properly characterise plant respiration in grapevines prior to selecting it as a target for improving WUE_{WP}.

5.8 INCREASING LEAF PHOTOSYNTHESIS

5.8.1 MESOPHYLL CONDUCTANCE TO CO_2: IMPROVING CO_2 AVAILABILITY FOR PHOTOSYNTHESIS

During photosynthesis, CO_2 moves from the atmosphere (C_a), surrounding the leaf to the sub-stomatal internal cavities (C_i) through stomata, and from there, to the site of carboxylation inside the chloroplast stroma (C_c) through the leaf mesophyll. The pathway through stomata is regulated by stomatal conductance (g_s), which affects both photosynthesis and transpiration so that increasing photosynthesis by increasing g_s results in reduced WUE_{leaf} (Figure 4). The internal leaf CO_2 diffusion component is determined by the so-called mesophyll conductance (g_m), which can be divided into at least three components, i.e. conductance through intercellular air spaces (g_{ias}), through cell wall (g_w) and through the liquid phase inside cells (g_{liq}). Concerning g_m, evidence has accumulated showing that, contrary to early thoughts, this is finite and variable, and that it can change as fast as g_s in response to environmental variables (reviewed in Flexas et al. 2008). While g_{ias} and g_w mostly depend on complex leaf structural traits that may not change in the short term, g_{liq} has been shown to partly depend on rapidly regulated proteins, particularly aquaporins (Hanba et al. 2004, Flexas et al. 2006b).

Several authors have already suggested that increasing g_m would increase WUE_{leaf} (Parry et al. 2005, Flexas et al. 2008), because it would result in increased C_c and, hence, photosynthesis (A_N) without any effect on leaf transpiration, i.e. it will displace the A_N–g_s relationship vertically (Figure 4). Moreover, one environmental variable typically inducing strong decreases of g_m is water stress (Flexas et al. 2002b, 2008). A recent study by Miyazawa et al. (2008) in tobacco showed that water stress-induced decreases of g_m could be mimicked by adding $HgCl_2$ to well-watered plants. Because $HgCl_2$ is an inhibitor of gating some aquaporins, and there was no evidence for water stress-induced changes in the amounts of aquaporins, it was suggested that water stress led to decreased g_m by means of aquaporin closure (Miyazawa et al. 2008). Therefore, we propose that increasing g_m at any given g_s (i.e. increasing the ratio g_m/g_s) will result in increased A_N/g_s and, hence, potentially increased WUE_{WP} (Figure 2).

A positive relationship between g_m/g_s and A_N/g_s under water stress occurs in field-grown grapevines, under the conditions described in Figure 5. With decreasing g_s because of progressive drought, both g_m/g_s and A_N/g_s increased simultaneously (Figure 7a), except at very severe water stress (i.e. the lowest g_s) in which A_N/g_s decreased because of metabolic impairment of photosynthesis (Flexas et al. 2002b) while g_m/g_s still increased. With the exception of this value, a high degree of correlation was observed between g_m/g_s and A_N/g_s (Figure 7b). Similar simultaneous increases in g_m/g_s and A_N/g_s were shown in water-stressed tobacco or in irrigated tobacco after mercurial addition (Miyazawa et al. 2008).

Therefore, it appears that g_m is a potential target for breeding and/or genetic engineering of WUE_{leaf}. That g_m has a genetic basis in grapevines was evidenced by Patakas et al. (2003a,b), who described important differences among cultivars in the leaf anatomical characteristics typically affecting g_{ias}, g_w and, perhaps, g_{liq} (i.e. g_m), such as palisade and spongy parenchyma thickness, the fraction of the intercellular air spaces or the surface of mesophyll cells exposed to the intercellular air spaces per unit leaf area. These anatomical differences were related to differences in photosynthetic efficiency, but g_m was not determined in that study.

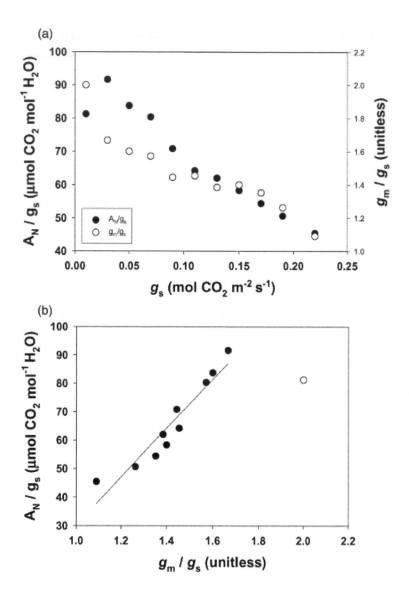

FIGURE 7: (a) The response of intrinsic WUE$_{leaf}$ (A$_N$/g$_s$, filled circles) and the ratio of mesophyll to stomatal conductance to CO$_2$ (g$_m$/g$_s$, empty circles) during progressive water stress in field-grown grapevines in Mallorca (Balearic Islands, Spain). Environmental conditions and original data are from Figure 5. (b) The relationship between A$_N$/g$_s$ and g$_m$/g$_s$ during the drought cycle (the outlier corresponds to the lowest absolute g$_s$, when metabolic impairment of photosynthesis occurred, see Flexas et al. 2002b). WUE, water use efficiency.

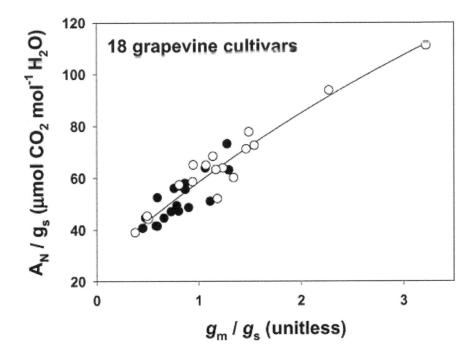

FIGURE 8: The relationship between A_N/g_s and g_m/g_s in 18 different cultivars of *Vitis vinifera* growing outdoors in pots during summer in Mallorca (Balearic Islands, Spain). Plants were either irrigated at field capacity daily (filled circles) or without water for 6 days (open circles, corresponding to a 40% decline in substrate water content). Data recalculated from Bota et al. (2001).

However, Bota et al. (2001) performed simultaneous measurements of gas exchange and chlorophyll fluorescence in up to 20 different grapevine cultivars, subjected to either full irrigation or water stress. From these experimental data, a preliminary, rough estimate of g_m was calculated using the variable chlorophyll fluorescence method already developed for grapevines (Flexas et al. 2002b). The estimations of g_m are defined as 'preliminary, rough' because several input parameters of this method (i.e. leaf absorptance and light distribution between photosystems I and II, Rubisco specificity factor and leaf respiration) were not determined specifically for each cultivar and treatment, but a common constant value was used after determinations in cultivars Tempranillo and Manto Negro (Flexas et al.

2002b). Although differences among cultivars in Rubisco specificity are probably low (Bota et al. 2002, see next section), slight inter-cultivar differences in leaf absorptance and respiration may bias results to some extent (indeed, of the 20 cultivars originally evaluated, only 18 were included in the present study, the other two displaying unreliable g_m values). Despite their preliminary character, the results are promising because significant differences in g_m/g_s were strongly and positively correlated with the already described differences in A_N/g_s, both considering irrigated plants only or including water-stressed plants (Figure 8).

In summary, differences in gm among grapevine cultivars exist, being associated with leaf anatomical differences (Patakas et al. 2003a,b), and, perhaps, with patterns of aquaporin expression or activity, which translate to differences in A_N/g_s, both under irrigation and water stress (Figure 8). While anatomical differences may depend on complex gene interactions, aquaporins depend on single genes, although their post-translational regulation may be more complex. As such, they are primary target candidates for attempts at genetic manipulation aimed to increase WUE_{leaf}. Studies are urgently needed to extend our knowledge on genotypic variations in gm and WUE_{leaf}, to evaluate whether these translate into whole-plant and yield-based differences in WUE and to identify specific aquaporins responsible for these differences.

5.8.2 RUBISCO: IMPROVING CO_2 AVAILABILITY FOR PHOTOSYNTHESIS

Because grapevine photosynthesis operates in the Rubisco-limited region and decreased gs and gm during water stress result in lowered chloroplast CO_2 availability (C_c), increasing Rubisco specificity for CO_2 may be more effective in increasing WUE_{leaf} in grapevines under Mediterranean conditions than increasing Rubisco abundance, activation state or catalytic rate (Parry et al. 2007). Nevertheless, increased Rubisco activity should not be totally ruled out as a way to improve WUE_{leaf}. In a recent study with the *Vitis* rootstock Richter-110 (Flexas et al. 2009), changes in the maximum velocity of carboxylation ($V_{c,max}$, which relates to Rubisco activity) occurred as a result of water stress and re-watering, and were related to intrinsic WUE_{leaf}. In these plants, $V_{c,max}$ decreased during water stress (while A_N/g_s increased as a consequence of increased g_m/g_s), but acclimation to water stress over 1 week resulted in increased $V_{c,max}$ above control values after re-watering, so that A_N/g_s in previously stressed plants was kept higher than in never-stressed plants for weeks. Therefore, although further studies are needed to verify this trend, increasing Rubisco capacity could be a means of increasing WUE_{leaf} during water stress and re-watering cycles in grapevines.

Rubisco specificity for CO_2 (τ) was determined in two grapevine cultivars, Tempranillo and Manto Negro. As expected, an identical value (100 mol/mol) was obtained (Bota et al. 2002). It is very unlikely that evolution under human selection resulted in relevant differences in the rbcL sequence among cultivars. Nevertheless, contrary to early thoughts, there is now evidence that substantial variability in τ exists even within C_3 plants, and values larger than those of grapevines (i.e. up to 110 mol/mol) have been described for the Mediterranean species *Limonium gibertii* (Galmés et al. 2005). Even higher values (up to 240 mol/mol) are found in some red algae, the highest known value corresponding to *Galdieria partita* (Uemura et al. 1997). Therefore, there are better forms of Rubisco outside grapevines, which may be used to improve carboxylation efficiency and WUE_{leaf}. Moreover, such differences have a clear genetic basis. For instance, the large subunits of Rubisco in spinach ($\tau = 80$), grapevines ($\tau = 100$) and *L. gibertii* ($\tau = 110$) present up to 39 non-conserved amino acids (Figure 9). Of this, *Vitis* (the species having intermediate specificity) shares 17 positions in common with *Limonium* (and not with spinach) and 14 in common with spinach (and not with *Limonium*). In contrast, the two species with extreme specificities only share five positions in common (and not with *Vitis*). Although this result may suggest that improvement of τ could be achieved by site-directed mutagenesis, this goal seems yet too far because of the lack of complete knowledge on the relationship between specific amino acids and Rubisco kinetics. Therefore, trying to replace a native Rubisco with a better foreign one seems to be the most likely option to succeed, at least in the short term.

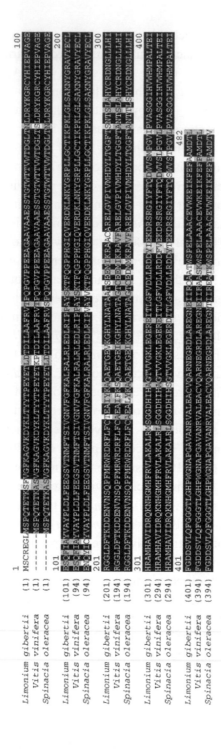

FIGURE 9: Alignment (Vector NTI; Invitrogen, Carlsbad, CA, USA) of the amino acid sequences for the Rubisco large subunit from *Limonium gibertii* (τ = 110, GenBank accession number CAH10354), *Vitis vinifera* (τ = 100, accession number YP567084) and *Spinacia oleracea* (τ = 80, accession number NP054944). Black boxes indicate fully conserved residues among species. Dark grey boxes indicate partially conserved residues, while the exceptions are indicated in pale grey (similar type of amino acids) or white (different type of amino acids).

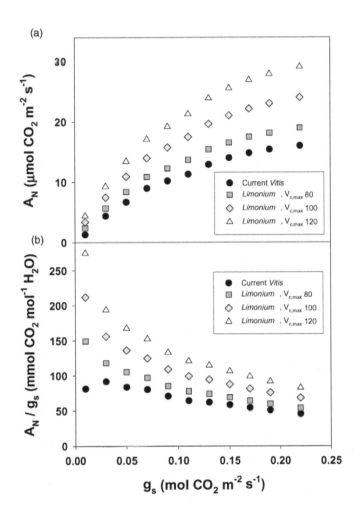

FIGURE 10: Simulation of the theoretical effects of replacing native Rubisco of grapevine with Rubisco from *Limonium gibertii* on (a) net photosynthesis (A_N) and (b) intrinsic WUE (A_N/g_s) at different water stress intensities (i.e. different stomatal conductance, gs). The environmental conditions and original data (black circles) are from Figure 3. Three different scenarios are considered, representing the achievement of a grapevine plant with a Rubisco specificity factor (τ) equal to that of *Limonium* but with a maximum velocity of carboxylation ($V_{c,max}$) of (i) 80 µmol/m²/s (dark grey squares), as described for *Vitis vinifera* by Schultz (2003b); (ii) 100 µmol/m²/s (pale grey rhomboids), as described for *Vitis vinifera* by Flexas et al. (2006a); or (iii) 120 µmol/m²/s (white triangles), as described for *Limonium gibertii* by Dr Jeroni Galmes et al. (unpublished data, 2007). All the $V_{c,max}$ values are scaled to a leaf temperature of 35°C using the temperature functions described by Bernacchi et al. (2002).

Advances in chloroplast transformation have led to successful replacement of the native large Rubisco subunit of tobacco by other higher plants versions (Kanevski et al. 1999, Whitney and Andrews 2001, Sharwood et al. 2008). So far, transformed plants present low amounts of Rubisco because of the lack of proper chaperones and other molecules necessary for correct transcription, translation and assembling of the enzyme (Sharwood et al. 2008). However, these seem to be minor problems, and there are good perspectives towards the possibility of obtaining improved plants by transferring Rubiscos within higher plants.

Bearing this in mind, a simulation of the theoretical effects of placing Rubisco from *L. gibertii* into grapevines is presented. Parameters from real conditions, such as those described in Figure 5 (i.e. assuming that transformation will not affect conductance of CO_2 and leaf temperature), and the model by Farquhar et al. (1980) have been used. Applying this model, values of A_N (Figure 10a) and of A_N/g_s (Figure 10b) are calculated for each water stress condition in Figure 5. We have considered three different scenarios: a grapevine plant with a Rubisco specificity factor (τ) equal to that of *Limonium* but with a maximum velocity of carboxylation ($V_{c,max}$) of (i) 80 $\mu mol/m^2/s$, as described for *V. vinifera* by Schultz (2003b); (ii) 100 $\mu mol/m^2/s$, as described for *V. vinifera* by Flexas et al. (2006a); or (iii) 120 $\mu mol/m^2/s$, as found for *L. gibertii* by Dr Jeroni Galmés et al. (unpublished data, 2007). Depending on the scenario considered, such transformation would result in potential increases of maximum net photosynthesis (A_N) by 15 to 80% as compared with current values (Figure 10a). This would result in similar increases of intrinsic WUE_{leaf} under irrigation, but these increases will become much larger as water stress intensifies, potentially reaching increases of up to 100% or higher (Figure 10b).

Therefore, the expected benefits of Rubisco engineering on WUE_{leaf} are high, and, hence, this should be a priority area of research for the immediate future. In particular, Rubiscos from grape cultivars should be fully characterised both molecularly (sequence) and biochemically (τ), and the first steps for chloroplast transformation in grapevines attempted.

TABLE 2: Summary of the different approaches to improving WUE, proposed in the present review, with their corresponding techniques and targets, including those achievable by agronomy/management and those achievable by biotechnology.

Way to improve WUE	Technique/target	Reference(s)
Agronomy/management		
Reducing soil evaporation and runoff	Mulching	Buckerfield and Webster (2001); Hatfield et al. (2001)
	Vineyard inter-cropping	Monteiro and Lopes (2007); Gulías et al. (2008)
Modifying root growth patterns + inducing partial stomatal closure and reduced plant transpiration	Regulated deficit irrigation	Cifre et al. (2005); Costa et al. (2007)
	Partial root drying	Dry et al. (2001); Chaves et al. (2007)
Optimising light interception by the canopy and radiation use efficiency	Training system	Carbonneau (1980); Escalona et al. (1999a, 2003)
	Pruning	Williams and Ayars (2005)
Biotechnology		
Reducing leaf (and canopy) transpiration	Reducing cuticular conductance	—
	Reducing night and/or day stomatal conductance (ABA genes)	Nilson and Assmann (2007); Matus et al. (2008)
Optimising water uptake, transport and transpiration	Aquaporins	Kaldenhoff et al. (2008); Vandeleur et al. (2009)
Reducing carbon losses in respiration	Reducing alternative oxidase	Loomis and Amthor (1999), present review
Increasing photosynthesis at any given rate of transpiration	Increasing CO_2 availability (aquaporin-mediated gm)	Flexas et al. (2008), present review
	Increasing carboxylation efficiency (Rubisco specificity factor for CO_2)	Parry et al. (2007), present review

For agronomic techniques, two references are given as examples of their application in grapevines, while for biotechnological techniques, the first reference is general and the second refers specifically to grapevines. ABA, abscisic acid; WUE, water use efficiency.

5.9 CONCLUDING REMARKS

Present and predicted future climate conditions indicate a reduction of water availability for both human and agricultural consumption, especially in grapevine growing areas. These changes enforce the need to improve WUE in grapevines in order to secure an environmentally friendly and sustainable viticulture.

This review presents several agronomic approaches to improve the WUE_C (summarised in Table 2), which, in some cases, could also lead to improved grape quality. Important progress can be achieved from agronomic practices aimed at reducing direct soil evaporation and runoff, including mulching, vegetative growth contention by spring competition for water with cover crops and green pruning. However, a precise evaluation of these in terms of contribution to water saving and WUE_C improvement in different environments is urgent.

Additional to these agronomic practices, the possibility of increasing WUE_{WP} through genetic changes in grapevines has been highlighted as promising (Table 2). These improvements would be based on a better control of water losses (reducing night transpiration, improvement of stomatal control), with the help of recent biotechnological developments achieved in other plant species. However, concomitant reductions in carbon gain should be taken into account when evaluating these results.

Alternatively, WUE_{WP} could be increased by genetically improving net plant carbon uptake, through either reducing carbon losses by respiration or improving gross photosynthesis. Recent reports on the role of mesophyll conductance in photosynthesis limitations and the relationships of CO_2 diffusion with the expression of aquaporins suggest that they may become a potential target through which to achieve such a goal. On the other hand, recent identification of plants with Rubisco of superior specificity factor, which could be transferred to crops such as grapevines, opens an exciting new approach for improving WUE.

As the present review shows, improving WUE is an unavoidable subject of research for sustainable viticulture, and offers fascinating areas of research, which merits to be explored in near future.

REFERENCES

1. Alsina, M.M., De Herralde, F., Aranda, X., Save, R. and Biel, C. (2007) Water relations and vulnerability to embolism are not related: experiments with eight grapevine cultivars. Vitis 46, 1–6.
2. Amthor, J.S. (2000) The McCree-de Wit-Penning de Vries-Thornley respiration paradigms: 30 years later. Annals of Botany 86, 1–20.
3. Antolín, M.C., Ayari, M. and Sánchez-Díaz, M. (2006) Effects of partial rootzone drying on yield, ripening and berry ABA in potted Tempranillo grapevines with split roots. Australian Journal of Grape and Wine Research 12, 13–20.
4. Antolín, M.C., Santesteban, H., Santamaría, E., Aguirreolea, J. and Sánchez-Díaz, M. (2008) Involvement of abscisic acid and polyamines in berry ripening of Vitis vinifera (L.) subjected to water deficit irrigation. Australian Journal of Grape and Wine Research 14, 123–133.
5. Araus, J.L. (2004) The problems of sustainable water use in the Mediterranean and research requirements for agriculture. Annals of Applied Biology 144, 259–272.
6. Baeza, P., Ruiz, C., Cuevas, E., Sotés, V. and Lisarrague, J.R. (2005) Ecophysiological and agronomic response of Tempranillo grapevines to four training systems. American Journal of Enology and Viticulture 56, 129–138.
7. Bartoli, C.G., Gómez, F., Gergoff, G., Guiamét, J.J. and Puntarulo, S. (2005) Upregulation of the mitochondrial alternative oxidase pathway enhances photosynthetic electron transport under drought conditions. Journal of Experimental Botany 56, 1269–1276.
8. Bernacchi, C.J., Portis, A.R., Nakano, H., Von Caemmerer, S. and Long, S.P. (2002) Temperature response of mesophyll conductance. Implications for the determination of Rubisco enzyme kinetics and for limitations to photosynthesis in vivo. Plant Physiology 130, 1992–1998.
9. Bindon, K., Dry, P. and Loveys, B. (2008) Influence of partial rootzone drying on the composition and accumulation of anthocyanins in grape berries (Vitis vinifera cv. Cabernet Sauvignon). Australian Journal of Grape and Wine Research 14, 91–103.
10. Bisson, L.F., Waterhouse, A.L., Ebeler, S.E., Walker, M.A. and Lapsley, J.T. (2002) The present and future of the international wine industry. Nature 418, 696–699.
11. CrossRef,PubMed,CAS,Web of Science® Times Cited: 32,ADS
12. Blum, A. (2005) Drought resistance, water-use efficiency, and yield potential – are they compatible, dissonant, or mutually exclusive? Australian Journal of Agricultural Research 56, 1159–1168.
13. Bota, J., Flexas, J. and Medrano, H. (2001) Genetic variability of photosynthesis and water use in Balearic grapevine cultivars. Annals of Applied Biology 138, 353–365.
14. Bota, J., Flexas, J., Keys, A.J., Loveland, J., Parry, M.A.J. and Medrano, H. (2002) CO2/O2 specificity factor of ribulose-1,5-bisphosphate carboxylase/oxygenase in grapevines (Vitis vinifera L.): first in vitro determination and comparison to in vivo estimations. Vitis 41, 163–168.

15. Bota, J., Stasyk, O., Flexas, J. and Medrano, H. (2004) Effect of water stress on partitioning of 14C-labelled photosynthates in Vitis vinifera. Functional Plant Biology 31, 697–708.

16. Boyer, J.S., Wong, S.C. and Farquhar, G.D. (1997) CO2 and water vapor exchange across leaf cuticle (epidermis) at various water potentials. Plant Physiology 114, 185–191.

17. Buckerfield, J. and Webster, K. (2001) Managing young vines. Responses to mulch continue – results from five years of field trials. The Australian and New Zealand Grapegrower and Winemaker 543, 71–78.

18. Caird, M.A., Richards, J.H. and Donovan, L.A. (2007) Nighttime stomatal conductance and transpiration in C3 and C4 plants. Plant Physiology 143, 4–10.

19. Carbonneau, A. (1980) Recherche sur les systèmes de conduite de la vigne: essai de maîtrise du microclimat et de la plante entière pour produire économiquement du raisin de qualité. PhD Thesis, Université du Bordeaux II, Léo-Saignat, Bordeaux, 235 pp.

20. Castellarin, S.D., Matthews, M.A., Di Gaspero, G. and Gambetta, G.A. (2007) Water deficits accelerate ripening and induce changes in gene expression regulating flavonoid biosynthesis in grape berries. Planta 227, 101–112.

21. Chaves, M.M., Santos, T.P., Souza, C.R., Ortuño, M.F., Rodrigues, M.L., Lopes, C.M., Maroco, J.P. and Pereira, J.S. (2007) Deficit irrigation in grapevine improves water-use-efficiency without controlling vigour and production quality. Annals of Applied Biology 150, 237–252.

22. Choné, X., Van Leeuwen, C., Dubourdie, D. and Gaudillére, J.P. (2001) Stem water potential is a sensitive indicator of grapevine water status. Annals of Botany 87, 477–483.

23. Cifre, J., Bota, J., Escalona, J.M., Medrano, H. and Flexas, J. (2005) Physiological tools for irrigation scheduling in grapevines (Vitis vinifera L.): an open gate to improve water-use efficiency? Agriculture, Ecosystems and Environment 106, 159–170.

24. Comas, L.H., Eissenstat, D.M. and Lakso, A.N. (2000) Assessing root death and root system dynamics in a study of grape canopy pruning. New Phytologist 147, 171–178.

25. Condon, A.G., Richards, R.A., Rebetzke, G.J. and Farquhar, G.D. (2004) Breeding for high water-use efficiency. Journal of Experimental Botany 55, 2447–2460.

26. Conejero, W., Alarcón, J.J., García-Orellana, Y., Nicolás, E. and Torrecillas, A. (2007) Evaluation of sap flow and trunk diameter sensors used for irrigation scheduling in early maturing peach trees. Tree Physiology 27, 1753–1759.

27. Correia, M.J., Pereira, J.S., Chaves, M.M., Rodrigues, M.L. and Pacheco, C.A. (1995) ABA xylem concentrations determine maximum daily leaf conductance of field-grown Vitis vinifera L. plants. Plant, Cell and Environment 18, 511–521.

28. Costa, J.M., Ortuño, M.F. and Chaves, M.M. (2007) Deficit irrigation as a strategy to save water: physiology and potential application to horticulture. Journal of Integrative Plant Biology 49, 1421–1434.

29. Cramer, G.R., Ergül, A., Grimplet, J., Tillett, R.L., Tattersall, E.A.R., Bohlman, M.C., Vincent, D., Sonderegger, J., Evans, J., Osborne, C., Quilici, D., Schlauch, K.A., Schooley, D.A. and Cushman, J.C. (2007) Water and salinity stress in grape-

vines: early and late changes in transcript and metabolite profiles. Functional & Integrative Genomics 7, 111–134.

30. Dobrowsky, S.Z., Pushnik, J.C., Zarco-Tejada, P.J. and Ustin, S.L. (2005) Simple reflectance indices track heat and water stress-induced changes in steady-state chlorophyll fluorescence at canopy scale. Remote Sensing of the Environment 97, 403–414.

31. Dry, P.R. and Loveys, B.R. (1998) Factors influencing grapevine vigour and the potential for control with partial rootzone drying. Australian Journal of Grape and Wine Research 4, 140–148.

32. Dry, P.R., Loveys, B.R., McCarthy, M.G. and Stoll, M. (2001) Strategic irrigation management in Australian vineyards. Journal International des Sciences de la Vigne et du Vin 35, 129–139.

33. Escalona, J.M., Flexas, J. and Medrano, H. (1999a) Contribution of different levels of plant canopy to total carbon assimilation and intrinsic water use efficiency of Manto Negro and Tempranillo grapevines. Acta Hotriculturae 493, 141–148.

34. Escalona, J.M., Flexas, J. and Medrano, H. (1999b) Stomatal and non-stomatal limitations of photosynthesis under water stress in field-grown grapevines. Australian Journal of Plant Physiology 26, 421–433.

35. Escalona, J.M., Flexas, J. and Medrano, H. (2002) Drought effects on water flow, photosynthesis and growth of potted grapevines. Vitis 41, 57–62.

36. Escalona, J.M., Flexas, J., Bota, J. and Medrano, H. (2003) From leaf photosynthesis to grape yield: influence of soil water availability. Vitis 42, 57–64.

37. Evain, S., Flexas, J. and Moya, I. (2004) A new instrument for passive remote sensing: 2. Measurement of leaf and canopy reflectance changes at 531 nm and their relationship with photosynthesis and chlorophyll fluorescence. Remote Sensing of Environment 91, 175–185.

38. Farquhar, G.D., Von Caemmerer, S. and Berry, J.A. (1980) A biochemical model of photosynthetic CO2 assimilation in leaves of C3 species. Planta 149, 78–90.

39. Feng, L., Han, Y., Liu, G., An, B., Yang, J., Yang, G., Li, Y. and Zhu, Y. (2007) Overexpression of sedoheptulose-1,7-bisphosphate enhances photosynthesis and growth under salt stress in transgenic rice plants. Functional Plant Biology 34, 822–834.

40. Fernández, J.E., Green, S.R., Caspari, H.W., Diaz-Espejo, A. and Cuevas, M.V. (2008) The use of sap flow measurements for scheduling irrigation in olive, apple and Asian pear trees and in grapevines. Plant and Soil 305, 91–104.

41. Flexas, J., Badger, M., Chow, W.S., Medrano, H. and Osmond, C.B. (1999a) Analysis of the relative increase in photosynthetic O2 uptake when photosynthesis in grapevine leaves is inhibited following low night temperatures and/or water stress. Plant Physiology 121, 675–684.

42. Flexas, J., Escalona, J.M. and Medrano, H. (1999b) Water stress induces different levels of photosynthesis and electron transport rate regulations in grapevines. Plant, Cell and Environment 22, 39–48.

43. Flexas, J., Escalona, J.M., Evain, S., Gulías, J., Moya, I., Osmond, C.B. and Medrano, H. (2002a) Steady-state chlorophyll fluorescence (Fs) measurements as a tool to follow variations of net CO2 assimilation and stomatal conductance during water-stress in C3 plants. Physiologia Plantarum 114, 231–240.

44. Flexas, J., Bota, J., Escalona, J.M., Sampol, B. and Medrano, H. (2002b) Effects of drought on photosynthesis in grapevines under field conditions: an evaluation of stomatal and mesophyll limitations. Functional Plant Biology 29, 461–471.

45. Flexas, J., Bota, J., Galmés, J., Medrano, H. and Ribas-Carbó, M. (2006a) Keeping a positive carbon balance under adverse conditions: responses of photosynthesis and respiration to water stress. Physiologia Plantarum 127, 343–352.

46. Flexas, J., Ribas-Carbó, M., Hanson, D.T., Bota, J., Otto, B., Cifre, J., McDowell, N., Medrano, H. and Kaldenhoff, R. (2006b) Tobacco aquaporin NtAQP1 is involved in mesophyll conductance to CO2 in vivo. The Plant Journal 48, 427–439.

47. Flexas, J., Ribas-Carbo, M., Diaz-Espejo, A., Galmés, J. and Medrano, H. (2008) Mesophyll conductance to CO2: current knowledge and future prospects. Plant, Cell and Environment 31, 602–631.

48. Flexas, J., Barón, M., Bota, J., Ducruet, J.-M., Gallé, A., Galmés, J., Jiménez, M., Pou, A., Ribas-Carbo, M., Sajnani, C., Tomás, M. and Medrano, H. (2009) Photosynthesis limitations during water stress acclimation and recovery in the drought-adapted Vitis hybrid Richter-110 (V. berlandieri × V. rupestris). Journal of Experimental Botany 60, 2361–2377.

49. Florez-Sarasa, I.F., Bouma, T.J., Medrano, H., Azcon-Bieto, J. and Ribas-Carbo, M. (2007) Contribution of the cytochrome and alternative pathways to growth respiration and maintenance respiration in Arabidopsis thaliana. Physiologia Plantarum 129, 143–151.

50. Galmés, J., Flexas, J., Keys, A.J., Cifre, J., Mitchell, R.A.C., Madgwick, P.J., Haslam, R.P., Medrano, H. and Parry, M.A.J. (2005) Rubisco specificity factor tends to be larger in plant species from drier habitats and in species with persistent leaves. Plant, Cell and Environment 28, 571–579.

51. Galmés, J., Pou, A., Alsina, M.M., Tomàs, M., Medrano, H. and Flexas, J. (2007) Aquaporin expression in response to different water stress intensities and recovery in Richter-110 (Vitis sp.): relationship with ecophysiological status. Planta 226, 671–681.

52. Gaudillère, J.P., Van Leeuwen, C. and Ollat, N. (2002) Carbon isotope composition of sugars in grapevine, and integrated indicator of vineyard water status. Journal of Experimental Botany 53, 757–763.

53. Gibberd, M.R., Walker, R.R., Blackmore, D.H. and Condon, A.G. (2001) Transpiration efficiency and carbon-isotope discrimination of grapevines grown under well-watered conditions in either glasshouse or vineyard. Australian Journal of Grape and Wine Research 7, 110–117.

54. Ginestar, C., Eastham, J., Gray, S. and Iland, P. (1998) Use of sap-flow sensors to schedule vineyard irrigation. I. Effects of post-veraison water deficit on water relations, vine growth, and yield of Shiraz grapevines. American Journal of Enology and Viticulture 49, 413–420.

55. Grant, O.M., Tronina, L., Jones, H.G. and Chaves, M.M. (2007) Exploring thermal imaging variables for the detection of stress responses in grapevine under different irrigation regimes. Journal of Experimental Botany 58, 815–825.

56. Gregory, P.J. (2004) Agronomic approaches to increasing water use efficiency. In: Water use efficiency in plant biology. Ed. M.A.Bacon (Blackwell Publishing Ltd.: Oxford) pp. 142–167.

57. Gulías, J., Cifre, J., Pou, A., Moreno, M.T., Tomàs, M. and Medrano, H. (2008) Vineyard cover crop effects on leaf water use efficiency and grape production and quality under Mediterranean conditions. VIII International symposium on grapevine physiology and biotechnology, Adelaide, Australia (Book of Abstracts) p. 172.

58. Hanba, Y.T., Shibasaka, M., Hayashi, Y., Hayakawa, T., Kasamo, K., Terashima, I. and Katsuhara, M. (2004) Overexpression of the barley aquaporin HvPIP2; 1 increases internal CO2 conductance and CO2 assimilation in the leaves of transgenic rice plants. Plant and Cell Physiology 45, 521–529.

59. Hatfield, J.L., Sauer, T.J. and Prueger, J.H. (2001) Managing soils to achieve greater water use efficiency: a review. Agronomy Journal 93, 271–280.

60. Huang, X., Lakso, A.N. and Eissenstat, D.M. (2005) Interactive effects of soil temperature and moisture on 'Concord' grape root physiology. Journal of Experimental Botany 56, 2651–2660.

61. Intrieri, C., Zerbi, G., Marchiol, L., Poni, S. and Caiado, T. (1995) Physiological response of grapevine leaves to lightflecks. Scientia Horticulturae 61, 47–59.

62. Intrigliolo, D.S. and Castel, J.R. (2008) Trunk diameter variations as water stress indicator in plum and grapevine. Acta Horticulturae 792, 363–369.

63. Jaillon, O., Aury, J.M., Noel, B., Policriti, A., Clepet, C., Casagrande, A., Choisne, N., Aubourg, S., Vitulo, N., Jubin, C., Vezzi, A., Legeai, F., Hugueney, P., Dasilva, C., Horner, D., Mica, E., Jublot, D., Poulain, J., Bruyère, C., Billault, A., Segurens, B., Gouyvenoux, M., Ugarte, E., Cattonaro, F., Anthouard, V., Vico, V., Del Fabbro, C., Alaux, M., Di Gaspero, G., Dumas, V., Felice, N., Paillard, S., Juman, I., Moroldo, M., Scalabrin, S., Canaguier, A., Le Clainche, I., Malacrida, G., Durand, E., Pesole, G., Laucou, V., Chatelet, P., Merdinoglu, D., Delledonne, M., Pezzotti, M., Lecharny, A., Scarpelli, C., Artiguenave, F., Pè, M.E., Valle, G., Morgante, M., Caboche, M., Adam-Blondon, A.-F., Weissenbach, J., Quétier, F. and Wincker, P. (2007) The grapevine genome sequence suggests ancestral hexaploidization in major angiosperm phyla. Nature 449, 463–467.

64. Juszczuk, I.M., Flexas, J., Szal, B., Dabrowska, Z., Ribas-Carbo, M. and Rychter, A.M. (2007) Effect of mitochondrial genome rearrangement on respiratory activity, photosynthesis, photorespiration, and energy status of MSC16 cucumber (Cucumis sativus L.) mutant. Physiologia Plantaurm 131, 527–541.

65. Kaldenhoff, R., Ribas-Carbo, M., Flexas, J., Lovisolo, C., Heckwolf, M. and Uehlein, N. (2008) Aquaporins and plant water balance. Plant, Cell and Environment 31, 658–666.

66. Kanevski, I., Maliga, P., Rhoades, D.F. and Gutteridge, S. (1999) Plastome engineering of ribulose-1,5-bisphosphate carboxylase/oxygenase in tobacco to form a sunflower large subunit and tobacco small subunit hybrid. Plant Physiology 119, 133–141.

67. Kebeish, R., Niessen, M., Thiruveedhi, K., Bari, R., Hirsch, H.-J., Rosenkranz, R., Stäbler, N., Schönfeld, B., Kreuzaler, F. and Peterhänsel, C. (2007) Chloroplastic photorespiratory bypass increases photosynthesis and biomass production in Arabidopsis thaliana. Nature Biotechnology 25, 593–599.

68. Kriedemann, P.E., Törökfalvy, E. and Smart, R.E. (1973) Natural occurrence and photosynthetic utilisation of sunflecks by grapevine leaves. Photosynthetica 7, 18–27.

69. Long, S.P., Zhu, X.-G., Naidu, S.L. and Ort, D.R. (2006) Can improvement in photosynthesis increase crop yields? Plant, Cell and Environment 29, 315–330.
70. Loomis, R.S. and Amthor, J.S. (1999) Yield potential, plant assimilatory capacity, and metabolic efficiencies. Crop Science 39, 1584–1596.
71. Loveys, B.R. and Kriedemann, P.E. (1974) Internal control of stomatal physiology and photosynthesis. I. Stomatal regulation and associated changes in endogenous levels of abscisic and phaseic acids. Australian Journal of Plant Physiology 1, 407–415.
72. Lovisolo, C., Hartung, W. and Schubert, A. (2002) Whole-plant hydraulic conductance and root-to-shoot flow of abscisic acid independently affected by water stress in grapevines. Functional Plant Biology 29, 1349–1356.
73. Lovisolo, C., Perrone, I., Hartung, W. and Schubert, A. (2008a) An abscisic acid-related reduced transpiration promotes gradual embolism repair when grapevines are rehydrated after drought. New Phytologist 180, 642–651.
74. Lovisolo, C., Tramontini, S., Flexas, J. and Schubert, A. (2008b) Mercurial inhibition of root hydraulic conductance in Vitis spp. rootstocks under water stress. Environmental and Experimental Botany 63, 178–182.
75. Macfarlane, C., Hansen, L., Florez-Sarasa, I. and Ribas-Carbo, M. (2009) Temperature has no effect on alternative oxidase activity in plant mitochondria. Plant, Cell and Environment 32, 585–591.
76. Martin, B., Nienhuis, J., King, G. and Schaefer, A. (1989) Restriction fragment length polymorphisms associated with water-use efficiency in tomato. Science 243, 1725–1728.
77. Masle, J., Gilmore, S.R. and Farquhar, G.D. (2005) The ERECTA gene regulates plant transpiration efficiency in Arabidopsis. Nature 436, 866–870.
78. Matthews, M.A., Ishii, R., Anderson, M.M. and O'Mahony, M. (1990) Dependence of wine sensory attributes on vine water status. Journal of the Science of Food and Agriculture 51, 321–335.
79. Matus, J.T., Galbiati, M., Cominelli, E., Francia, P., Tonelli, C., Cañón, P., Medina, C. and Arce-Johnson, P. (2008) Functional analysis of a putative MYB60 orthologue: a candidate gene to increase drought tolerance in grapevine (Vitis vinifera). VIII International symposium on grapevine physiology and biotechnology, Adelaide, Australia (Book of Abstracts) p. 169.
80. Medrano, H., Bota, J., Abadía, A., Sampol, B., Escalona, J.M. and Flexas, J. (2002) Drought effects on light energy dissipation in high light-acclimated, field-grown grapevines. Functional Plant Biology 29, 1197–1207.
81. Medrano, H., Escalona, J.M., Cifre, J., Bota, J. and Flexas, J. (2003) A ten-year study on the physiology of two Spanish grapevine cultivars under field conditions: effects of water availability from leaf photosynthesis to grape yield and quality. Functional Plant Biology 30, 607–619.
82. Miyazawa, S.-I., Yoshimura, S., Shinazaki, Y., Maeshima, M. and Miyake, C. (2008) Deactivation of aquaporins decreases internal conductance to CO_2 diffusion in tobacco leaves grown under long-term drought. Functional Plant Biology 35, 553–564.
83. Monteiro, A. and Lopes, C.M. (2007) Influence of cover crop on water use and performance of vineyard in Mediterranean Portugal. Agriculture, Ecosystems and Environment 121, 336–342.

84. Morison, J.I.L., Baker, N.R., Mullineaux, P.M. and Davies, W.J. (2008) Improving water use in crop production. Philosophical Transactions of the Royal Society of London Series B 363, 639–658.

85. Murchie, E.H., Pinto, M. and Horton, P. (2009) Agriculture and the new challenges for photosynthesis research. New Phytologist 181, 532–552.

86. Nelson, D.E., Repetti, P.P., Adams, T.R., Creelman, R.A., Wu, J., Warner, D.C., Anstrom, D.C., Bensen, R.J., Castiglioni, P.P., Donnarummo, M.G., Hinchey, B.S., Kumimoto, R.W., Maszle, D.R., Canales, R.D., Kroliwoski, K.A., Dotson, S.B., Gutterson, N., Ratcliffe, O.J. and Heard, J.E. (2007) Plant nuclear factor Y (NF-Y) B subunits confer drought tolerance and lead to improved corn yields on water-limited acres. Proceedings of the National Academy of Sciences of the United States of America 104, 16450–16455.

87. Nilson, S.E. and Assmann, S.M. (2007) The control of transpiration. Insights from Arabidopsis. Plant Physiology 143, 19–27.

88. Parry, M.A.J., Flexas, J. and Medrano, H. (2005) Prospects for crop production under drought: research priorities and future directions. Annals of Applied Biology 147, 211–226.

89. Parry, M.A.J., Madgwick, P.J., Carvalho, J.F.C. and Andralojc, P.J. (2007) Prospects for increasing photosynthesis by overcoming the limitations of Rubisco. Journal of Agricultural Science 145, 31–43.

90. Patakas, A., Kofidis, G. and Bosabalidis, A.M. (2003a) The relationship between CO_2 transfer mesophyll resistance and photosynthetic efficiency in grapevine cultivars. Scientia Horticulturae 97, 255–263.

91. Patakas, A., Stavrakas, D. and Fisarakis, I. (2003b) Relationship between CO_2 assimilation and leaf anatomical characteristics in two grapevine cultivars. Agronomie 23, 293–296.

92. Patakas, A., Noitsakis, B. and Chouzouri, A. (2005) Optimization of irrigation water use in grapevines using the relationship between transpiration and plant water status. Agriculture, Ecosystems and Environment 106, 253–259.

93. Peterhansel, C., Niessen, M. and Kebeish, R.M. (2008) Metabolic engineering towards the enhancement of photosynthesis. Photochemistry and Photobiology 84, 1317–1323.

94. Pou, A., Flexas, J., Alsina, M.M., Bota, J., Carambula, C., De Herralde, F., Galmés, J., Lovisolo, C., Jiménez, M., Ribas-Carbo, M., Rusjan, D., Secchi, F., Tomàs, M., Zsófi, Z. and Medrano, H. (2008) Adjustments of water-use efficiency by stomatal regulation during drought and recovery in the drought-adapted Vitis hybrid Richter-110 (V. berlandieri × V. rupestris). Physiologia Plantarum 134, 313–323.

95. Ribas-Carbo, M., Taylor, N.L., Giles, L., Busquets, S., Finnegan, P.M., Day, D.A., Lambers, H., Medrano, H., Berry, J.A. and Flexas, J. (2005) Effects of water stress on respiration in soybean (Glycine max. L.) leaves. Plant Physiology 139, 466–473.

96. Rivero, R.M., Kojima, M., Gepstein, A., Sakakibara, H., Mittler, R., Gepstein, S. and Blumwald, E. (2007) Delayed leaf senescence induces extreme drought tolerance in a flowering plant. Proceedings of the National Academy of Sciences of the United States of America 104, 19631–19636.

97. Rodrigues, M.L., Chaves, M.M., Wendler, R., David, M.M., Quick, W.P., Leegood, R.C., Stitt, M. and Pereira, J.S. (1993) Osmotic adjustment in water stressed grape-

vine leaves in relation to carbon assimilation. Australian Journal of Plant Physiology 20, 309–321.

98. Rodrigues, M.L., Santos, T.P., Rodrigues, A.P., De Souza, C.R., Lopes, C.M., Maroco, J.P., Pereira, J.S. and Chaves, M.M. (2008) Hydraulic and chemical signalling in the regulation of stomatal conductance and plant water use in field grapevines growing under deficit irrigation. Funtional Plant Biology 35, 565–579.

99. Rodríguez-Pérez, J.R., Riaño, D., Carlisle, E., Ustin, S. and Smart, D.R. (2007) Evaluation of hyperspectral reflectance indexes to detect grapevine water status in vineyards. American Journal of Enology and Viticulture 58, 302–317.

100. Saez, A., Robert, N., Maktabi, M.H., Schroeder, J.I., Serrano, R. and Rodriguez, P.L. (2006) Enhancement of abscisic acid sensitivity and reduction of water consumption in Arabidopsis by combined inactivation of the protein phosphatases type 2C ABI1 and HAB1. Plant Physiology 141, 1389–1399.

101. Salón, J.L., Chirivella, C. and Castel, J.R. (2005) Response of cv. Bobal to timing of deficit irrigation in Requena, Spain: water relations, yield, and wine quality. American Journal of Enology and Viticulture 56, 1–8.

102. Satisha, J., Prakash, G.S. and Venugopalan, R. (2006) Statistical modeling of the effect of physio-biochemical parameters on water use efficiency of grape varieties, rootstocks and their stionic combinations under moisture stress conditions. Turkish Journal of Agriculture and Forestry 30, 261–271.

103. Schroeder, J.I., Kwak, J.M. and Allen, G.J. (2001) Guard cell abscisic acid signaling and engineering drought hardiness in plants. Nature 410, 327–330.

104. Schultz, H.R. (2000) Climate change and viticulture: a European perspective on climatology, carbon dioxide and UV-B effects. Australian Journal of Grape and Wine Research 6, 2–12.

105. Schultz, H.R. (2003a) Differences in hydraulic architecture account for near-isohydric and anisohydric behaviour of two field-grown Vitis vinifera L. cultivars during drought. Plant, Cell and Environment 26, 1393–1405.

106. Schultz, H.R. (2003b) Extension of a Farquhar model for limitations of leaf photosynthesis induced by light environment, phenology and leaf age in grapevines (Vitis vinifera L. cvv. White Riesling and Zinfandel). Functional Plant Biology 30, 673–687.

107. Sharma-Natu, P. and Ghildiyal, M.C. (2005) Potential targets for improving photosynthesis and crop yield. Current Science 88, 1918–1928.

108. Sharwood, R.E., Von Caemmerer, S., Maliga, P. and Whitney, S.M. (2008) The catalytic properties of hybrid Rubisco comprising tobacco small and sunflower large subunits mirror the kinetically equivalent source Rubiscos and can support tobacco growth. Plant Physiology 146, 83–96.

109. Smart, R.E. (1974) Photosynthesis by grapevine canopies. Journal of Applied Ecology 11, 997–1006.

110. Soar, C.J., Speirs, J., Maffei, S.M. and Loveys, B.R. (2004) Gradients in stomatal conductance, xylem sap ABA and bulk leaf ABA along canes of Vitis vinifera cv. Shiraz: molecular and physiological studies investigating their source. Functional Plant Biology 31, 659–669.

111. Soar, C.J., Speirs, J., Maffei, S.M., Penrose, A.B., McCarthy, M.G. and Loveys, B.R. (2006) Grape vine varieties Shiraz and Grenache differ in their stomatal response to

VPD: apparent links with ABA physiology and gene expression in leaf tissue. Australian Journal of Grape and Wine Research 12, 2–12.

112. De Souza, C.R., Maroco, J.P., Dos Santos, T.P., Rodrigues, M.L., Lopes, C.M., Pereira, J.S. and Chaves, M.M. (2003) Partial root zone drying: regulation of stomatal aperture and carbon assimilation in field-grown grapevines (Vitis vinifera cv. Moscatel). Functional Plant Biology 30, 653–662.

113. De Souza, C.R., Maroco, J.P., Dos Santos, T.P., Rodrigues, M.L., Lopes, C.M., Pereira, J.S. and Chaves, M.M. (2005) Control of stomatal aperture and carbon uptake by deficit irrigation in two grapevine cultivars. Agriculture, Ecosystems and Environment 106, 261–274.

114. Stoll, M., Loveys, B. and Dry, P. (2000) Hormonal changes induced by partial root-zone drying of irrigated grapevine. Journal of Experimental Botany 51, 1627–1634.

115. Tarara, J.M. and Ferguson, J.C. (2006) Two algorithms for variable power control of heat-balance sap flow gauges under high flow rates. Agronomy Journal 98, 830–838.

116. Troggio, M., Vezzulli, S., Pindo, M., Malacarne, G., Fontana, P., Moreira, F.M., Costantini, L., Grando, M.S., Viola, R. and Velasco, R. (2008) Beyond the genome, opportunities for a modern viticulture: a research overview. American Journal of Enology and Viticulture 59, 117–127.

117. Uemura, K., Anwaruzzaman, M.S. and Yokota, A. (1997) Ribulose-1,5-bisphosphate carboxylase/oxygenase from thermophilic red algae with a strong specificity for CO2 fixation. Biochemical and Biophysical Research Communications 233, 568–571.

118. Vandeleur, R.K., Mayo, G., Shelden, M.C., Gilliham, M., Kaiser, B.N. and Tyerman, S.D. (2009) The role of plasma membrane intrinsic protein aquaporins in water transport through roots: diurnal and drought stress responses reveal different strategies between isohydric and anisohydric cultivars of grapevine. Plant Physiology 149, 445–460.

119. Verslues, P.E. and Bray, E.A. (2006) Role of abscisic acid (ABA) and Arabidopsis thaliana ABA-insensitive loci in low water potential-induced ABA and proline accumulation. Journal of Experimental Botany 57, 201–212.

120. Vincent, D., Ergül, A., Bohlman, M.C., Tattersall, E.A.R., Tillett, R.L., Wheatley, M.D., Woolsey, R., Quilici, D.R., Joets, J., Schlauch, K., Schooley, D.A., Cushman, J.C. and Cramer, G.R. (2007) Proteomic analysis reveals differences between Vitis vinifera L. cv. Chardonnay and cv. Cabernet Sauvignon and their responses to water deficit and salinity. Journal of Experimental Botany 58, 1873–1892.

121. Vivier, M.A. and Pretorius, I.S. (2002) Genetically tailored grapevines for the wine industry. Trends in Biotechnology 20, 472–478.

122. Whitney, S.M. and Andrews, T.J. (2001) The gene for the ribulose-1,5-bisphosphate carboxylase/oxygenase (Rubisco) small subunit relocated to the plastid genome of tobacco directs the synthesis of small subunits that assemble into Rubisco. Plant Cell 13, 193–205.

123. Wilkinson, S. and Davies, W.J. (2008) Manipulation of the apoplastic pH of intact plants mimics stomatal and growth responses to water availability and microclimatic variation. Journal of Experimental Botany 59, 619–631.

124. Williams, L.E. and Ayars, J.E. (2005) Grapevine water use and the crop coefficient are linear functions of the shaded area measured beneath the canopy. Agricultural and Forest Meteorology 132, 201–211.

125. Yunusa, I.A.M., Walker, R.R., Loveys, B.R. and Blackmore, D.H. (2000) Determination of transpiration in irrigated grapevines: comparison of the heat-pulse technique with gravimetric and micrometeorological methods. Irrigation Science 20, 1–8.

126. Zhang, X., Wollenweber, B., Jiang, D., Liu, F. and Zhao, J. (2008) Water deficits and heat shock effects on photosynthesis of a transgenic Arabidopsis thaliana constitutively expressing ABP9, a bZIP transcription factor. Journal of Experimental Botany 59, 839–848.

127. Zsófi, Z., Gál, L., Szilágy, Z., Szücs, E., Marschall, M., Nagy, Z. and Bálo, B. (2009) Use of stomatal conductance and pre-dawn water potential to classify terroir for the grape variety Kékfrankos. Australian Journal of Grape and Wine Research 15, 36–47.

128. Zufferey, V., Murisier, F. and Schultz, H.R. (2000) A model analysis of the photosynthetic response of Vitis vinifera L. cvs Riesling and Chasselas leaves in the field: I. Interactions of age, light and temperature. Vitis 39, 19–26.

CHAPTER 6

MANAGEMENT INTENSITY AND TOPOGRAPHY DETERMINED PLANT DIVERSITY IN VINEYARDS

JURI NASCIMBENE, LORENZO MARINI, DIEGO IVAN, AND MICHELA ZOTTINI

6.1 INTRODUCTION

In the last decades, intensively cultivated areas have faced a severe loss of biodiversity [1]. However, the maintenance and improvement of biodiversity in agricultural landscapes is progressively more recognized as a key issue for improving human life-quality, promoting the cultural value of anthropogenic landscapes, and in general enhancing the provision of several ecosystem services [2].

Vineyards are amongst the most intensive forms of agriculture often resulting in simplified landscapes where semi-natural vegetation is restricted to small scattered patches. However, a recent trend among wine producers is to increasingly promote the cultural value of this landscape, recognizing the importance of coupling wine production with environmental quality. This tendency toward a more biodiversity friendly management is likely to reflect a general change in the mentality of vineyards owners (see e.g.

Management Intensity and Topography Determined Plant Diversity in Vineyards. © Nascimbene J, Marini L, Ivan D, and Zottini M. PLoS ONE, *8,10 (2013), doi:10.1371/journal.pone.0076167. Licensed under Creative Commons Attribution 4.0 International License, http://creativecommons.org/licenses/by/4.0/.*

3) and is stimulating research on biodiversity in these poorly investigated agro-ecosystems [4,5]. The improvement of wild plant diversity in vineyards may sustain higher landscape biodiversity, providing refuge and food source for several vertebrates and arthropods, including those that are beneficial for pest control [3,6].

Management intensity is expected to influence plant diversity negatively although little research has been done in vineyards. On the other hand, some simple changes in farming activities, which are compatible with grape production, may potentially yield positive effects on the diversity of several taxonomic groups. It is therefore crucial to evaluate the role of those farming activities that are likely to influence plant diversity, such as the use of herbicides treatments, the frequency of mowing [7,8], mechanization and the supply of nitrogen fertilizers [9].

However, in hilly landscapes also topographic factors (e.g. slope, altitude, aspect) may influence plant diversity, even overriding the effect of management intensity and should therefore be taken into account [10]. Steep slopes tend to form shallow soils with lower nutrient and water availability [11]. If phosphorus and/or water limitation help to maintain a species-rich sward in chalk grassland, it would be expected that, at sites with varied topography, steeper slopes would be more resistant to change in vegetation composition than flatter areas. However, it is still unknown the potential interaction between slope and other management practices such as mowing and herbicide treatments.

Hence, the main aim of this study was to test the effect on plant diversity of management intensity and topography in vineyards located in a homogenous intensive hilly landscape. Specifically, this study will evaluate the role of slope, mowing and herbicide treatments frequency, and nitrogen supply in shaping plant diversity and composition of life-history traits. We tested the effect of these factors on the abundance of rosulate and reptant species that are expected to be advantaged in more intensively managed sites. Specifically, these species are expected to be more resistant to mowing frequency [7,12,13] and this should be reflected by a higher relative cover in more frequently mowed vineyards. Moreover, this management-related pattern is expected to have a negative effect on community evenness since the dominance of rosulate and reptant species should hinder the establishment and development of less competitive plants.

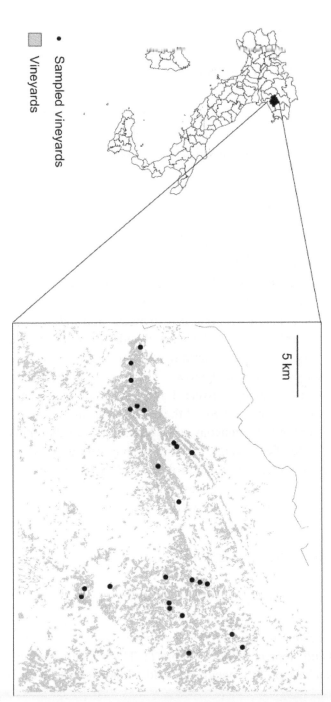

FIGURE 1: Study area and distribution of the 25 vineyards in the Treviso province (NE Italy).

6.2 METHODS

6.2.1 ETHICS STATEMENT

All necessary permits were obtained for the described field studies. In particular, we thank the owners of the vineyards who kindly allowed us to work in their fields and provided information on management practices.

6.2.2 STUDY AREA

The study was carried out in the area of the Conegliano-Valdobbiadene DOCG including 6100 ha of vineyards in the northern part of the province of Treviso (Veneto, NE Italy, N 45°52'40", E 12°17'5"; Figure 1). This area is characterized by a hilly landscape where altitude ranges between 70 and 450 m, annual precipitation is between 900 and 1000 mm and mean annual temperature is 11°C. This hilly landscape is intensively cultivated with vineyards while semi-natural vegetation is restricted to small scattered forest patches or hedgerows. The cultivated area is composed by small vineyards, usually between 1 and 2 ha, belonging to several owners. This fragmented arrangement of the ownerships implies that management practices are not homogeneous at the landscape scale, strongly depending on the attitude of each single owner, and may vary even between adjacent vineyards. In particular, the control of weeds may have different intensity depending on mowing frequency and the use of herbicides. Also nitrogen supply is not constant, even if nitrogen input is generally low.

6.2.3 SAMPLING DESIGN

Twenty-five vineyards belonging to different owners were selected to represent the whole geographical range of the Conegliano-Valdobbiadene DOCG and the gradient of management intensity (from intensive to extensive) and slope conditions (Table 1). The selection process was based on the database of the Conegliano-Valdobbiadene DOCG Consortium and information retrieved through farmers' interviews. We interviewed c. 300 farmers to explore the management practices applied in the region. Of these only c. 60 agreed to share information with us and among them we selected the 25 vineyards to test four factors: mowing frequency, use of herbicide, nitrogen supply, and slope. These factors were selected based on the current knowledge on the main drivers of herbaceous plant diversity. Since we were interested in disentangling the effects of different management practices (mowing, use of herbicides, and nitrogen supply) and slope, vineyards were selected to keep as much as possible low collinearity between our predictors (all correlation <0.4). This allowed us to clearly separate the effect of each factor and test interactions. Mowing techniques included both manual devices such as string trimmer and mechanical devices mounted on tractors such as mowing bars and grass choppers. On average the vineyards manually mown present a significant (P<0.01) steeper slope (n=6, mean slope=75%, SE=5.22%) than the vineyard mown with the tractor (n=19, mean slope=22%, SE=4.78%). A potential effect of slope could therefore include also an additional effect of the mowing technique. The strong correlation between slope and mowing techniques did not allow selecting vineyards in order to keep the two factors statistically independent. Moreover, vineyards were selected to reduce difference in altitudes between the sites. Further criteria to include a vineyard in the study were an age above 10 years and a minimum area of 1 ha. All the vineyards have spontaneous vegetation and farmers did not sow any seed mix for at least 10 years.

TABLE 1: Average and range values of the main topographical and management related factors characterizing the 25 vineyards included in this study.

Topographical factors	Mean±SD	Range (Min-Max)
Altitude (m)	190±58	100-350
Slope (%)	34±30	0-90
Aspect (°)	185±74	76-340
Management factors		
N of mowing treatments yr⁻¹	3.6±1	2-6
N of herbicide treatments yr⁻¹	0.8±0,9	0-3
Total N (Kg ha⁻¹)	18±25	0-70

TABLE 2: Results of the multiple regression models testing the effect of vineyard slope, mowing frequency, herbicide treatment and fertilization on the four measures of diversity.

	(b) Gamma			(a) Alpha			(c) Beta (%)			(d) Evar		
	b	SE	P	b	SE	P	b	SE	P	b	SE	P
Intercept	33.396	1.310	<0.01	13.46	0.44	<0.01	60.23	0.91	<0.01	0.39	0.01	<0.01
Slope	6.813	1.429	<0.01	0.99	0.49	0.05	4.24	0.93	<0.01	0.037	0.01	<0.01
Mowing	-4.319	1.501	<0.01	-1.96	0.49	<0.01	-	-	-	-	-	-
Herbicide	-	-	-	-	-	-	-	-	-	-0.024	0.01	0.02
Total N	-	-	-	-0.97	0.45	0.04	-	-	-	-	-	-
Slope x Mowing	-4.851	1.625	<0.01	-	-	-	-	-	-	-	-	-

Vascular plants were sampled once between April 2nd and 23th, 2012 before any management interventions. In each vineyard, 10 1 m x 1 m plots were randomly placed in the central part of the cultivated area in the field between grape rows. Within each plot, all vascular plants were recorded and for each species the abundance was visually estimated using 5% cover classes. Herbaceous species were further classified according to [14] as follows: perennial reptant and rosulate species, perennial species with erect stem, and summer annuals which overwinter by means of generative diasporas.

6.2.4 DATA ANALYSIS

As response variables we considered the additive components of diversity: alpha (α), beta (β) and gamma (γ) [15-17]. In this work, γ represents the total number of species found in vineyard, α represents the mean number of species at the plot level in each vineyard, and β was calculated as γ-α, giving estimates of the heterogeneity of the plant biota within each vineyard. β-diversity was expressed as a proportion of the total number of species (β/γ). We also computed community evenness using the E_{var} index [18]. This index has been chose due to its mathematical independence from species richness. We also tested the effect of our environmental predictors in selecting plant life forms indicative of management intensity (rosulate and reptant species).

To test the effect of management and topography on the above mentioned response variables we used linear multiple regression. The models initially included vineyard slope, mowing frequency, number of herbicide treatments and N fertilization. For the vineyard with slope steeper than 5% we tested in preliminary analyses the effect of aspect. As aspect was never associated with our response variables we omitted this predictor from the analyses presented here. All the predictors were centred to mean 0 and standard deviation 1. The slopes can therefore be used to evaluate the relative strength of the effect of the single variables. The models were simplified using a backward deletion procedure with P<0.05. The multiple regression was estimated using the lm() function in R, version 2.12.1 [19]. We could use P-values and traditional hypothesis testing on the slope due to the low collinearity between the predictors. A preliminary analysis using an information-theoretic approach based on AICc yielded very similar results and was therefore not presented here [20] (see electronic Text S1, Table S1).

6.3 RESULTS

A total of 141 species were found (see electronic Table S2). The mean number of species per vineyard (γ-diversity) was 35.2 ± 12.7 (range: 19-69), the mean number of species per plot (α-diversity) was 13.4 ± 3.3 (range: 8.8-21.3), and the mean heterogeneity (β-diversity) was 21.7 ± 10 (range: 10.2-50). The mean value of evenness was 0.39 ± 0.06, ranging between 0.27 and 0.55.

Both management intensity and topography had a significant effect on plant diversity (Table 2). In particular, α-diversity was negatively influenced by mowing frequency and fertilization while we found a marginal positive effect of slope. Vineyard species richness (γ-diversity) was negatively affected by mowing frequency and positively affected by slope. We also found a significant interaction between slope and mowing frequency, i.e. the negative effect of mowing was evident only on the steep slopes (Figure 2). Proportional β-diversity was also enhanced by steep slopes while no clear effect of management was found. Finally, evenness increased with increasing slope and declined with the number of herbicide treatments.

The models were simplified with a backward deletion procedure ($P<0.05$). The interactions Slope x Herbicide and Slope x Total N were never retained in the models.

The multiple regression analyses testing the effect of management intensity and slope on cover of rosulate and reptant species yielded significant results. In particular, the cover of rosulate and reptant species increased in more frequently mowed vineyards (b=8.34, SE=2.50, P<0.01), while their relative cover decreased with increasing slope (b=-8.80, SE=2.50, P<0.01). The dominance of rosulate and reptant species in more intensively managed and flat sites had a negative influence on the evenness of plant communities (Figure 3).

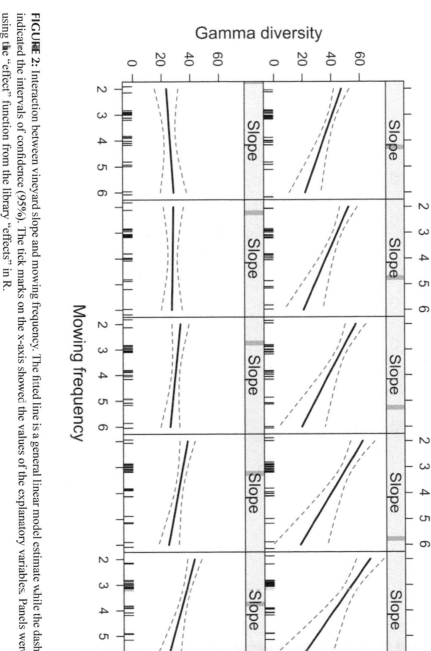

FIGURE 2: Interaction between vineyard slope and mowing frequency. The fitted line is a general linear model estimate while the dashed lines indicated the intervals of confidence (95%). The tick marks on the x-axis showed the values of the explanatory variables. Panels were drawn using the "effect" function from the library "effects" in R.

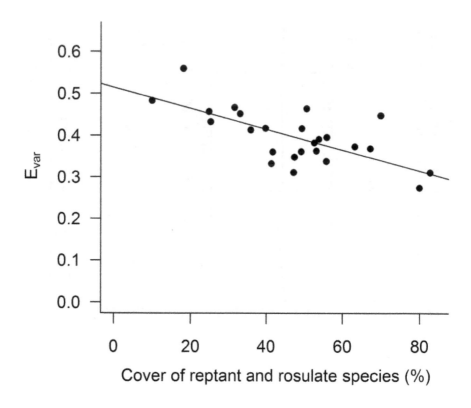

FIGURE 3: Relationship between the cover of rosulate and reptant species and species evenness. The line indicates a significant linear regression (P<0.01, R²=47.9).

6.4 DISCUSSION

Management intensity and topography are both relevant drivers of plant species diversity patterns in our vineyards, confirming results already available for other types of agro-ecosystems [9].

The two most important factors are slope and mowing frequency that respectively yielded positive and negative effects on different measures of plant diversity. At our best knowledge, this is the first time that a significant interaction between these two factors in determining local plant diversity is also demonstrated, providing new insights for effective management practices to promote plant diversity. In particular, this result warns against the detrimental effects of increasing mowing intensity on steep slope (slope higher than 40%) where plant communities are more diverse. On the contrary, increasing mowing intensity may be not detrimental to plant diversity in flat sites where the species pool is poorer. These results predict that the maintenance of high plant diversity in our study area is mainly related to the management intensity applied to vineyards on steep slopes where low mowing frequency is recommended. As the steep slopes were more often mown with manual devices than flat areas, the lower impact of this mowing technique could have contributed to explain the interaction between slope and mowing frequency. Steeper slopes might be buffered to some extent against invasion by more competitive species, probably due to edaphic factors including low phosphorus availability [10].

The analysis of plant traits clarifies the mechanism that is behind the observed effect of mowing frequency on plant diversity. The response of plant communities to mowing frequency is mediated by a process of selection of resistant growth forms, such in the case of rosulate and reptant species [7,12,13,21]. These species tend to cover a large amount of the available surface hindering the establishment of plants that are less tolerant to mowing. This process is also reflected by the negative effect of the increasing cover of rosultae and reptant species on the evenness of plant communities, indicating the tendency of these resistant species to dominate in more disturbed sites. The decrease of community evenness is therefore a suitable indicator to evaluate the effects of mowing. However, results also suggest that in vineyards on steep slopes this mechanism is likely to be less efficient, preserving these vineyards from a severe homogenization of plant communities in terms of plant traits. This pattern is probably related to the fact that plants on steep slopes are subjected to more limiting factors other than management [10,22] that prevent the selection of a few growth forms and the dominance of few species.

The other two management-related factors tested in this study, number of herbicide treatments and N fertilization, are less influential on plant diversity in our vineyards. The influence of herbicide treatments is significant only for community evenness, suggesting a similar mechanism of plant selection as for mowing frequency, resulting in a simplified community dominated by resistant species. The absence of a significant effect of herbicide treatments on the additive components of diversity (alpha, beta and gamma) is likely to reflect the fact that in our study area herbicides are only applied to a restricted zone under the grape rows (c. 50-60 cm), while the fields between the rows, where our plots were placed, are not directly sprayed.

Nitrogen fertilization has a negative effect only on the mean number of species per plot (alpha-diversity), but does not affect the other components of diversity. In particular it does not cause negative effect on the total number of species at the vineyard level, supporting the idea that the need to decrease nitrogen supply is not a priority for effectively enhancing plant diversity in this context. Interestingly, in other agro-ecosystems, this factor is among the main management-related drivers of plant diversity [22]. In these cases, however, nitrogen inputs are higher by orders of magnitude compared with our vineyards where nitrogen supply is generally low and even not constant along time in a given site. The management of nitrogen fertilization is indeed highly variable even within a single vineyard and is not considered a priority, as in the case of mowing, by the owners.

In general, our study corroborates the idea that some simple changes in farming activities, which are compatible with grape production, should be encouraged for improving the natural and cultural value of the area of the Conegliano-Valdobbiadene DOCG, by maintaining and improving wild plant diversity. In particular, mowing frequency should be low (e.g. 2-3 times per year), especially on steep slopes where plant communities are more diverse. Despite its marginal effect on plant diversity, also the use of herbicides should be reduced, contributing to improve the aesthetic value of the landscape. These measures are likely to reduce management costs and to yield positive effects on the diversity of several taxonomic groups, including organisms that benefit ecosystem services such as pollination, biological control and grape resistance to pathogens.

REFERENCES

1. Tscharntke T, Tylianakis JM, Rand TA, Didham RK, Fahrig L et al. (2012) Landscape moderation of biodiversity patterns and processes - eight hypotheses. Biol Rev 87: 661-685. doi:10.1111/j.1469-185X.2011.00216.x. PubMed: 22272640.
2. Gillespie M, Wratten SD (2012) The importance of viticultural landscape features and ecosystem service enhancement for native butterflies in New Zealand vineyards. J Insect Conserv 16: 13-23. doi:10.1007/s10841-011-9390-y.
3. Arlettaz R, Maurer ML, Mosimann-Kampe P, Nusslé S, Abadi F et al. (2012) New vineyard cultivation practices create patchy ground vegetation, favouring Woodlarks. J Ornithol 153: 229–238. doi:10.1007/s10336-011-0737-7.
4. Brugisser OT, Schmidt-Entling MH, Bacher S (2010) Effects of vineyards management on biodiversity at three trophic levels. Biol Conserv 143: 1521-1528. doi:10.1016/j.biocon.2010.03.034.
5. Nascimbene J, Marini L, Paoletti MG (2012) Organic farming benefits local plant diversity in vineyard farms located in intensive agricultural landscapes. Environ Manage 49: 1054-1060. doi:10.1007/s00267-012-9834-5. PubMed: 22411057.
6. Sanguankeo PP, León RG (2011) Weed management practices determine plant and arthropod diversity and seed predation in vineyards. Weed Res 51: 404–412. doi:10.1111/j.1365-3180.2011.00853.x.
7. Kramberger B, Kaligaric M (2008) Semi-natural grasslands: the effects of cutting frequency on long-term changes of floristic composition. Pol J Ecol 56: 33-43.
8. Zechmeister HG, Schmitzberger I, Steurer B, Peterseil J, Wrbka T (2003) The influence of land-use practices and economics on plant species richness in meadows. Biol Conserv 114: 165-177. doi:10.1016/S0006-3207(03)00020-X.
9. Marini L, Fontana P, Scotton M, Klimek S (2008) Vascular plant and Orthoptera diversity in relation to grassland management and landscape composition in the European Alps. J Appl Ecol 45: 361-370. doi: 10.1111/j.1365-2664.2007.01402.x
10. Bennie J, Hill M, Baxter R, Huntley B (2006) Influence of slope and aspect on long-term vegetation change in British chalk grasslands. J Ecol 94: 355-368. doi:10.1111/j.1365-2745.2006.01104.x.
11. Balme OE (1953) Edaphic and vegetational zoning on the Carboniferous limestone of the Derbyshire Dales. J Ecol 41: 331-344. doi:10.2307/2257045
12. Catorci A, Ottaviani G, Ballelli S, Cesaretti S (2011) Functional differentiation of central Apennine grasslands under mowing and grazing disturbance regimes. Pol J Ecol 59: 115-128.
13. Drobnik J, Romermann C, Bernhardt-Romermann M, Poschlod P (2011) Adaptation of plant functional group composition to management changes in calcareous grassland. Agric Ecosyst Environ 145: 29-37. doi:10.1016/j.agee.2010.12.021.

14. Pignatti S (1982) Flora d'Italia. Bologna Edagricole, 1-3: 2302.
15. Allan JD (1975) Components of diversity. Oecologia 18: 359-367. doi:10.1007/BF00345855.
16. Lande R (1996) Statistics and partitioning of species diversity, and similarity among multiple communities. Oikos 76: 5-13. doi:10.2307/3545743.
17. Roschewitz I, Gabriel D, Tscharntke T, Thies C (2005) The effects of landscape complexity on arable weed species diversity in organic and conventional farming. J Appl Ecol 42: 873-882. doi:10.1111/j.1365-2664.2005.01072.x.
18. Smith B, Wilson JB (1996) A consumer's guide to evenness indices. Oikos 76: 70-82. doi:10.2307/3545749.
19. R Development Core Team (2011) R: A language and environment for statistical computing. Vienna: R Foundation for Statistical Computing.
20. Burnham KP, Anderson DR (2002) Model selection and multimodel inference. A practical information–theoretic approach. Berlin: Springer Verlag, Verlag.
21. Gago P, Cabaleiro C, García J (2007) Preliminary study of the effect of soil management system on the adventitious flora of a vineyard in northwestern Spain. Crop Protect 26: 584-591. doi:10.1016/j.cropro.2006.05.012.
22. Marini L, Scotton M, Klimek S, Pecile A (2008) Patterns of plant species richness in Alpine hay meadows: Local vs. landscape controls. Basic Appl Ecol 9: 365-372. doi:10.1016/j.baae.2007.06.011.

There are several supplemental files that are not available in this version of the article. To view this additional information, please use the citation on the first page of this chapter.

CHAPTER 7

ADVANCED TECHNOLOGIES FOR THE IMPROVEMENT OF SPRAY APPLICATION TECHNIQUES IN SPANISH VITICULTURE: AN OVERVIEW

EMILIO GIL, JAUME ARNY, JORDI LLORENS, RICARDO SANZ, JORDI LLOP, JOAN R. ROSELL-POLO, MONTSERRAT GALLART, AND ALEXANDRE ESCOLA

7.1 INTRODUCTION

Crop protection is a key issue in farm management. It involves dealing with important risks and expensive pesticide products. Most of these products are to be applied by means of sprayers. Spraying techniques have been continuously evolving in recent decades. However, it is not only the sprayer itself, but everything related to the application of plant protection products (PPP) that needs to be taken into account in order to improve the results obtained. What is the canopy like? What is the canopy volume? What is the leaf area to be sprayed? What technologies are available to

Advanced Technologies for the Improvement of Spray Application Techniques in Spanish Viticulture: An Overview. © 2014 by the authors; licensee MDPI, Basel, Switzerland. Sensors, *14 (2014),* doi:10.3390/s140100691. *Licensed under Creative Commons Attribution 3.0 Unported License,* http://creativecommons.org/licenses/by/3.0/.

help growers improve their spray applications in a more effective, efficient and sustainable way?

In the 1980s, the term "precision agriculture" was used for the first time. Sensors, agricultural and statistical techniques and technology were combined to improve farm management techniques. One of the contexts where precision agriculture has been widely implemented is vineyards, in what is known as "precision viticulture". After some decades of rapid development, in this day and age when people are carrying around extremely powerful microprocessors and GPS receivers in their pockets, it seems reasonable to consider using these technologies to turn spray applications into a more sustainable operation.

In the late 1990s, researchers from the *Universitat Politècnica de Catalunya* and the *Universitat de Lleida* focused their work on improving the spray application of PPP in vineyards and orchards. Some interesting results have been obtained with public funding coming from the Spanish Government and the coordination with other researchers from the Department of Agriculture of the Catalan Government, the *Universitat Politècnica de València* and the *Institut Valencià d'Investigacions Agràries*. In this paper, part of that work regarding the use of sensors to retrieve data from the vineyard canopies and to monitor the drift produced during the spray applications is reviewed. Some methods and geostatistical procedures for mapping vineyard parameters are proposed, and the development of a variable rate sprayer is described. After about two decades of development efforts, it is time now to put everything together and at the disposal of growers with the objective of improving the spray applications in Spanish viticulture.

7.2 SENSORS FOR CANOPY CHARACTERIZATION

The first step towards a crop-adapted application of PPP is the characterization of the vegetation to be treated. The information on geometrical and structural characteristics (height, width, volume, leaf density, leaf area, etc.) of plantations can help to optimize many agricultural tasks, such as irrigation and fertilization, as well as pruning and crop training techniques, among others. As far as pesticide applications is concerned, knowledge of the geometrical characteristics of the crop allows optimizing the dose of the applied product, and adjusting it to the characteristics of the plants, thereby obtaining a reduction of its environmental and economic impact [1].

Structural and geometric parameters of trees have been traditionally obtained using time-consuming and costly manual measurements, which are also destructive in the case leaves are removed to measure foliar surfaces. In recent years, different instruments and measurement systems for characterizing the canopy in a non-destructive, contactless, fast, accurate and repeatable way have appeared as alternatives to manual methods. The main systems used for the characterization of plants involve the use of electronic data acquisition systems associated with sensors based on different physical measuring principles. Other than ultrasound-based systems, most sensors are based on the use of electromagnetic radiation in certain spectral ranges or bands (visible, infrared, etc.). Ultrasonic sensors themselves, digital photography techniques, as well as stereoscopic vision and laser scanning sensors (Light Detection and Ranging—LIDAR scanners), are probably the most widely used and promising techniques for characterizing tree crops [1]. Among the above, both stereoscopic vision systems and LIDAR sensors stand out for the possibility of providing three-dimensional models of plantations [1–3]. Digital cameras are low-cost, easy-to-use and popular instruments suitable for estimating some plant characteristics such as height, volume and leaf area index (LAI) with reasonable accuracy. However, at present, they do not allow accurate 3D characterization of plantations in real time because of the complex and time-consuming post-processing algorithms required. Stereoscopic systems provide 3D models by combining two monocular images taken simultaneously with a binocular digital camera by means of computational algorithms. These systems lose effectiveness under certain environmental conditions, especially in variable lighting situations. Meanwhile, both ultrasonic sensors and LIDAR systems base their operation on the measurement of the distance between an emitter and a target object using pulsed sound waves or laser light, respectively. The distance is determined by measuring the time taken for the pulse to travel the distance from the emission point to the detection point after reflecting off the object (ultrasonic sensors and time-of-flight LIDAR) or, alternatively, by measuring the phase difference between the incident and the reflected waves (phase-shift LIDAR). The main disadvantages of ultrasonic sensors come from their low spatial resolution, highly divergent sonic cones, lack of scanning systems and potential interference between proximal sensors [4]. On the other hand, due to their high speed of measurement, resolution and accuracy, LIDAR systems are becoming one of the most used sensors for characterizing vegetation. If, in addition, the measurements of these sensors are synchronized with their spatial coordinates obtained by geo-referencing systems (e.g., GPS), it is possible to obtain maps of the parameters of interest of the plots analysed: canopy volumes, LAI, crop foliage density, 3D crop model, etc. Figure 1 shows a 3D point cloud obtained in a vineyard by displacing a LIDAR sensor along an alley between two adjacent rows.

However, it is still necessary to continue reducing the costs of sensors and associated electronics. It is also essential to develop suitable software programs for the post-processing of data, to increase both the usability and speed of calculation to assist decision making. This will contribute to the expansion of the use of systems for characterization of vegetation in many agricultural practices, being the application of pesticides one of which more advances and improvements may incorporate.

7. 2.1 MEASUREMENT OF VEGETATIVE VOLUME

The use of terrestrial LIDAR technology for the geometric characterization of vineyards has mainly been concentrated in the last 10 years [2,3,5–9]. Two-dimensional terrestrial laser-LIDAR scanners (2D TLS) make two-dimensional sweeps in just one measuring plane. The additional third dimension can be obtained by moving the LIDAR in a direction perpendicular to the scanning plane. Although 2D TLS systems are normally simpler and more affordable than 3D TLS systems, they tend to be less accurate, and it can be difficult to properly control the movement of the LIDAR when collecting the data. Currently, the measurement systems based on 2D TLS allow three-dimensional scanning of vines with horizontal and vertical distances between points (size of the scan mesh) below 5 cm, and distance measurement accuracies of about 1 cm. To fully scan a row of

FIGURE 1: Point cloud obtained in a vineyard by the displacement of a LIDAR sensor along an alley between 2 rows (different colors correspond to different beam orientation angles).

vines, it is necessary to scan both sides of the row and unify both point clouds into a single coordinate system [3].

The 3D point clouds obtained provide a huge amount of data. These, once processed into appropriately formatted information, can be very useful in the precise application of PPP. An important variable is the volume occupied by the point cloud (Figure 2), the so-called tree row LIDAR-volume (TRLV) [4]. The TRLV depends on: (i) the real size of the vegetation; (ii) the shape and size of the scan mesh; and (iii) the position(s) of the sensor with respect to the vineyard.

Since the main function of plants is photosynthesis, the distribution and position of the leaves is directly related to the availability of light. For this reason, the preferred position of leaves is normally on the outer layer of the crown. With this precedent, in [6] the hypothesis that there must be a

FIGURE 2: Resulting volume after processing the point cloud obtained with a 2D TLS dynamic measurement system in a vineyard.

non-linear relationship between the TRLV and the leaf area density (LAD) is presented.

For the calculation of the LAD in a section of the vineyard, one must know the TRLV and the leaf surface area. The latter is calculated by manually defoliating and measuring the surface of the leaves.

The study reported in [6] reveals a good logarithmic fit between the TRLV and the LAD. According to the results, the LAD can be estimated from the TRLV. If the LAD is multiplied by the TRLV, the leaf area of the vegetation under study can be obtained. It is therefore concluded that by using the information provided by the LIDAR 3D Dynamic Measurement System, a good estimation of the leaf area in hedgerow vineyards can be obtained.

Improvements in obtaining parameters related to geometric characterization such as TRLV, LAD or leaf area represent a major advance in improving the accurate application of PPP.

7.2.2 LEAF AREA INDEX ESTIMATION BY USING THE TREE AREA INDEX PARAMETER

The LAI is defined as the one-side leaf area per unit of ground area and is
probably the most widely used index for characterizing grapevine vigour.
Currently, the LAI can be estimated using different types of sensors. Among
the proposed technologies, vineyard leaf area can be indirectly estimated
using ground-based laser sensors (or LIDAR systems) to obtain information
about the geometry of the canopy [2,9,10]. Specifically, canopy volume is
measured to subsequently obtain the total leaf area by an allometric relationship between both parameters. Faced with this procedure, [5] proposed to
use the tree area index (TAI) to estimate the LAI in vineyards.

The process for obtaining the TAI was explained in detail in [11] and
was later adapted to vineyards by [5]. Adopting the reference system
shown in Figure 3, the Oz axis (not shown) is parallel to the ground and
in the direction in which the tractor-mounted LIDAR sensor moves. Thus,
several scans can be obtained along the vine row, each of which results in a
vertical semicircle in the plane Oxy. Finally, all interception points within
the canopy are projected relative to the Oz axis onto a two-dimensional
grid of polar cells in the Oxy plane (Figure 3). In particular, the overall
projected cross-section of the canopy volume is divided into cells with
equal angular increments of $\Delta\theta = 3°$ and equal radial increments of $\Delta r =$
100 mm. For each of the cells (k, j), it is possible to calculate the number
of laser beams reaching the entrance side of the polar cell, $n_{k,j}$, and the
number of interceptions, $\Delta n_{k,j}$, within the cell. The TAI is finally calculated
using Equation (1):

$$TAI = -\frac{\Delta\theta}{W}\sum_{k=1}^{K}\sum_{j=1}^{J_K} r_j \delta_{k,j} \ln\left(1 - \frac{\Delta n_{k,j}}{n_{k,j}}\right) \qquad (1)$$

where, apart from the parameters $\Delta\theta$, $\Delta n_{k,j}$ and $n_{k,j}$ described above, W (m) is the distance between the vine rows, r_j is the radial distance between the polar cell and the sensor position and $\delta_{k,j}$ is a binary variable that represents the presence or absence of foliage in each cell ($\delta_{k,j} = 1$ when the coefficient $\Delta n_{k,j}/n_{k,j}$ is greater than or equal to 0.01, and $\delta_{k,j} = 0$ when the coefficient is less than 0.01). In fact, the TAI is formulated as the ratio between the crop area detected by the LIDAR sensor and the ground area. In their calculation, it is also assumed that the probability of the laser beam's transmission within the vines could be approximated by the Poisson probability model when sufficiently small distances (Δr) and a random spatial distribution of the leaves are considered. In any case, this approach allows the laser beam interception through the canopy to be described using an extinction prob-

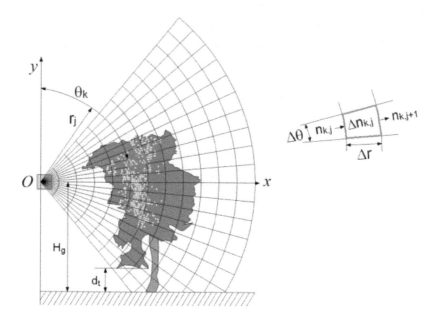

FIGURE 3: Two-dimensional grid of polar cells for calculating the tree area index Each polar cell is defined by two coordinates (r_j, θ_k). The first, r_j, is the distance from the reference origin (LIDAR sensor), and the second, θ_k, is the angle between the Oy axis and the radial direction (clockwise). Hg is the height of the LIDAR sensor with respect to the ground (approximately constant, 1.65 m), and d_t is the distance used to exclude intercepted points at ground and trunk level (with permission of the editor of *Precision Agriculture*).

ability model, in a manner similar to how the light extinction that occurs within vines is described.

To validate this method, the LMS-200 laser sensor (SICK AG, Waldkirch, Germany) has been used. More detailed information on this sensor can be found in [5]. The operation of the sensor is based on using the time-of-flight (TOF) principle to estimate the distances to the leaves within the canopy. The vines are scanned from one side of the row, and the data provided by the sensor are the polar coordinates of each interception point, i.e., the radial distance (with an accuracy of ±15 mm in a single-shot measurement and 5 mm standard deviation in a range up to 8 m) and the angle of the laser beam (with an angular resolution of 1°). Data transfer from the sensor to a laptop is done via the RS-232 protocol using a MATLAB-based program for sensor control and data acquisition. The specific research conducted in a vineyard (cv. Merlot) in Raimat (Lleida, Spain) has made it possible to establish a model (Equation (2)) to estimate the leaf area using the TAI:

$$LAI = 1.2646TAI - 0.1935 \tag{2}$$

where LAI is the leaf area index (m^2/m^2) and TAI is the tree area index (dimensionless). The model can be applied to row lengths of 1 m, 2 m and 4 m, and has its main advantage in its applicability regardless of the side of the row from which the LIDAR reading is performed. Additionally, this method is non-destructive and does not require the use of allometric relationships. The disadvantages to this method are the large amount of data provided and the complexity of the algorithm used to calculate the TAI.

7.2.3 ELECTRONIC VERSUS MANUAL CANOPY CHARACTERIZATION

The accuracy of electronic measurements has been widely evaluated, and several field tests have been developed to compare electronic canopy estimations with manual measurements. The authors of [12] compared ultrasonic and laser measurements of citrus canopy volumes with manual measurement methods. They concluded that laser measurements provided a better prediction of canopy volume than the ultrasonic system because of the inherent higher resolution, but in any case they recommended the use of both ultrasonic and laser sensors for automatic mapping and quantification of the canopy volume of citrus trees. Arnó et al. used a LIDAR sensor to evaluate the leaf area index in vineyards, and the results were compared with manual measurements [13]. They found a good correlation between both values, which allowed the creation of canopy maps for subsequent applications. Based on that evidence, and with the aim to evaluate the different electronic alternatives for canopy characterization in vineyards, a comparative research study was carried out by [9]. The overall goal of that study was to evaluate the applicability of ultrasonic and LIDAR sensors for mapping canopy structures of different varieties and crop stages in vineyards, and to correlate the measurements of the canopy characteristics using manual methods, LIDAR and ultrasonic sensors.

The three different canopy parameters were manually measured in each field test: crop height, crop width and leaf area index. The measuring procedure was arranged according to [14], where the total canopy height was divided into three parts. For each of those parts, height and canopy width values were obtained. A partial leaf area corresponding to each of the three height levels was determined by applying the weight–area ratio obtained for every variety and crop stage. This ratio was determined by measuring the weight and surface area of 50 leaf samples collected from the bottom, middle and upper parts of the canopy in a randomized procedure, following the method described in [15]. The leaf surface (one side) was measured with a LI-COR LI 3100C electronic planimeter (LI-COR, Lincoln, NE, USA).

The obtained average values for the most important parameters used to define the canopy structure, such as crop height, crop width and crop volume obtained with the three investigated methods (manual, ultrasonic and LIDAR sensors) were analysed. A preliminary evaluation indicates that the values of canopy height measured manually (C_{HM}) and with the LIDAR sensor (C_{HL}) were relatively close. The values of crop width measured manually (C_{WM}) were in all cases greater that those obtained with the ultrasonic sensor (C_{WU}). The crop width obtained with the LIDAR sensor

(C_{WL}) results in the lowest values, probably due to it being the most precise scanning method and its greater ability to detect gaps in the canopy. Then, as a consequence of the observed tendency of those parameters, the measurements and estimations of crop volume present the same ranking, going from the highest values, obtained manually (C_{VM}), to the lowest, obtained with the LIDAR sensor (C_{VL}).

The comparative assessment included not only the above described and most important canopy structure characteristics but also others, obtained either from manual measurements in the field such as the leaf area index or numerical values derived from the use of the sensors, such as percentage of zero values measured on the crop with ultrasonic or LIDAR sensors, Z_U and Z_L, respectively, or IL (impacts·m^{-1}) defined as the number of impacts (points where the laser beam detected the canopy). A good example of those relationships is shown in Figure 4, where the LAI can be predicted from the number of impacts obtained with LIDAR in the canopy or from the values of the canopy height measured with the same sensor, respectively.

The leaf wall area (LWA) is one of the proposed parameters to be used during pesticide spray applications in fruit crops, and is for that reason widely analysed in this field of research. Crop height values obtained with a LIDAR sensor (C_{HL}) allow one to calculate the total leaf wall area on one side of a row canopy. The comparison of the manually estimated leaf wall area (LWA$_M$) with the values measured with the LIDAR sensor (LWA$_L$) indicates that in most cases the manual estimation of this parameter exceeds that obtained with the LIDAR sensor by about 30%, except in some particular cases. Those differences can be related to the total row length, L, with average values of 0.29 m^2·m^{-1}. Those differences can substantially affect the calculation of the total amount of pesticide applied to a target area, leading to unnecessary overdose.

One of the most important problems during the electronic process of canopy characterization using electronic devices comes from the tedious process of data analysis. In order to facilitate this process, a specific tool, PROTOLIDAR v1.0, was developed [16] to use data specifically from the LidarScan v.1® program [17], with the aim of helping in the process of analysing the data generated by LIDAR measurements. PROTOLIDAR [18] was created to run in R® and was released under the GPL 2 license. It consists of a package with a set of functions that helps to characterize the canopy of the grapevine (height, width and front view) from the LIDAR scan, performs statistical analysis on the outputs, plots and calculates the LAI, LWA and Tree Row Volume (TRV).

7.3 VINEYARD MAPPING

7.3.1 MAPPING THE LEAF AREA INDEX USING A GROUND-BASED LIDAR SCANNER

Precision viticulture (PV) is a concept that is beginning to have an impact on the wine-growing sector [19]. Grape-yield maps and remote sensing tools are used with varying success, the management of the spatial variability of grape quality being the remaining challenge. Furthermore, the application of plant protection products also raises controversy about the risk of contamination by drift and residue in grapes due to overdosing. The use of vegetation maps (LAI maps) might be an interesting way to assist in defining zones of different leaf surfaces. Then these zones could be managed differentially by applying the proper dose and avoiding environmental problems.

So far, there are few references in relation to obtaining vineyard LAI maps using LIDAR sensors. In any case, to facilitate the subsequent zoning according to leaf surface, it is advisable to generate LAI raster maps using a specific protocol associated with LIDAR sensor data. This idea has been discussed in [20], by mapping six rows of vines (cv. Syrah) in Raimat (Lleida, Spain), occupying an area of about 0.70 ha. The plot is ideal for generating a vegetation map due to the presence of a considerable spatial variation in the vine vigour along the rows, probably due to changes in topography within the plot. The LIDAR sensor used was again the SICK LMS-200, together with an inertial sensor (IMU, Inertial Measurement Unit) and a GX 1230 GG model GPS + RTK system (Leica Geosystems AG, St. Gallen, Switzerland) for later georeferencing and mapping the acquired information. Considering LIDAR readings from both sides of the row, the LAI per meter of row length can be estimated by calculating the left and right side areas (leaf wall areas including gaps) and the area which encloses the top of the row [2]. LAI values can then be georeferenced at points that are equally spaced by 1 m and are placed along the line of the grapevine trunks. The information generated

(X, Y, LAI) can be displayed as a point vector map, and then converted to a raster map by geostatistical interpolation (Figure 5). Specifically, the raster map can be obtained by using the VESPER software [21], in this case, using a block kriging and a grid of 2 m for the projection of the interpolated data.

The spatial variability of the LAI is evident (Figure 5), and, more importantly, the pattern of variation seems to be highly structured, allowing the delimitation of well-defined and compact areas within the field. Adopting the approach suggested by [22], an unsupervised classification algorithm (fuzzy c-means) can be applied to the interpolated data to classify the LAI and then generate LAI zone maps (two or three zones, Figure 5). Ultimately, it is possible to optimize the applied doses of pesticides according to the amount of vegetation in each area and according to the benefits in product savings and drift reduction. There are also difficulties: The post processing is somewhat complex since the LIDAR sensor and GPS receiver normally have different frequencies, and the LIDAR data needs to be adjusted to match the LIDAR readings from both sides of the row.

7.3.2 GEOREFERENCED LIDAR 3D VINE PLANTATION MAP GENERATION

Generation of canopy maps and their further use in the improvement of different agronomy procedures in fruit and wine plantations has been the objective of several research groups. In the same way, canopy map generation has enhanced the quality of information obtained using LIDAR sensors, which has improved the knowledge of vegetation parameters. In addition, during the last decade, the systems for geographic positioning have helped to link all data obtained in the field into an accurate global geographic position. This work was conducted to generate a georeferenced canopy map of measured vine plantations using LIDAR measurements [8]. This process used to georeference every single point obtained with the LIDAR sensor is defined in [8]. Data obtained with LIDAR sensors while driving a tractor along a crop row can be managed and transformed into canopy impact density maps by evaluating the frequency of LIDAR returns. The readings have been made using the precise position information provided by a DGPS (Differential Global Positioning System) sensor.

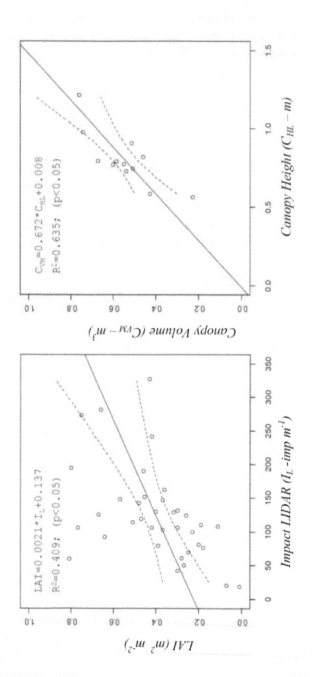

FIGURE 4: Relation between laser impacts obtained with LIDAR and LAI (Left). The correlation between canopy heights calculated with LIDAR and the canopy volume manually measured (Right) (published in [8], with permission of the editor of Sensors journal).

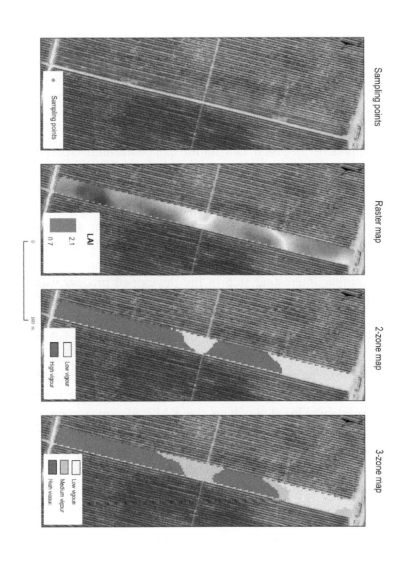

FIGURE 5: Sampling points. LAI raster map and maps of the LAI zones. LAI 2.1 0.7 Low vigour High vigour Low vigour Medium vigour High vigour Sampling points Sampling points Raster map 2-zone map 3-zone map

In that research, the team followed the same scan technique discussed in Section 2.3 of this paper, a technique that permits the scanning of the vineyard crop from both sides (Figure 6a). This scan provides a point cloud that defines the crop with high accuracy (Figure 6b). This point cloud can be georeferenced using the geographic position provided by the DGPS system; when this information is processed, it is possible to obtain an impact density map (Figure 6c). The impact density represents the foliar density of the canopy that has been scanned. This methodology was applied and tested in different vine varieties, and the results show an accurate definition of some crop parameters; this is for the case of the LAI, which in its conventional implementation is a destructive measurement.

This same process of calculation can be useful to create maps of plantations using different sensors or analysing different parameters from this LIDAR sensor, like height, width or foliage density of the crop. These maps could be used either in real time to instantaneously modify the working parameters of the sprayer in sprayer applications, or in a post-process way as a base of information for DSS (decision support systems) tools. [13,23–25].

Another interesting step in the developed process was the process to share and visualize the obtained impact density maps. Once the impact density map was obtained, the UTM coordinates of each pixel were defined. Applying a conversion process to this image led to the production of a new file that was saved as a Google Earth® compatible file (a KMZ file). The use of Virtual Globes, such as Google Earth® and NASA World Wind representations of scientific data, have been thoroughly reviewed by [26]. Launching the KMZ file on a computer in which Google Earth® had previously been installed enables the proposed application of the obtained maps and allows the drawing of the generated density map exactly over the orthophotograph of the measured parcel (Figure 7). Identification of gaps, leaf accumulation zones or other related aspects affecting canopy development can then be used in the crop management process.

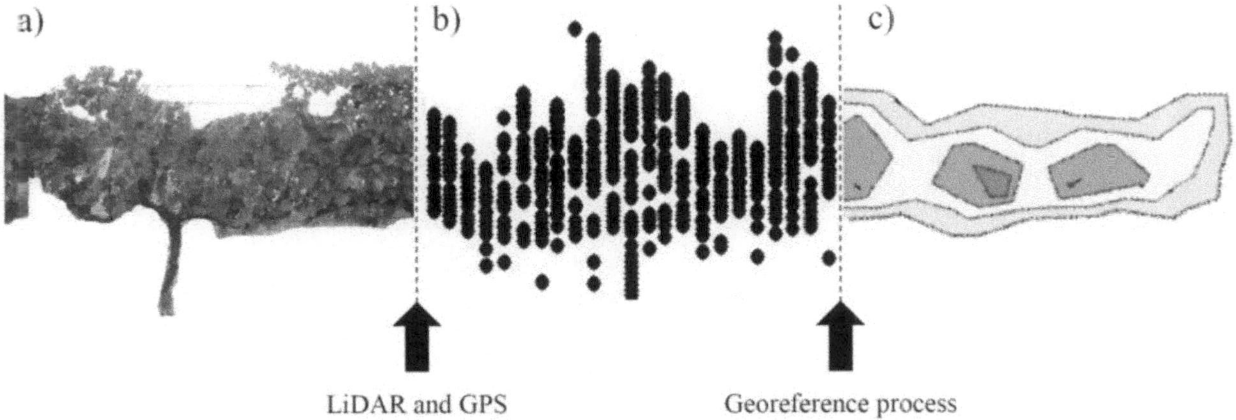

FIGURE 6: Graphical process to georeference LiDAR sensor data using the information provided by the DGPS system. The final results provide maps of the impact density in vineyards. (a) Image of real vines, lateral view; (b) Point cloud obtained, lateral view; (c) Impact density map generated, top view.

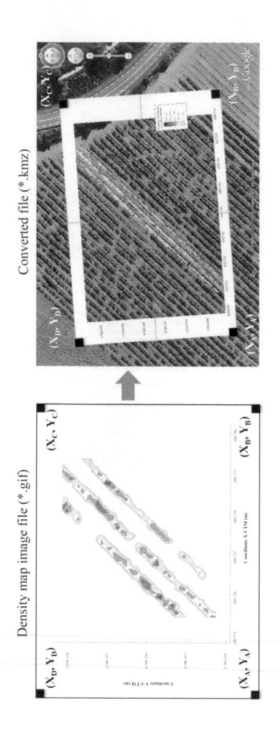

FIGURE 7: Procedure for conversion of an image file to a KMZ file. This proposed method superimposes the density map on the image of the field (published in [8] with permission of the editor of Sensors journal).

7.4 VARIABLE RATE APPLICATION OF PLANT PROTECTION PRODUCTS

Electronic canopy characterization allows the implementation of variable application rate techniques in fruit and vineyard crops, whereby pesticide application rates are modified according to crop characteristics [15,23,24,27–33]. In all cases, relevant benefits in terms of dose reduction, drift control and uniform deposition were achieved by all of the proposed methods. In the specific case of vineyards, the research group of Universitat Politècnica de Catalunya has developed a sprayer prototype that can apply a variable amount of liquid according to the canopy variability along the crop row [33]. The control algorithm is based on the measurement of the canopy width and its variations along the crop line. Once that parameter is electronically determined, information about the forward speed of the tractor along the row and the canopy height of every single measured section is added; the algorithm was developed in order to calculate the canopy volume to be sprayed per unit time, which is expressed in $m^3 \cdot min^{-1}$. Equation (3) indicates the relationship applied for this process:

$$C_{V,j} = -\frac{1,000}{60} \times C_{W,j} \times C_{H,j} \times v \tag{3}$$

where C_{Vj} is the unit canopy volume to be sprayed per unit time at section j $(m^3 \cdot min^{-1})$; C_{Wj}, the canopy width at a certain position at section j (m); C_{Hj}, the canopy height at section j (m); and v, the forward speed of the tractor $(km \cdot h^{-1})$.

The main objective of the algorithm was to modify the emitted nozzle flow rate based on the measurements of the canopy volume along the crop row and its variations in order to maintain a constant objective application coefficient of 0.095 $L \cdot m^{-3}$, selected according to previous research [34,35]. The prototype was developed to be capable of a variable application rate according to the canopy variations along the crop line by proper modification of the nozzle flow rate.

In order to obtain continuous data about the canopy characteristics, three ultrasonic sensors were fitted to a stainless steel mast placed on the left side of a conventional air-blast orchard sprayer. A GPS antenna was also installed on top of the mast so that a GPS receiver could be used to evaluate the uniformity of the forward speed along the track and to record geographical coordinates. The sensors continuously estimated the canopy width from only the left side of the sprayer. All the sensors were connected to a controller placed in a waterproof box located on the right rear side of the sprayer. The controller was a Compact Field Point (National Instruments, Austin, TX, USA) equipped with analogue and digital input/output modules. A rugged computer and wireless router were also connected to remotely monitor and control the system. A box containing three sets of electrovalves (proportional and on–off), an electronic flow meter and a general pressure sensor were installed on top of the sprayer at the rear. Individual pressure sensors were also placed on every manifold.

The results indicated an average potential saving of 21.9%. There was a higher savings potential in the narrow canopy zones where the canopy width was below 0.22 m, which had an average savings of 31.4%. This value dropped to a 12.5% average for zones with a canopy width of over 0.22 m. These results indicated a similar response by the prototype that was independent of the canopy variation; instead, it was influenced by the crop stage and sensor position. In general, these estimated savings values are similar to the results of previous research [15,36] and can be directly related to more precise and safe use of plant protection products in accordance with the new European Directive for the sustainable use of pesticides [37].

7.5 USE OF LIDAR SENSOR IN DRIFT MEASUREMENTS

In general, the arrangement of field tests for drift measurement is very difficult and expensive. The ISO 22866:2005 standard defines the procedure to quantify drift during field tests, but this method is complex, time-consuming and heavily depends on external conditions such as wind speed and direction, which makes such tests difficult to use and may result in poor repeatability of results. As an alternative process, using sensor technology is another interesting option for drift evaluation purposes. Several studies [7,38–40] were carried out using LIDAR technology to measure drift.

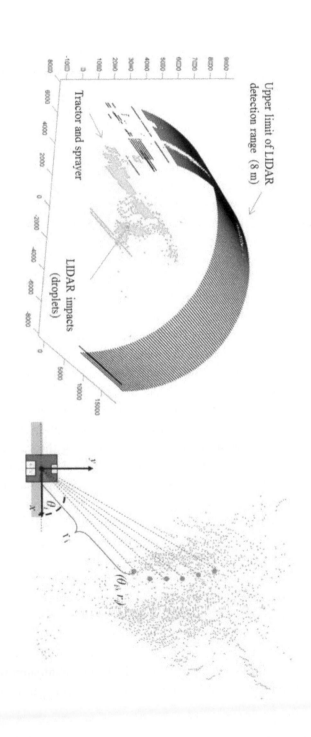

FIGURE 8: Example of LIDAR data plotted. 3D view of the drift cloud escaping the canopy (Left). Measurement process determining the values of angular position (θi) and radial distance (rᵢ) of every LIDAR impact (Right) (published in [7] with permission of the editor of Sensors journal).

The specific scenario of spray processes in orchards is one of the most risky activities from the environmental point of view. In [7], a LIDAR system was used to measure the concentration of small droplets in the air above an orange orchard canopy during and after the sprayer operation. The authors of [38] developed a model to predict airborne drift based on the target structure. The model utilizes LIDAR measurements of optical transmission to predict the characteristics of airborne drift of PPPs leaving the target orchard at different growth stages and modified the drift characteristic for different methods of dose adjustment. Good agreement was demonstrated between the measurements and predictions of the drift from a semi-dwarf apple orchard at full-dose application rates.

Based on the above-described work, the research group arranged a research project [7] with the aims of (1) verifying the use of a LIDAR sensor to measure the drift cloud during pesticide application in a vineyard and (2) studying the effect of different working parameters (nozzle type, sprayer characteristics and air settings) on the total amount of liquid exceeding the target canopy. Drift measurements were made using a LIDAR sensor located at a distance 4 m from the last sprayed canopy row, oriented so that it could measure the cloud drift on a perpendicular plane relative to the canopy row (Figure 8).

The LIDAR scanner used in this work was a low-cost general-purpose Sick LMS-200, adjusted to have an angular resolution of 1° and a scanning angle of 180°. The spray drift cloud exceeding the canopy was scanned for an average of 40 s (the total time of LIDAR scanning in a single test) during the spray track along the row, for 20 s before the sprayer passed in front of the LIDAR and for 20 s afterward, representing a total measurement distance of 50 m. In order to compare the LIDAR measurements, spray deposit over 20 m away from the last row (perpendicular lane) was also collected and quantified using tartrazine as a tracer.

The obtained results indicate that, in general, the use of the LIDAR sensor represents an interesting and easy technique for establishing the potential drift of a specific set of sprayer settings and environmental conditions. The LIDAR system provides an idealized optical view of spray droplets escaping the canopy and its distribution in position with respect to the target. Furthermore, it allows drift to be evaluated with less labour, cost and time than other current methods. In general, a good correlation has been observed between the measured drift cloud with LIDAR and the deposition distribution obtained with the artificial collectors placed in the test bench. However, it seems that drift measurements using LIDAR can be affected by droplet size. The proposed use of LIDAR will help users to configure an adequate deposition over the whole canopy according to the specifications of the treatment and could be used as a drift prediction tool depending on the target geometry, also in accordance with [38].

As an alternative approach, the research group is developing a LIDAR-based system that analyses the temporal characteristics of the backscattered signal received when the laser beam passes through the drift cloud (range-resolved LIDAR) [40]. Such LIDAR devices, which are widely used in atmospheric studies, are based on different types of interaction between the electromagnetic radiation near the visible range and atmospheric molecules, elastic backscatter being the most commonly used technique. These systems allow the characterization and visualization, in near real-time, of the drift plume section in terms of the density of PPP droplets present in the air and hence allow the quantification of the amount of PPP lost. The LIDAR instrument being developed, specifically designed for the characterization of the PPP drift, benefits from the experience gained by the group as a result of various experimental trials with an ultraviolet LIDAR for atmospheric studies, which was employed to monitor PPP drift in orchards [41].

7.6 SOFTWARE TO DETERMINE THE OPTIMAL VOLUME RATE IN VINEYARDS

In the context of advanced technologies in vineyard spraying, several decision support systems (DSS) have been developed in the last decade for improving and making more sustainable pesticide applications (e.g., [42,43]). In this context, DOSAVIÑA [44,45] is a DSS used to determine the optimal volume rate for spraying of vineyards based on achieving the optimal coverage (impacts·cm^{-2}) according to the characteristics of the crop canopy. It also considers the establishment of an efficiency factor that depends on aspects such as the type of sprayer used, nozzle type and size, operational parameters and weather conditions. Several field experiments showed that the use of this DSS reduced the volume rate by an average of 39.9% and reduced pesticide use by 53% in comparison with conventional applications, maintaining the deposition, uniformity and disease control [45]. DOSAVIÑA establishes the applied volume rate according to the characteristics of the canopy. This fact implies an important reduction in pesticide use in comparison with traditional application values. The results obtained demonstrated the amount of this reduction and also the corresponding application efficiency. Applications made according to the DOSAVIÑA recommendations resulted in generally higher values of recovery relative to the total applied amount that was retained on the leaves.

This reduction in pesticide use can be directly related to the intended global reduction in the use of pesticide proposed by the EU [37]. However, those values could also be linked to those proposed by [46], which stated that for agricultural applications, dose has little to do with efficacy because there is already sufficient pesticide to kill all the pests in the field many times over.

7.7 CONCLUSIONS/OUTLOOK

All the previous research on the use of new technologies to improve the pesticide application process in viticulture has demonstrated interesting results in terms of the reduction of the total amount of pesticides used, reduction of the risk of environmental problems and the possibility to adapt the applied amount of PPP to the canopy characteristics. A further analysis of all those aspects, either individually or as a whole, indicates how all the new developments and their applications, in this case in viticulture, represent tools to implement the policies in the new and very restrictive European legal frame in all aspects concerning the use of plant protection products.

Since new legislation has applied efforts to the use-phase of pesticides, it is now time to integrate all the disposable tools that previous research have demonstrated to be interesting devices. However, there is still another key element that is absolutely needed to achieve success in the whole process: an adequate training of all the professionals involved in the process, which also according to the mandate established in [37] represents, as demonstrated by different researchers, the key factor for the whole integration. Only when adequate training has been achieved in all of the European territory will the system be able to implement the policies in the legal framework and to produce in a better and more sustainable way.

REFERENCES

1. Rosell, J.R.; Sanz, R. A review of methods and applications of the geometric characterization of tree crops in agricultural activities. Comput. Electron. Agric. 2012, 81, 124–141.
2. Rosell-Polo, J.R.; Sanz, R.; Llorens, J.; Arnó, J.; Escolà, A.; Ribes-Dasi, M.; Masip, J.; Camp, F.; Gràcia, F.; Solanelles, F.; et al. A tractor-mounted scanning LIDAR for the non-destructive measurement of vegetative volume and surface area of tree-row plantations: A comparison with conventional destructive measurements. Biosyst. Eng. 2009, 102, 128–134.
3. Rosell, J.R.; Llorens, J.; Sanz, R.; Arnó, J.; Ribes-Dasi, M.; Masip, J.; Escolà, A.; Camp, F.; Solanelles, F.; Gràcia, F.; et al. Obtaining the three-dimensional structure of tree orchards from remote 2D terrestrial LIDAR scanning. Agric. Forest Meteorol. 2009, 149, 1505–1515.
4. Escolà, A.; Planas, S.; Rosell, J.R.; Pomar, J.; Camp, F.; Solanelles, F.; Gràcia, F.; Llorens, J.; Gil, E. Performance of an ultrasonic ranging sensor in apple tree canopies. Sensors 2011, 11, 2459–2477.
5. Arnó, J.; Escolà, A.; Vallès, J.M.; Llorens, J.; Sanz, R.; Masip, J.; Palacín, J.; Rosell-Polo, J.R. Leaf area index estimation in vineyards using a ground-based LIDAR scanner. Precision Agric. 2013, 14, 290–306.
6. Sanz, R.; Rosell, J.R.; Llorens, J.; Gil, E.; Planas, S. Relationship between tree row LIDAR-volume and leaf area density for fruit orchards and vineyards obtained with a LIDAR 3D Dynamic Measurement System. Agric. Forest Meteorol. 2013, 171, 153–162.
7. Gil, E.; Llorens, J.; Llop, J.; Fàbregas, X.; Gallart, M. Use of a terrestrial lidar sensor for drift detection in vineyard spraying. Sensors 2013, 13, 516–534.
8. Llorens, J.; Gil, E.; Llop, J.; Queraltó, M. Georeferenced LiDAR 3D vine plantation map generation. Sensors 2011, 11, 6237–6256.
9. Llorens, J.; Gil, E.; Llop, J.; Escolà, A. Ultrasonic and lidar sensors for electronic canopy characterization in vineyards: advances to improve pesticide application methods. Sensors 2011, 11, 2177–2194.
10. López-Lozano, R.; Baret, F.; García de Cortázar-Atauri, I.; Bertrand, N.; Casterad, M.A. Optimal geometric configuration and algorithms for LAI indirect estimates under row canopies: The case of vineyards. Agric. Forest Meteorol. 2009, 149, 1307–1316.
11. Walklate, P.J.; Cross, J.V.; Richardson, G.M.; Murray, R.A.; Baker, D.E. Comparison of different spray volume deposition models using LIDAR measurements of apple orchards. Biosys. Eng. 2002, 82, 253–267.

12. Tumbo, S.D.; Salyani, M.; Whitney, J.D.; Wheaton, T.A.; Miller, W.M. Investigation of laser and ultrasonic ranging sensors for measurements of citrus canopy volume. Appl. Eng. Agric. 2002, 18, 367–372.

13. Arnó, J.; Vallès, J.M.; Llorens, J.; Blanco, R.; Palacín, J.; Sanz, R.; Masip, J.; Ribes-Dasi, M.; Rosell, J.R. Ground Laser Scanner Data Analysis for LAI Prediction in Orchards and Vineyards. In Proceedings of International Conference on Agricultural Engineering 2006, Bonn, Germany, 3–6 September 2006.

14. Manktelow, D.W.L.; Praat, J.P. The Tree-Row-Volume Spraying System and Its Potential Use in New Zealand. In Proceedings of the NZ Plant Protection Conference, Lincoln, New Zealand, 18–21 August 1997; pp. 119–124.

15. Llorens, J.; Gil, E.; Llop, J.; Escolà, A. Variable rate dosing in precision viticulture: Use of electronic devices to improve application efficiency. Crop Prot. 2010, 29, 239–248.

16. Rinaldi, M.; Llorens, J.; Gil, E. Electronic Characterization of the Phenological Stages of Grapevine Using A LIDAR Sensor. In Proceedings of the 9th European Conference on Precision Agriculture, Lleida, Spain, 8–11 July 2013.

17. LidarScan v.1 Software; Universitat de Lleida: Lleida, Spain, 2007.

18. Frascati, F. Create Packages for R under Windows XP. Available online: http://cran.r-project.org/ doc/contrib/Frascati-Rpackages.pdf (accessed on 20 November 2013).

19. Arnó, J.; Martínez-Casasnovas, J.A.; Ribes-Dasi, M.; Rosell, J.R. Review. Precision viticulture. Research topics, challenges and opportunities in site-specific vineyard management. Spanish J. Agric. Res. 2009, 7, 779–790.

20. Arnó, J.; Del Moral, I.; Escolà, A.; Company, J.; Masip, J.; Sanz, R.; Rosell, J.R.; Martínez-Casasnovas, J.A. Mapping the Leaf Area Index in Vineyard Using A Ground-Based Lidar Scanner. In Proceedings of the 11th International Conference on Precision Agriculture, Indianapolis, IN, USA, 15–18 July 2012.

21. Precision Agriculture Laboratory. Available online: http://www.sydney.edu.au/agriculture/pal/ (accessed on 19 June 2013).

22. Fridgen, J.J.; Kitchen, N.R.; Sudduth, K.A.; Drummond, S.T.; Wiebold, W.J.; Fraisse, C.W. Management zone analyst (MZA): Software for subfield management zone delineation. Agronomy J. 2004, 96, 100–108.

23. Balsari, P.; Doruchowski, G.; Marucco, P.; Tamagnone, M.; van de Zande, J.C.; Wenneker, M. A system for adjusting the spray application to the target characteristics. Agr. Eng. Int. CIGR Ejournal 2008, X, 1–11.

24. Gil, E.; Escolà, A. Variable Rate Application of Plant Protection Products in Vineyard Using Ultrasonic Sensors. In Proceedings of the 9th Workshop on Sustainable Plant Protection Techniques in Fruit Growing, Alnarp, Sweden, 11–14 September 2007; pp. 61–62.

25. Shimborsky, E. Digital Tree Mapping and Its Applications. In Proceedings of the 4th European Conference on Precision Agriculture, Berlin, Germany, 15–19 June 2003; pp. 645–650.

26. Ballagh, L.M.; Raup, B.H.; Duerr, R.E.; Khalsa, S.J.S.; Helm, C.; Fowler, D.; Gupte, A. Representing scientific data sets in KML: Methods and challenges. Comput. Geosci. 2011, 37, 57–64.

27. Escolà, A. Method for Real-Time Variable Rate Application of Plant Protection Products in Precision Horticulture/Fructiculture. Ph.D. Thesis, Universitat de Lleida: Lleida, Spain, 2010.

28. Doruchowski, G.; Balsari, P.; Van de Zande, J.C. Development of A Crop Adapted Spray Application System for Sustainable Plant Protection in Fruit Growing. In Proceedings of International Symposium on Application of Precision Agriculture for Fruits and Vegetables, Orlando, FL, USA, 1 April 2009.

29. Zaman, Q.U.; Salyani, M. Effects of foliage density and ground speed on ultrasonic measurement of citrus tree volume. Appl. Eng. Agric. 2004, 20, 173–178.

30. Balsari, P.; Tamagnone, M. An Ultrasonic Airblast Sprayer. In Proceedings of the International Conference on Agricultural Engineering, Oslo, Norway, 24–27 August 1998; pp. 585–586.

31. Solanelles, F.; Escolà, A.; Planas, S.; Rosell, J.R.; Camp, F.; Gracia, F. An electronic control system for pesticide application proportional to the canopy width of tree crops. Crop Prot. 2006, 95, 473–481.

32. Escolà, A.; Rosell-Polo, J.R.; Planas, S.; Gil, E.; Pomar, J.; Camp, F.; Llorens, J.; Solanelles, F. Variable rate sprayer. Part 1—Orchard prototype: Design, implementation and validation. Comput. Electron. Agric. 2013, 95, 122–135.

33. Gil, E.; Llorens, J.; Llop, J.; Fàbregas, X.; Escolà, A.; Rosell-Polo, J.R. Variable rate sprayer. Part 2—Vineyard prototype: Design, implementation, and validation. Comput. Electron. Agric. 2013, 95, 136–150.

34. Byers, R.E.; Hickey, K.D.; Hill, C.H. Base gallonage per acre. Va. Fruit 1971, 60, 19–23.

35. Gil, E. Metodología y Criterios Para la Selección y Evaluación de Equipos de Aplicación de Fitosanitarios Para la Viña. Ph.D. Thesis, Universitat de Lleida: Lleida, Spain, 2001, in press.

36. Escolà, A.; Camp, F.; Solanelles, F.; Llorens, J.; Planas, S.; Rosell, J.R.; Gràcia, F.; Gil, E. Variable Dose Rate Sprayer Prototype for Tree Crops Based on Sensor Measured Canopy Characteristics. In Proceedings of Precision Agriculture 2007, Skiathos, Greece, 3–6 June 2007; pp. 563–571.

37. Directive 2009/128/EC of the European Parliament and of the Council of 21 October 2009 Establishing a Framework for Community Action to Achieve the Sustainable Use of Pesticides; 2009/128/EC; European Parliament: Bruxelles, Belgium, 2009.

38. Walklate, P.J. Modelling Canopy Interactions for Drift Mitigation. In Proceedings of the International Conference on Pesticide Application for Drift Management, Waikoloa, III, USA, 27–29 October 2004; pp. 370–377.

39. Solanelles, F.; Gregorio, E.; Sanz, R.; Rosell, J.R.; Arnó, J.; Planas, S.; Escolà, A.; Masip, J.; Ribes-Dasi, M.; Gràcia, F.; et al. Spray Drift Measurements in Tree Crops Using a Lidar System. In Proceedings of the 10th Workshop on Spray Application Techniques in Fruit Growing, Wageningen, The Netherlands, 30 September–2 October 2009; pp. 40–41.

40. Gregorio, E.; Solanelles, F.; Rocadenbosch, F.; Rosell, J.R.; Sanz, R. Airborne Spray Drift Measurement Using Passive Collectors and Lidar Systems. In Remote Sensing for Agriculture, Ecosystems, and Hydrology XIII, Prague, Czech Republic, 19 September 2011.

41. Gregorio, E.; Rosell-Polo, J.R.; Sanz, R.; Rocadenbosch, F.; Solanelles, F.; Garcerá, C.; Chueca, P.; Arnó, J.; del Moral, I.; Masip, J.; et al. LIDAR as an alternative to passive collectors to measure pesticide spray drift. Atmos. Environ. 2013, doi:10.1016/j.atmosenv.2013.09.028.

42. Kuflik, T.; Prodorutti, D.; Frizzi, A.; Gafni, Y.; Simon, S.; Pertot, I. Optimization of copper treatments in organic viticulture by using a web-based decision support system. Comput. Electron. Agric. 2009, 68, 36–43.

43. Walklate, P.J.; Cross, J.V. A webpage calculator for dose rate adjustment of orchard spraying products. Aspects Appl. Biol. Int. Adv. Appl. 2010, 99, 359–366.

44. Gil, E.; Escolà, A. Design of a decision support method to determine volume rate for vineyard spraying. Appl. Eng. Agric. 2009, 25, 145–151.

45. Gil, E.; Llorens, J.; Landers, A.J.; Llop, J.; Giralt, L. Field validation of dosaviña, a decision support system to determine the optimal volume rate for pesticide application in vineyards. Eur. J. Agronomy 2011, 35, 33–46.

46. Ebert, T.A.; Downer, R.A. A different look at experiments on pesticide distribution. Crop Prot. 2006, 25, 299–309.

CHAPTER 8

SOME CRITICAL ISSUES IN ENVIRONMENTAL PHYSIOLOGY OF GRAPEVINES: FUTURE CHALLENGES AND CURRENT LIMITATIONS

H.R. SCHULTZ AND M. STOLL

8.1 INTRODUCTION

Grapevines are cultivated in six out of seven continents, between latitudes 4° and 51° in the Northern Hemisphere (NH) and between 6° and 45° in the Southern Hemisphere (SH) across a large diversity of climates (oceanic, warm oceanic, transition temperate, continental, cold continental, Mediterranean, subtropical, attenuated tropical, arid and hyperarid climates) (Peguy 1970, Tonietto and Carbonneau 2004). Accordingly, the range and magnitude of environmental factors differ considerably from region to region and so do the principal environmental constraints for grape production. Problems of low winter temperatures have limited grape cultivation in the past in areas with continental climates in Eastern Europe, Asia and North America. Low temperatures during the growing season have prevented the extension of grape-growing in regions approximately be-

Schultz HR and Stoll M. Some Critical Issues in Environmental Physiology of Grapevines: Future Challenges and Current Limitations. Australian Journal of Grape and Wine Research, *16,s1 (2010), DOI: 10.1111/j.1755-0238.2009.00074.x. Reprinted with permission from the authors.*

yond the 12°C temperature isotherm (April–October (NH), October–April (SH)) (Jones et al. 2005a). The effects of hot temperatures, on the contrary, are less clear with respect to the distribution of grapevine cultivation areas. In general, the 22°C temperature isotherm is considered limiting for wine grape production (Jones 2007a, Schultz and Jones 2008), but many areas in the tropics are much warmer than this (Tonietto and Carbonneau 2004) and detrimental effects of high temperatures may be largely mitigated if water supply is sufficient and/or if humidity is high. Within the existing production areas, water shortage is probably the most dominant environmental constraint (Williams and Matthews 1990), and even in moderate temperate climates, grapevines often face some degree of drought stress during the growing season (Morlat et al. 1992, van Leeuwen and Seguin 1994, Gaudillère et al. 2002, Gruber and Schultz 2009).

The primary and global challenge for the grape and wine industry of the future will be climate change because its direct (temperature, precipitation, CO_2 concentration, etc.) and indirect consequences (resource management, energy efficiency, sustainability in production and consumer acceptance, etc.) will affect all facets of the industry. Climate change has already caused significant warming in most grape-growing areas of the world during the last approximately 55 years (Jones et al. 2005a,b). The degree of warming varies strongly between regions, with large observed differences within each climate maturity grouping (cool, intermediate, warm, hot), ranging form 0.1°C during the growing season for Burgundy, to about 4°C for the Northern Rhone Valley (Jones et al. 2005a). Warming will continue, and depending on the baseline scenario (Intergovernmental Panel on Climate Change (IPCC) 2008) and the model used to make regional predictions up to 2050, temperatures may increase between somewhat less than 1°C for the growing season in South Africa, to near 3°C in Eastern Washington, Southern Portugal (Jones et al. 2005a) and some areas in Australia (Webb et al. 2007, 2008).

Predictions of the total annual amount of precipitation and its annual and regional distribution are much more uncertain (IPCC 2008). However, according to many experts, water and its availability and quality will be the main pressures on, and issues for, societies and the environment (IPCC

2008). Because of rising temperatures and solar radiation in many places, and decreasing and/or more irregular precipitation patterns, climate change will exacerbate soil degradation and desertification (IPCC 2008). Desertification is often accompanied by soil salinization which today affects 7% of the global total land area and 20–50% of the global irrigated farmland (IPCC 2008). Irrigation in agriculture already accounts for about 70% of the total water use worldwide, and the irrigated surface area has increased linearly since 1960. Driven by apparent changes in the climate conditions in viticultural areas previously entirely rain-fed, there is already an increasing interest in irrigation. However, population growth is predicted to reach between 8.7 billion (by 2050) in the most conservative estimation to about 15 billion (by 2100) in a A2 'worst case' scenario (IPCC baseline scenarios 2007). This will cause a general increase in water demand on a global scale, and will become a problem for agricultural water use in light of sharp increases in water consumption of the urban, industrial and environmental sectors (Fereres and Soriano 2007). Some fresh water basins in the world termed 'water-stressed' by the IPCC (2008), that is water availability decreases below 1000 m^3/capita/yr or withdrawal to average run-off increases above a ratio of 0.4, are partly congruent to areas where grapes are currently cultivated on a larger scale (e.g. the Murray–Darling River basin in Australia). These developments will put enormous pressure on irrigated land not directly devoted to food production with the combined consequences (temperature and water) that grape cultivation will be partly displaced from traditional areas (Schultz and Jones 2008) and will be forced to use more marginal land under environmental conditions previously termed less suitable.

Therefore, more than ever before, there is a need to understand the physiological mechanisms and the genetic background underlying the interactions between plants and the environment, and to pay attention to research fields which will be pivotal for the development of sustainable concepts under changing conditions. Because of the large array of possible issues under this rather wide topic, this review will not be able to address every aspect in adequate detail. Focus will be on some key subjects and on the challenges and limitations the research community is currently facing.

8.2 EFFICIENT WATER USE

8.2.1 WATER USE EFFICIENCY (WUE)

Regional estimates on changes in water availability under future climate conditions suggest that for most grape-growing regions, the propensity for water deficiency will increase during the growing season (IPCC 2008). Together with increasing competition for irrigation water, WUE in viticulture will be an important issue of the future. Therefore, to secure a sustainable and effective use of water, more information on the physiological and genetic basis of WUE is needed (Chaves and Oliveira 2004). There is evidence for variation in WUE among species and cultivars in grapevines (Düring and Scienza 1980, Eibach and Alleweldt 1984, Chaves et al. 1987, Bota et al. 2001, Schultz 2003a, Flexas et al. 2004, Souza et al. 2005b, Soar et al. 2006a, Flexas et al. 2008) and some of the aspects of genetic variation of WUE have been elucidated in *Arabidopsis thaliana* as a model plant (Masle et al. 2005, Nilson and Assmann 2007, Lake and Woodward 2008). Despite demonstrated interest in WUE, it can be defined in different ways and the physiological basis of its regulation is not fully understood. WUE depends on complex interactions between environmental factors and physiological mechanisms such as stomatal behaviour, photosynthetic capacity, and leaf and plant anatomy (for a review see Bacon 2004). Additionally, all mechanisms that tend to maintain plant survival or a certain productivity under conditions of limited water supply or high evaporative demand (or adverse conditions in general) come at certain 'costs', which, from an agricultural point of view, will either reduce dry matter production or competitive ability (Jones 1992). In that sense, high WUE genotypes may not be the ideal compromise between drought tolerance and economic performance.

WUE: challenges and limitations of measurements and data interpretation. WUE is often determined from single leaf gas-exchange measurements, relating net CO_2 assimilation rate (A) either to stomatal conductance for water vapour (g), termed intrinsic water use efficiency (WUE_i) (Osmond et al. 1980) (Eqn 1),

$$WUE_i = \frac{A}{g} \tag{1}$$

or A to leaf transpiration rate (E), termed instantaneous water use efficiency WUEinst (Eqn 2).

$$WUE_{inst} = \frac{A}{E} \tag{2}$$

The former (Eqn 1) is a way to largely exclude the effects of changing evaporative demand on water flux out of the leaf (Bierhuizen and Slatyer 1965) and has been predominantly used in studies of water stress effects on grapevines (i.e. Düring 1987, Schultz 1996, Flexas et al. 1998, Escalona et al. 1999, Bota et al. 2001, Souza et al. 2005b, Chaves et al. 2007, Pou et al. 2008, Zsófi et al. 2009). In general, an increase in WUE_i under drought or deficit irrigation strategies has been observed in these studies, but it is difficult to integrate WUE_i over time. Diurnal changes in g and environmental conditions may preclude a close relationship to long-term WUE measurements such as, for instance, leaf carbon isotope composition described below.

Less frequently used as WUE_i is WUE_{inst} or the integrated WUE (W_p) (Farquhar et al. 1989) which includes losses of carbon beause of respiration at night or from non-photosynthetic organs such as roots and water loss at night because of incomplete stomatal closure or high cuticular conductance (Eqn 3). Data like these are most relevant but are confined to pot studies and are reflected in ratios such as dry matter produced per unit of water lost (i.e. Düring and Scienza 1980, Gibberd et al. 2001).

$$W_p = \frac{A(1 - \Phi_C)}{E(1 - \Phi_E)} \tag{3}$$

where Φ_C = carbon loss via respiration and where Φ_E = unproductive water loss

More recently, the analyses of ^{13}C carbon isotope discrimination ($\Delta^{13}C$) has become popular and widespread for the evaluation of water use in plants and grapevines under field conditions (Gaudillère et al. 2002, Medrano et al. 2005, Souza et al. 2005b, Chaves et al. 2007). In contrast to gas-exchange techniques that provide measurements of photosynthesis rates at a single point in time, leaf carbon isotope signature ($\delta^{13}C$) integrates the ratio of intercellular (C_i) to atmospheric CO_2 concentration (C_a) for longer periods of time (Condon et al. 2004). The basis of the biochemical discrimination (Δ) against ^{13}C in C_3 plants lies within the primary carboxylating enzyme, ribulose-1,5-bisphosphate carboxylase-oxygenase (Rubisco) which discriminates against ^{13}C because of the intrinsically lower reactivity of ^{13}C compared with ^{12}C (Farquhar et al. 1982).

Variations in $\Delta^{13}C$ are used to analyse temporal and spatial trends in plant–carbon water relations. In the most frequently used version of the Farquhar et al. (1982) model – the linear (or reduced) form relates $\Delta^{13}C$ linearly to C_i/C_a (Eqn 4).

$$\Delta^{13}C = -a + (b' - a)\frac{C_i}{C_a} \qquad (4)$$

With a and b' being isotopic fractionation coefficients (a = 4.4‰, fractionation during CO_2 diffusion through stomata, O'Leary 1981; b' = 27‰, fractionation associated with reactions by Rubisco and phosphoenol pyruvate (PEP)-carboxylase, Farquhar and Richards 1984). The linear relationship between C_i/C_a and $\Delta^{13}C$ can be used to calculate WUE_i (A/g) because C_i/C_a reflects the balance between A and stomatal conductance for CO_2 (gc) and because gc and stomatal conductance for water vapour (g) are related by a constant factor (g = 1.6 g_c) (Farquhar et al. 1980) (Eqn 5)

$$WUE_i = \frac{C_a}{1.6}\frac{(b' - \Delta^{13}C)}{(b' - a)} \qquad (5)$$

Inherent in Equations 4 and 5 are, that if C_i/C_a decreases, then $\Delta^{13}C$ decreases and WUE_i will increase. If we relate the linear model of carbon isotope discrimination in an analogous manner to WUE_{inst}, where leaf transpiration rate, E, is a function of g and the leaf to air vapour pressure deficit (LAVPD) (kPa), (Eqn 6) (Farquhar and Richards 1984)

$$E = gLAVPD \qquad (6)$$

then,

$$WUE_{inst} = \frac{C_a}{(1.6\ LAVPD)} \frac{(b' - \Delta^{13}C)}{(b' - a)} \qquad (7)$$

which implies that WUE_{inst} will vary with C_i (affecting $\Delta^{13}C$) and LAVPD because C_a is essentially constant. It also implies that because evaporative demand is considered in WUE_{inst}, but not in WUE_i, both can show different trends (opposite directions) while these can still be consistent with the same trend in $\Delta^{13}C$ (Seibt et al. 2008).

8.2.2 VPD RESPONSE AND SIGNALLING MECHANISMS

Research has shown that stomatal conductance of grapevines is sensitive to the vapour pressure deficit of the air (VPD) (kPa), (i.e. Düring 1987, Soar et al. 2006a,b, Poni et al. 2009) closely related to LAVPD, which will directly affect WUE_{inst} (g↓, E↑]) and WUE_i (g↓). Figure 1 shows the relationships between WUE_{inst} and WUE_i with LAVPD for four grapevine varieties grown in pots (30L, 3-year old vines) under a wide range of climatic conditions (photosynthetic photon flux density, PPFD 750–1750 μmol/m²/s; air temperature, Tair, 19°C–37°C) during different drought and recovery experiments (pre-dawn water potential ψ_{PD}, −0.15 to −1.45 MPa) over a period of several weeks (Chouzouri and Schultz, unpublished data). When all data are pooled, there is a clear relationship of WUE_{inst} to LAVPD, with WUE_{inst} decreasing with increasing LAVPD, while there is none for WUE_i. This is consistent with the optimisation theory (Cowan 1977) and is important because (i) it indicates that LAVPD negatively affects WUE_{inst} over a wide range of plant water status; and (ii) because this is not apparent in WUE_i, measurements conducted under high LAVPD and water deficit may show opposite tendencies in these two parameters. This becomes clear from diurnal gas-exchange data of a field experiment with the variety Syrah (Shiraz), where LAVPD was high during a continuous dry down over several months (Figure 2). As expected, WUE_i was higher for stressed than for irrigated plants throughout most of the day at any level of water deficit (Figure 2(a),(b),(c)). However, WUE_{inst} was either not different between irrigated and non-irrigated plants (Figure 2(d)) or lower in the stressed plants when water deficit became more severe (Figure 2(e),(f)) and stomatal closure caused LAVPD to rise substantially (Figure 2(j)–(l)) because of increases in leaf temperature (Figure 2(g),(i)).

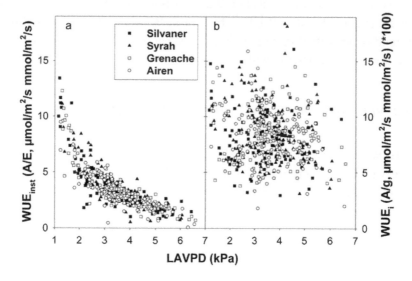

FIGURE 1: Response of instantaneous water use efficiency (WUEinst) (a) and intrinsic water use efficiency (WUEi) (b) to leaf to air vapour pressure deficit (LAVPD) of four grapevine varieties (Silvaner, Syrah, Grenache, Airen) over a wide range of climate conditions (photosynthetic photon flux density (PPFD) 750–1750 µmol/m2/s; air temperature (Tair), 19°C–37°C) during different drought and recovery experiments (predawn water potential (ψPD), −0.15 to −1.45 MPa) over a period of several weeks. Plants were grown outdoors in 30 L pots in 2003 (Chouzouri and Schultz unpublished data).

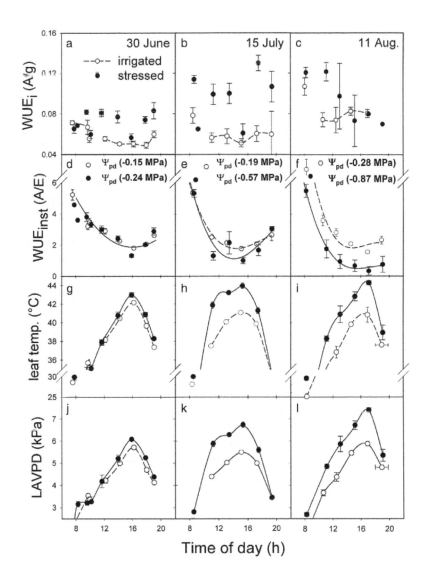

FIGURE 2: Diurnal time courses in intrinsic water use efficiency (WUEi) ((a)–(c)), instantaneous water use efficiency (WUEinst) ((d)–(f)), leaf temperature ((g)–(i)) and leaf to air vapour pressure deficit (LAVPD) ((j)–(l)) of irrigated (open symbols) and non-irrigated (stressed) (closed symbols) field-grown Syrah grapevines in a commercial vineyard near Montpellier, France on 30 June (2 weeks post bloom), 15 July and 11 August (veraison). The development in ψPD is indicated in (d)–(f). For details on experimental set-up and materials and methods see Schultz (2003a).

There is uncertainty whether a pronounced stomatal response to VPD during water deficit is beneficial for improving WUE, but there are clear differences in this response between varieties (Düring 1987, Soar et al. 2006b) and rootstock/scion combinations (Soar et al. 2006a, Koundouras et al. 2008). Water stress intensified the VPD response in some rootstock/scion combinations but had the opposite effect on others (Soar et al. 2006a). Irrespective of the irrigation strategy applied, g decreased in response to rising VPD under partial root zone drying (PRD) (Loveys et al. 2004) and regulated deficit irrigation (RDI) (Patakas et al. 2005, Pou et al. 2008) which would allow the plant to better control E. However, depending on the intensity of the closure response and the extent of the subsequent rise in temperature and LAVPD, WUE_{inst} may be reduced (as in Figure 2), which has been observed in other studies on grapevines (Naor et al. 1994, Naor and Bravdo 2000, Koundouras et al. 2008) and trees (Jifon and Syvertsen 2003, Baldocchi and Bowling 2003). In this context, it is important to realise, that a reduction in WUE_{inst} may occur despite $\Delta^{13}C$ data suggesting the opposite (Koundouras et al. 2008, Seibt et al. 2008). Figure 3 illustrates this using different scenarios and combining Equations 4, 5 and 7. At the same carbon isotope signature indicating an increase in WUE_i with decreasing $\Delta^{13}C$, WUE_{inst} may increase, stay constant or even decrease (Figure 3(b)) depending on the development in LAVPD (Figure 3(a)). It is recognised that this model ignores the contribution of changes in mesophyll conductance with drought and/or temperature increase and photorespiration (Farquhar et al. 1982, Flexas et al. 2002, 2007) or even decreased activities of photosynthetic enzymes under severe stress (Flexas et al. 2006). Incorporating these aspects into a more comprehensive model can yield similar results (Baldocchi and Bowling 2003, Seibt et al. 2008). Added in Figure 3 are some data from Gibberd et al. (2001) obtained in a pot study of 19 different grapevine genotypes, where different groups were separated depending on obtained relationships between WUE on the whole plant level (dry matter produced/amount of water transpired, e.g. close to Eqn 3) and $\Delta^{13}C$ under well-watered conditions. Functional groups which exhibited a greater decrease in g had lower $\Delta^{13}C$ values but gained little in WUE (Figure 3B), probably because increasing leaf temperature and LAVPD increased E and did offset decreases in g. This effect will strengthen with decreasing wind-speed, decreasing water availability, increasing leaf size, increasing ambient temperature and denser canopies, where the latter four could be expected to be more important in the future because of warming and rising ambient CO_2 concentrations. All the listed factors and their impact on leaf and canopy temperature are also playing a role in the use of new proxy and remote sensing techniques for stress monitoring and irrigation management (Jones et al. 2002, Möller et al. 2007, see section Canopy (leaf) temperature and imaging technologies).

Differences in the responsiveness of stomata to environmental factors, such as VPD and/or water deficit, seem to be modulated by abscisic acid (ABA), either through the increased ability of the xylem to supply ABA (Loveys 1984a, Loveys et al. 2004), increases in xylem pH affecting ABA ionization and general metabolism (Stoll et al. 2000) or through differential gene expression in the ABA biosynthetic pathway involved in regulating varietal responses to VPD (Soar et al. 2006a). However, absolute control of g by ABA is disputed, and hydraulic signalling may also play a role (Schultz 2003a, Rodrigues et al. 2008) and both factors may actually partly depend on each other in the regulation and recovery of g during and after a water stress (Lovisolo et al. 2008a). Depending on the degree of stress, the rate and degree of recovery may be very different. Flexas et al. (2004, 2006) and Galmés et al. (2007) found that grapevines subjected to a mild deficit (i.e. maximum g above 100 mmol/m^2/s) fully recovered within a day after rewatering, whereas more severely stressed plants recovered slowly during the following week and did not reach pre-stress levels in A and g.

Nitric oxide (NO) may be another signalling molecule (aside of or together with ABA) involved in stomatal responses to drought. NO has been shown to be involved in many biological events (Quan et al. 2008), and the synthesis of NO by nitric oxide synthase and nitrate reductase in response to drought may elicit the increase of anti-oxidant enzymes which can scavenge water stress-induced hydrogen peroxide (Sang et al. 2008) or may mediate ABA-induced closure of stomata (Neil et al. 2003). Little is known about the relationships between drought and NO accumulation in grapevines, but Patakas and Zotos (2005) found a rapid increase in NO-induced fluorescence close to the stomates in leaves of stressed plants.

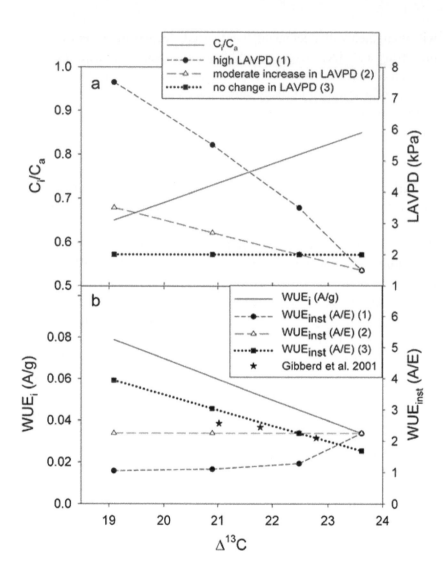

FIGURE 3: Modelled changes in intrinsic water use efficiency (WUEi) and instantaneous water use efficiency (WUEinst) at a given Ci/Ca ratio and three cases of possible changes in leaf to air vapour pressure deficit (LAVPD) (a) as a function of $\Delta13C$ (b). The model is based on Equations 4, 5 and 7 and the conceptual analysis presented by Seibt et al. (2008). Some data from Gibberd et al. (2001) obtained in a pot study with 19 different grapevine genotypes are added. In this study, different groups were separated depending on obtained relationships between water use efficiency (WUE) on the whole plant level (dry matter produced/amount of water transpired, e.g. close to Eqn 3) and $\Delta13C$ under well-watered conditions.

8.2.3 ISOTOPIC SIGNATURES IN LEAVES AND FRUITS

For grapevine leaves from plants exposed to different irrigation regimes, even the relationships between WUEi and $\Delta^{13}C$ may be weak (Souza et al. 2005b, Chaves et al. 2007), but there are reasonable relationships of WUEi with $\Delta^{13}C$ of the fruit (Medrano et al. 2005, Souza et al. 2005b, Chaves et al. 2007). These differences have been interpreted as being related to the fact that sugar accumulation starts late in the season and results from current leaf photosynthates, which are produced during the water deficit as compared with the isotopic signature of leaves which are formed earlier during the season (Gaudillère et al. 2002). There could also be more post-photosynthetic fractionation processes (namely respiration) in berries, which might result in differences in the carbon isotope composition of the two organs (Badeck et al. 2005). Despite good correlations to plant water status during berry ripening (i.e. van Leeuwen et al. 2001, Gaudillère et al. 2002), so far the final evidence is lacking that fruit $\Delta^{13}C$ is an indicator of crop scale water use efficiency (i.e biomass produced per unit water used, Jones (1992), because correlations to WUEi may not reflect relationships on the whole-canopy level.

Aside from the isotopic discrimination of carbon during photosynthesis and respiration, it may also be useful in future studies to measure the isotopic enrichment of oxygen ($^{18}O‰$) occurring in leaves during the evaporation of water. The enrichment will increase with a decrease in g, an increase in leaf temperature, a decrease in relative humidity (or an increase in VPD and LAVPD) (Figure 4), a decrease in E, and as such may allow the distinction between effects of soil water deficit and evaporative demand on WUE (Farquhar et al. 2007). This response to relative humidity on the leaf level can also be retraced in grape juice (Tardaguila et al. 1997) and wines coming from areas with different evaporative demand where isotopic enrichment can be related to the prevailing conditions 30 days before harvest (Figure 4) (Hermann and Voerkelius 2008).

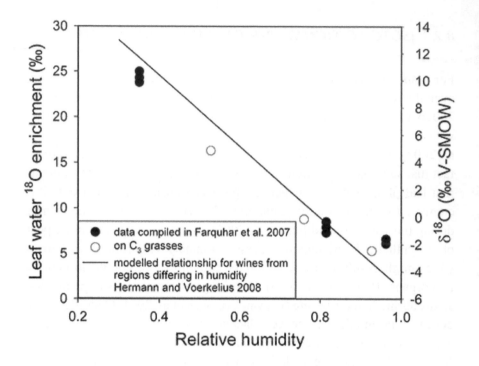

FIGURE 4: Enrichment of leaf water versus relative humidity for five C_3 grass species. Original data from Helliker and Ehleringer (2000a, closed symbols, 2000b, open symbols) redrawn in Farquhar et al. (2007). Copyright American Society of Plant Biologists used with permission (http://www.plantphysiol.org). The line represents the published modelled correlation between the isotopic oxygen value in wine as a function of relative humidity of the region of origin from Hermann and Voerkelius (2008). More details can be found in these references. V-SMOW, Vienna Standard Mean Ocean Water (the standard isotopic reference).

8.2.4 VARIETAL DIFFERENCES IN STRESS RESPONSES (ISOHYDRIC VERSUS ANISOHYDRIC)

There is clear evidence that varieties differ in their physiological response to stress in their response velocity (Düring and Scienza 1980), adaptive mechanisms (i.e. Düring and Loveys 1982, Loveys 1984b, Schultz 1996, Bota et al. 2001), and effectiveness in regulating gas-exchange (i.e.

Chaves et al. 1987, Escalona et al. 1999, Flexas et al. 2002, Schultz 2003a, Souza et al. 2005a,b). A fortunate choice of varieties to study has led to the illustration of these differences in a field experiment with the varieties Grenache and Syrah, whose stomatal behaviour under water deficit led to a classification termed 'near-isohydric' and 'anisohydric' in reference to Stocker (1956) (Schultz 2003a). Subsequent studies have shown, that these differences in g were also triggered in response to VPD (Soar et al. 2006a), and correlated with differential expression of key genes in the ABA biosynthetic pathway in leaves but not roots (Soar et al. 2006a). The isohydric–anisohydric classification leaves sufficient space to test other varieties, whether they can fit these categories or can enlarge the existing spectrum in order to quantify a general type of WUE, which may be important in irrigated viticulture in the future. However, recent findings in an outdoor 'heating experiment' (40°C for three consecutive days) showed that anisohydric Syrah vines up-regulated g, E, and A in response to temperature (Soar et al. 2009) at a common VPD. This may be a positive adaptation in terms of heat tolerance through increased evaporative cooling at the expense of long-term transpiration efficiency (Soar et al. 2009).

The continuum between classical isohydry and anisohydry was well illustrated in a study by Turner et al. (1984) with nine woody and herbaceous species, where the response of E to VPD varied largely. Despite of the correlation to ABA physiology (Soar et al. 2006a), the underlying mechanisms for isohydric or anisohydric behaviour are little understood, and there is evidence that plant hydraulic conductance may play a decisive role (Franks et al. 2007). This conclusion was based on the fact that even in plants capable of strongly down-regulating g as a reaction to water deficit (Franks et al. 2007) it could not be prevented that midday leaf water potential fell to levels where xylem air embolisms may have become the dominant factor in determining water flow and where a functional correlation between embolism formation and leaf gas-exchange existed (Domec et al. 2006, Maherali et al. 2006). Nevertheless, Alsina et al. (2007) did not find a correlation between drought tolerance and vulnerability to xylem embolisms of different grapevine varieties, whereas other data from pot and field experiments showed a correlation (Schultz 2003a, Chouzouri and Schultz 2005, Lovisolo et al. 2008b). The solution to the dilemma in terms of experimental set up may be to test varieties in their combined response of gas-exchange to environmental stresses using the Ball-Woodrow-Berry model (Ball et al. 1987) (Figure 5(a),(b)) and to accompany its application with both ABA and hydraulic conductance measurements to gain clarity. The advantage of the model is that it incorporates the composite sensitivity of stomata to assimilation rate, CO_2 concentration, humidity and temperature (Ball et al. 1987), is applicable to grapevines (Schultz et al. 1999, Schultz 2003b), and gives an indication of overall stomatal sensitivity to these factors (Figure 5(a)) (Eqn 8):

$$g = g_0 \kappa A \frac{rh}{C_a} \tag{8}$$

where g_0 is a residual stomatal conductance to H_2O vapour (A$\rightarrow$0 when light intensity approaches 0), and k is a constant representing stomatal sensitivity (Tenhunen et al. 1990, Harley and Tenhunen 1991) (Figure 5(a)). Measurements of relative humidity, rh, and CO_2 partial pressure (C_a) outside the leaf boundary layer can be used to drive the model. Because this relationship is purely empirical, k and (A $[rh/C_a]$) are dimensionless (Ball et al. 1987). While the use of rh has been criticised as being non-mechanistic (Aphalo and Jarvis 1993), the model has been successfully coupled to assimilation models of individual perennial crops (Katul et al. 2000), including grapevines (Schultz 2003b) or entire terrestrial ecosystems (Lloyd et al. 1995, Wang et al. 2007, Mo et al. 2008). The constant representing stomatal sensitivity, k, i.e. the slope of the relationship between g and (A $[rh/C_a]$) is sensitive to water deficit (Figure 5(a)) and can be used to distinguish the sensitivities of g of different varieties (Figure 5(b)). Note, the smaller k, the higher is the stomatal sensitivity (Harley and Tenhunen 1991), thus Figure 5(b) shows that stomates of Grenache exhibit a higher sensitivity (smaller k) than Syrah at similar water potentials.

8.3 SPECIFIC PHYSIOLOGICAL ASPECTS

8.3.1 AQUAPORINS

The differences between varieties in the response scheme to drought and/or VPD may be mediated by aquaporins (AQP) (Sade et al. 2009, Vandeleur et al. 2009). AQP are members of the mayor membrane intrinsic protein family, can act as water channels and can regulate cell-to-cell water transport (Maurel et al. 2008). So far, 35 AQP have been identified in Arabidopsis (Johanson et al. 2001) and 28 in the Vitis genome (Fouquet et al. 2008). Under short-term water deficit, gene expression for AQP in grapevines was stronger in roots than in leaves, particularly under moderate stress levels (Galmés et al. 2007). This may be an indication that the function of some AQP is more important in roots than in leaves, which would be consistent with the idea that they facilitate water transport (Luu and Maurel 2005), although a relation of leaf hydraulic conductance to AQP expression has also been found (Cochard et al. 2007). However, the role of AQP in regulating plant water status is a complex issue, because different subfamilies and subclasses may be up- or down-regulated or remain unchanged depending on the degree of water deficit and/or the time during the stress period (Galmés et al. 2007). Lovisolo and Schubert (2006) showed that AQP may be involved in the recovery of grapevine shoot water relations after embolisms were formed during water stress. However, it remains unclear if this effect was more important in the regulation of water flow in roots or shoots.

A recent study exploring the nature of the isohydric and anisohydric response pattern of different grapevine cultivars suggested that physiological and anatomical differences in the roots played a major role in water transport (Vandeleur et al. 2009). In this study, substantial differences in root hydraulic conductance under water deficit and recovery could be related to the differential expression of the two most highly expressed plasma membrane intrinsic protein (PIP) AQP (*VvPIP1;1* and *VvPIP2;2*). In the anisohydric cultivar (in this case Chardonnay), root hydraulic conductance was up-regulated under water deficit and so was *VvPIP1*;1, whereas in the isohydric variety (Grenache) there was no up-regulation indicating that water transport across roots was regulated to match transpirational demand (Vandeleur et al. 2009) and possibly stomatal conductance (Galmés et al. 2007). In a functional study on AQP in different *Vitis* spp. used as rootstocks, Lovisolo et al. (2008b) also found differences in their role in root hydraulic conductance and susceptibility to embolism formation. However, expression patterns in AQP genes not necessarily correlate always with physiological parameters, such as hydraulic conductance (Galmés et al. 2007), because the complexity of responses on the metabolic, cellular, organ or whole plant level may mask such a correlation. Nevertheless, the involvement of AQP in the regulation of isohydric/anisohydric behaviour has also been demonstrated for different lines of tomatoes (*Solanum lycopersicum*) (Sade et al. 2009).

8.3.2 MESOPHYLL CONDUCTANCE

A decisive factor in the determination of WUE at the leaf level, apart from g is the mesophyll conductance, gi, (Flexas et al. 2002, Flexas et al. 2006), because the role of CO_2 transfer from the substomatal cavity to the photosynthetic enzyme(s) is not reflected in C_i (i.e. Flexas et al. 2002, Flexas et al. 2006), but will be in $\Delta^{13}C$ (Seibt et al. 2008). Recently, it was shown that both water (Flexas et al. 2002, 2007) and salt stress (Geissler et al. 2009) can reduce g_i and in the latter case also reduce WUE_{inst}.

There may be differences in this component between genotypes because of differences in leaf anatomy currently unknown or in the response of gi to factors such as variations in C_i (Flexas et al. 2007) and temperature (Bernacchi et al. 2002). If confirmed, they may offer a way to improve WUE in the future (Flexas et al. 2009, this issue). Because g_i also changes with CO_2 concentration, it is important to investigate possible differences between genotypes, because g_i will play a crucial role in the photosynthetic performance under future CO_2 concentrations (Flexas et al. 2007, Flexas et al. 2008). The possibility of genetic differences was already indicated by variations in the CO_2 specificity factor of Rubisco in the absence and presence of water deficit (Bota et al. 2001).

8.3.3 NIGHT-TIME WATER USE AND RESPIRATION

Night-time water use and dark respiration also affect daily WUE (see Eqn 3), and changes in plant water status will alter both, yet there is only little information available on both processes for grapevines.

Incomplete stomatal closure during the night has been observed in a diverse range of C_3 and C_4 species, among these are many horticultural and crop species (for a review see Caird et al. 2007). This can lead to substantial night-time transpirational water loss even in dessert species were water is a strongly limiting factor (Snyder et al. 2003). The magnitude of water transpired during the night depends on the cuticular conductance, g, and VPD. It can account for E values between 5 and 15% of daytime rates, sometimes even more, and has been detected using a variety of techniques, from single leaf gas exchange measurements to whole plant sap flow, and field scale lysimeters (Caird et al. 2007). Green et al. (1989) reported that 20–30% of total daily transpiration could occur at night in kiwi and apple orchards in New Zealand under certain conditions. There may be a genetic attribute involved, because successions of *Arabidopsis thaliana* showed different night g in a common environment which correlated with the average VPD of their native habitat (Caird et al. 2007). While E during the night responds positively to VPD, night-time g may show a similar negative response as observed during the day (Bucci et al. 2004). There may be some benefits to the plant by not fully closing its stomata including better nutrient availability because of continuous mass flow or a better carbon balance because of high early morning g when VPD and temperature are still low (Caird et al. 2007). There is little information available on night-time water losses of grapevines, but it could potentially be important for regions with frequently high wind velocities (i.e southern France, California coast) (Chu et al. 2009) and high night-time temperatures and VPD. In fact, Schmid (1997) observed that high wind velocities correlated with high night-time sap flow rates of field-grown Riesling grapevines in Germany. This effect on whole plant E was much stronger later during the season (September, Northern Hemisphere) than during mid-summer (July–August). The seasonal difference in this study was attributed to a partial loss of stomatal control and seemed unrelated to soil water content (Schmid 1997). However, most studies have shown that water deficit and salt stress will cause night-time g to decrease (Caird et al. 2007).

Data presented in Figure 6, are among the few examples available where the night-time flux of water and its dependence on temperature (Figure 6(a)) and VPD (Figure 6(b)) has been measured for grapevines (Schmid 1997). The strong response to both environmental factors indicates that these phenomena should be studied specifically in warmer and dryer areas. The relationship to temperature was very similar to that of leaf respiration measured concomitantly in a gas exchange cuvette (Figure 6(a), Schmid 1997).

Apart from photosynthesis, plant respiration is the other basic component of plant productivity and it is remarkable how little information on respiratory responses of grapevines to environmental factors there is. Dark respiration rate (R) has been shown to change with temperature (Figure 6(a)), leaf age and the temperature response coefficient, Q_{10}, and depends on phenology and canopy position (sun and shade leaves) (Schultz 1991). Leaves are generally responsible for most of whole plant above-ground respiration in grapevines (Palliotti et al. 2004). Under water stress, the response seems to depend on the plant species and the plant organ investigated (Flexas et al. 2005) and on the degree of stress exposed to (Flexas et al. 2006). In a recent review on this subject, Atkin and Macherel (2009) reported that root and whole plant respiration were almost always reduced, whereas the response of mature leaves was different, with about two-thirds of the reviewed studies showing a reduction in R and most of the remainder showing no differences with a few reports on increasing R under severe water deficit. For grapevines, results are also not clear. Escalona et al. (1999) did not find significant effects of water stress on leaf respiration for the varieties Manto Negro and Tempranillo, whereas Gómez del Campo et al. (2004) observed a decrease in R (leaves) and some differences between varieties. They also found that the relationship of respiration rate to photosynthesis first decreased before it increased when water stress became severe. However, because the measurement temperature was not constant in that study, these observations are difficult to compare with others. Figure 7 shows an example from a field trial, where R was measured at a leaf temperature of 20°C before or after determining ψ_{PD} for the varieties Grenache and Syrah during a progressive dry-down. R decreased for both varieties but somewhat faster for Grenache than for Syrah (Figure 7(a)), whereas the ratio of R/A_{max} was similar for both declining at first and increasing at more severe water deficits (Figure 7(b)) corroborating the results of Gómez del Campo et al. (2004). These results are not totally in line with the general view that A is much more sensitive than R to water stress irrespective of the degree of water deficit (Atkin and Macherel 2009). It would be interesting to determine the response of R of different varieties to increasing temperatures in order to determine the acclimation potential for future climate developments because substantial differences have been observed with other plants (Atkin et al. 2008).

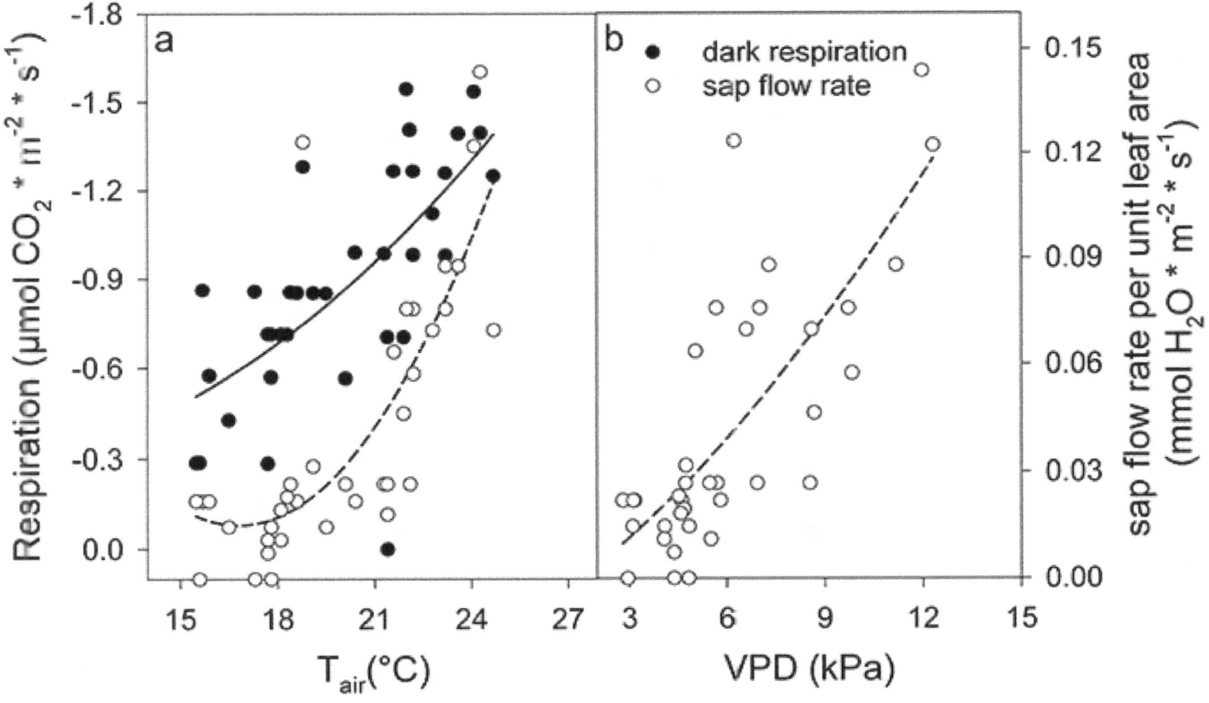

FIGURE 6: Night-time water flux through field-grown grapevines (cv. Riesling) determined with sap flow measurements (Granier method) as a function of temperature (a) and vapour pressure deficit (VPD) (b). Experiments were conducted on two consecutive nights (24 and 25 July 1994) in Geisenheim, Germany. Respiration rates (A) were concomitantly measured on an individual leaf enclosed in a gas exchange cuvette on one of the plants used for sap flow determination (adapted from Schmid 1997).

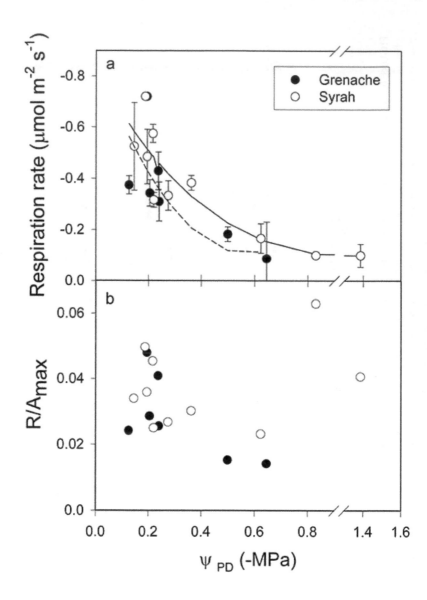

FIGURE 7: Respiration rates of mature leaves at 20°C (a) and the ratio of respiration rate (R) to the maximum photosynthetic rate, Amax (b) for the varieties Grenache and Syrah during a continuous water deficit in the field. R was measured immediately before or after measurements of ψ_{PD}. Amax was determined during the day following measurements of R. Data comprise a period of about 3 months (beginning of June to end of August). Data in (a) are means ± standard deviation (n = 3–5) (Schultz unpublished data).

8.4 MANAGEMENT ASPECTS

8.4.1 IRRIGATION SYSTEMS

Irrigation systems are crucial in translating genetic and physiological information on how WUE is responding to the environment into suitable application systems. There are various deficit irrigation techniques currently investigated, such as RDI (i.e. Matthews et al. 1987, Bravdo and Naor 1996, McCarthy et al. 2002, Girona et al. 2006), PRD (i.e. Dry et al. 1996, Loveys et al. 2004, Santos et al. 2005, Rodrigues et al. 2008) or sustained deficit irrigation (SDI) (Fereres and Soriano 2007) to achieve a more efficient use of water and better crop quality. A recent survey on the efficiency of these techniques stated that there was no significant advantage of one over the other (Sadras 2009), but the current challenge remains to find the best suited parameters for scheduling irrigation irrespective of the system used. In most cases, a certain percentage of potential (reference) crop evapo-transpiration is used, but this can result in severe water stress during short time periods of temperature extremes (Goodwin and Jerie 1992), so some form of knowledge about soil and/or plant water status is necessary.

Soil water monitoring has the inherent problem, that root distribution is mostly unknown and that the relationship between soil water content and physiological indicators of plant water status, such as pre-dawn (ψ_{PD})-, mid-day (ψ_{M})-, and stem water potential (ψ_{ST}) vary with soil type (Soar and Loveys 2007), the hydraulic resistances within the soil-plant system, the evaporative demand of the atmosphere (Kramer and Boyer 1995) and will additionally be mediated by factors such as both scion and rootstock, phenological stage or plant age. The use of ψ_{PD} for evaluating plant water status and consequently for making decisions about irrigation is promising because the daily maximum rates of A and g as well as vegetative growth are related to ψ_{PD} (Schultz 1996, Escalona et al. 1999, Rodrigues et al. 2008). The main disadvantage in terms of practicability is the time of measurement. Various proposals to substitute direct measurements of ψ_{PD} have been made for perennial plants, using vegetative growth components (Pellegrino et al. 2005), gas exchange parameters like g (Cifre et al. 2005), continuous measurements of sap flow (Ginestar et al. 1998, Patakas et al. 2005, Conejero et al. 2007) or trunk diameter fluctuations (Goldhamer and Fereres 2004, Ortuño et al. 2006) and, seemingly most related, midday measurements of ψ_M and ψ_{ST} (Choné et al. 2001, Williams and Araujo 2002, Girona et al. 2006).

Because ψ_{PD} is the parameter least affected by diurnal variations in weather conditions, this parameter seems best suited for cool and variable climates with intermittent summer rainfall (Gruber and Schultz 2009), whereas ψ_{ST} and ψ_M seem suitable for more stable climates (Naor 1998, Choné et al. 2001, Williams and Araujo 2002, Möller et al. 2007). Nevertheless, both respond directly to the prevailing environmental conditions (solar radiation, temperature, VPD) and to phenological stage (Matthews et al. 1987, Möller et al. 2007, Olivo et al. 2009), so it seems difficult to determine suitable thresholds for irrigation. Recent trials on the use of thermal and visible imagery of grapevines using the crop water stress index (CWSI) for irrigation management showed that while the relationship CWSI to g remained unchanged throughout the season, CWSI to ψ_{ST} varied (Möller et al. 2007).

There is a need to develop more dynamic scheduling systems and to fine tune further the balance between water use, mineral nutrient uptake and fruit quality (Keller 2005). There is also an obvious tendency to use red varieties in many irrigation trials related to wine quality, possibly because of more obvious responses in fruit composition (i.e. phenolic compounds in general), and consequently a lack of understanding with respect to white varieties. However, the future will force many regions to apply deficit irrigation strategies irrespective of the colour of their grapes.

8.4.2 CANOPY (LEAF) TEMPERATURE AND IMAGING TECHNOLOGIES

The usefulness of canopy temperature as a measure of 'crop water stress' was recognised in the 1960s (Tanner 1963, Gates 1964) and suggested to be useful in irrigation management. Several types of CWSI have been derived since then, relating canopy temperature (measured using infrared thermometry) to either a 'non water stressed' baseline, referring to the temperature of well-watered plants (i.e. Idso et al. 1981), or to wet and dry reference surfaces (T_{wet} and T_{dry}), for example wetted or fully transpiring leaves (Jones et al. 2002), canopy sections (Jones 1999), artificial leaf replica (Jones et al. 2002) or wet and dry reference plants (Grant et al. 2007). The CWSI is a surrogate for g (Idso 1982, Jones 1999) and can be used as an indicator of grapevine water availability (Jones et al. 2002, Grant et al. 2007, Stoll and Jones 2007).

The increase in leaf temperature with increasing water deficit is a response which may be somewhat problematic in measurements on individual leaves. For instance, in order to measure WUE_i and WUE_{inst} in the field, leaf measurements need to be conducted under saturating light to make them comparable between treatments, thus leaves will be exposed to the sun. Under these conditions the leaf energy balance will change, increasing leaf temperature and LAVPD (Fuchs 1990) and decreasing g which may drive WUE_{inst} and WUE_i in opposite directions (i.e. Schulze and Hall 1982). Such an 'indirect' contribution may have also influenced the data presented in Figure 2 and many other studies. The sensitivity of leaf temperature to changes in g, and hence the utility of thermal imaging, depends on the absorbed radiation, boundary layer conductance (leaf size or canopy density and wind speed) and air humidity. Figure 8, adapted from Jones et al. (2002), illustrates how the modelled difference between wet and dry leaves varies as a function of absorbed radiation and wind speed. It is evident that the sensitivity, i.e. the temperature difference, increases with radiation absorbed and with decreasing wind speed.

It is uncertain how important the variability in leaf temperature across a canopy is with respect to the assessment of CWSI in an application for irrigation management (Jones et al. 2002, Möller et al. 2007). The variability in g increases substantially below g values of about 100 mmol/m²/s and should be reflected in leaf temperature if no adjustment in leaf angle occurs (Möller et al. 2007). Although restricted CO_2 diffusion across leaves is the most dominant cause for decreased photosynthetic rates under water stress, metabolic impairment of about 15% of total A limitation becomes apparent at or below 100 mmol/m²/s, irrespective of the species tested (Flexas et al. 2006). In non-irrigated field-grown grapevines in a production situation, g largely resides below 100 mmol/m²/s during extended periods of the growing season in different canopy sections and throughout the day when leaves are measured in their natural position (for examples see Schultz et al. 2001, Schultz 2003b). Although this type of measurement is not a common practice, the results seem more relevant to whole canopies, and may be more compatible with the use of imaging technologies.

It is also uncertain how estimates of CSWI relate to how efficiently water is used in a whole canopy situation, because leaves are exposed to a wide array of sunlight intensity at any time during the day and WUE of entire plants may differ from the WUE of individual leaves. A recent model analyses on different sections of tree canopies has shown that WUE_i was highest and WUE_{inst} was lowest where VPD and light exposure and subsequent temperature were highest (Seibt et al. 2008). If integrated over the whole canopy, these effects can lead to a reduction in plant WUE during water deficit at an unchanged $\Delta^{13}C$ even at relative moderate water deficits (ψ_{PD} near −0.4 MPa) as recently shown for grapevines with measurements of whole plant gas-exchange (Poni et al. 2009). Thus, assessing the variability in leaf temperature across a canopy and the relative change in temperature caused by different water supply is an important area of current research to exploit this response for the use of deficit irrigation strategies (Grant et al. 2007, Möller et al. 2007).

8.4.3 REMOTE SENSING

Measuring the physiological status of a plant using non-invasive remote sensing techniques is challenging and a radiation-footprint is central to many biochemical and biophysical properties of a grapevine. Thus, radiation interaction is measured for a wide range of purposes, and remote sensing data are recorded both from satellite or aircraft sensors and from ground based measurements. The techniques available are used to track for vineyard variability (Bramley and Hamilton 2004), to monitor vineyard canopy density (Johnson et al. 1996, Hall et al. 2002, Zarco-Tejada et al. 2005), to trace nutrient deficiencies (Johnson 2001, Martin et al. 2007) or to act as a management support system prior to harvest (Johnson et al. 2001). Furthermore, simultaneous signals for different physiological indicators have been introduced to predict berry phenolics and colour (Lamb et al. 2004, Agati et al. 2007) to estimate the photosynthetic light use efficiency (Cerovic et al. 2002, Rascher and Pieruschka 2008) or plant water status (Seelig et al. 2008). These approaches already show great potential to monitor the physiological status non-invasively. Nevertheless, scaling up from leaf-level to a complex three dimensional canopy structure will remain challenging in future.

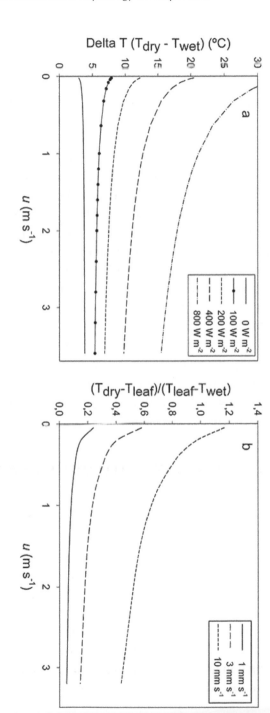

FIGURE 8: (a) Variation in temperature of a dry reference surface (Tdry) – temperature of a wet reference surface (Twet) as a function of wind speed (u, m s⁻¹) and net radiation absorbed (W m⁻²) for leaves with a 10-cm characteristic dimension and at an air temperature of 20°C and a relative humidity of 50%. (b) Corresponding response of one possible version of a crop water stress index (CWSI), Ig = ((Tdry – Tleaf)/(Tleaf – Twet)) (which is proportional to g) for leaf conductances of 1, 3, 10 mm s⁻¹ (note that 1 mm s⁻¹ = 25 mmol m–2 s⁻¹) (Jones et al. 2002, Journal of Experimental Botany, reproduced with permission of Oxford University Press). Tleaf, leaf temperature.

8.4.4 COVER CROPS

Aside from the vines themselves, cover crops also play a role in the carbon and water balance of a vineyard, yet their impact is largely neglected although they have an additional effect on the turnover rate of organic matter in the soil, another possible source of CO_2 during climate warming (Litton and Giardina 2008). It seems important to somewhat refocus research on cover crops because, depending on the region/climate type combination, these plants will have a substantial effect on sustainability in both vineyard mechanisation and their additional role in affecting the carbon and water balance of vineyards. Figure 9 shows some recent results on comparing C_3 and C_4 species in studies on the suitability of cover crops in terms of improving vineyard WUE (Uliarte and Schultz unpublished). This research area will gain importance because open soil cultivation will increase CO_2 release because of organic matter degradation and may not be a sustainable solution in light of increased precipitation intensity and increased risk for erosion (IPCC 2008). Additionally, competition for resources such as water between cover crop and vines will increase in the future and thus will need to be minimised by carefully selecting adapted species (Olmstead et al. 2001) and by considering both carbon and water resource efficiency (Lopes et al. 2004).

8.5 LIMITATIONS OF OUR EXPERIMENTAL SYSTEMS

8.5.1 WATER STATUS AND ITS MEASUREMENTS

There are limitations and contradictions in some experimental systems we use. It is sometimes difficult to relate results obtained on potted plants to data obtained in the field, but pot studies are a necessary compromise to investigate certain phenomena. However, there are some observations which should be addressed in the future because they may hold some relevant physiological answers. For example, several studies on the factors controlling g in grapevines have reported substantial effects at ψ_{PD}'s in the vicinity of -0.2 MPa (i.e. Lovisolo et al. 2002, Pellegrino 2003, Pou et al.

2008, and some results from our group), which is considered no stress for field-grown grapevines. In such cases, there is a need to investigate why these changes in physiology can occur at these deficit levels in pots while there is not even a response in growth or g under field conditions (Escalona et al. 1999, Schultz 2003a). If water potential is equilibrating with the wettest soil layer, and moisture distribution in pots is not uniform, a potential disequilibrium between soil and plant water potential can occur resulting in large variations of potential plant responses (Donovan et al. 2003). In these cases, enclosing two tubes consisting of porous tissue and containing sand into the pot medium and placing them at opposite sides ensures homogeneous water distribution (Pellegrino 2003) and good relations between plant and soil water potential (Chouzouri and Schultz 2005).

In addition to these problems, Jones (2007b), in a recent review, outlined the common lack of application of adequate measures of plant and soil water status in many physiological and agronomic studies on drought tolerance, breeding and the management of irrigation systems. He particularly criticised the omission of necessary water-status measurements in the majority of published molecular studies dealing with effects of drought on gene expression or effects of transgenes on the performance under water stress in general. Although his survey was not on studies related to grapevines, his suggestion to 'enhance inputs from environmental plant physiologists to improve the value of molecular studies' would certainly benefit both sides (Jones 2007b).

8.5.2 STOMATAL PATCHINESS

Some aspects of physiological research on grapevines have been ignored in the recent past, although they may intuitively be important. For instance, heterogeneous stomatal closure across leaves under water deficit, salt stress, high light and low humidity leads to erroneous estimates of C_i. This has been known to occur for a long time (Laisk 1983, Downton et al. 1988, Mott and Buckley 1998, Mott and Peak 2007) and there was a period when this phenomenon was actively researched and demonstrated to occur in grapevine leaves (i.e. Downton et al. 1988, Downton et al. 1990, Düring 1992, Düring and Loveys 1996) and where C_i values derived from gas-exchange measurements could not be published as a consequence. However, despite its demonstrated occurrence in hundreds of species (Beyschlag and Eckstein 1998) and the fact that patchiness appears to reduce WUE, the absence of any apparent physiological function has caused this to remain a little studied phenomenon (Mott and Peak 2007). In more recent studies on the importance of mesophyll conductance in the limitation of photosynthesis under water stress, patchiness was measured (Flexas et al. 2002) but was considered unimportant, and others have doubted that there is any significance to it (Lawlor and Tezera 2009). Yet, significant patchiness has been shown to occur in grapevine leaves below g values of about 120 mmol m^{-2} s^{-1} with a linear decrease in the number of open patches with decreasing g (Düring and Loveys 1996). Thus, it should not be ignored that there may be substantial variability between varieties in the occurrence of patchiness which could also be related to their drought or salt resistance. A recent hypothesis states that patchy patterns of stomatal opening and closing are similar to structures and behaviours in locally connected networks of dynamic units that perform complex tasks and may as such be a form of communication across leaves (Mott and Peak 2007). Combining image analysing technologies and programes such as thermal and fluorescence imagery may allow the study of these phenomenon on a finer scale in the future (Mott and Peak 2007).

8.5.3 STRUCTURAL AND PHYSIOLOGICAL MODELS

One possibility to integrate physiological responses on a single leaf level into canopy or even stand scale responses are coupled structural-functional models, which have been developed for annual species (maize, Fournier and Andrieu 1999), trees (peach, Allen et al. 2005) and recently grapevines (Grenache and Syrah, Louarn et al. 2008). These models can integrate structural components of a canopy, such as shape, orientation and location of plant organs which influence light interception and thus canopy energy balance with functional properties such as stomatal aperture, photosynthetic capacity and photomorphogenesis or other metabolic processes. Because vineyard canopy structure, functioning and management are important in the formation of yield and quality (i.e. Smart et al. 1982, Reynolds

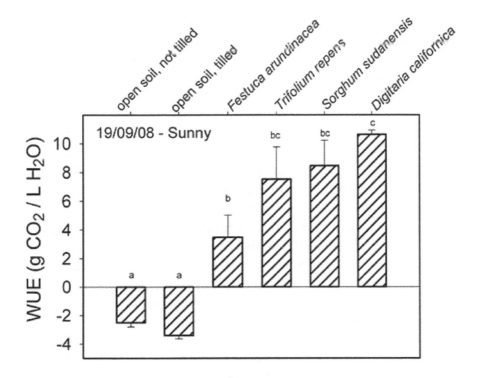

FIGURE 9: Response of average daily water use efficiency (WUE) on a per m 2 vineyard surface area basis for different soil treatments (open soil, tilled, not tilled) and C3 and C4 cover crop species on 19 September 2008 in the field in Geisenheim, Germany. Measurements were conducted throughout the day with six large custom-made chambers (0.5 m basal diameter, 0.5 m height) connected to two EGM four gas-exchange analysers (PP-systems, Hutchin, England). Plots had been seeded with Arizona cottoncrop (Digitaria californica, C4), Sudan grass (Sorghum sudanensis, C4), white clover (Trifolium repense, C3), and tall fescue (Festuca arundinacea, C3) at the beginning of the season (April). (Uliarte and Schultz, unpublished data).

and Wardle 1989, Carbonneau and Cargnello 2003, Gladstone and Dokoo-zlian 2003, Downey et al. 2004), these type of models, in the future, may serve to deduce management decisions with respect to yield and quality production, and go well beyond simple descriptions of canopy architec-ture (leaf area density distribution) and light harvesting (Schultz 1995). Louarn et al. (2008) have used a combination of approaches to construct virtual canopies of two varieties, Grenache and Syrah, with four common spur-pruned canopy systems (Gobelet, bilateral free cordon system, and two bilateral cordon system with vertical shoot orientation differing in the number of catch wires). They employed a limited number of parameters to describe the volume occupied by a shoot (turbid-medium-like envelope) and combined this with results from random samplings for the position of individual shoot organs (leaves as discrete geometric polygons) within this volume to generate individual shoots with individual leaf positions and orientations. Coupled to a set of descriptors of plant architecture, bud location and shoot orientation and angle, complex three dimensional canopies were regenerated (Figure 10(a)–(d)).

A particular advantage of this more statistical approach as compared to earlier three dimensional descriptions of grapevine canopies (Mabrouk et al. 1997, Mabrouk and Sinoquet 1998) was the improved integration of inter-plant variability and the implementation of varietal specific parameters (Louarn et al. 2008). This type of model allows for a more accurate simulation of light interception and can be coupled to the mechanistic gas-exchange model of Farquhar et al. (1980) (Louarn et al. 2005) which has been fully parameterised for grapevines (Schultz 2003b) (Figure 10(e)). If research development continues, in the future, this type of approach will allow the simulation of the response of entire vineyards to changes in environmental conditions such as water deficit or salt stress and may be able to give some answers to the impact of climate change and the mitigation possibilities in terms of canopy structure and management.

8.5.4 THE CO_2 PROBLEM

In a recent editorial for the New Phytologist titled 'An Inconvenient Truth' with reference to the Academy award for the best documentary film by former US Vice President Al Gore, Woodward (2007) described and analysed the dilemma between practical experiments with elevated CO_2 concentrations and the need to understand and predict the future responses of plants in the field. Aside from the fact that increasing CO_2 concentrations will impact on global temperature, CO_2 itself is generally beneficial to plant growth, although the response strongly varies between species (Long et al. 2004). Because stomata are sensitive to CO_2 but photosynthesis increases in response to it, increased biomass production at reduced water losses is expected (Ainsworth and Rogers 2007, Woodward 2007). Additionally, even a reduced plant sensitivity to the pollutant ozone may be a consequence (Morgan et al. 2006). However, Woodward (2007) continued that CO_2 enrichment experiments usually do not mimic the gradual increase in CO_2 plants are experiencing in the field but rather follow a step-up approach, and possible differences in plant responses to these approaches are unknown. Additionally, CO_2 enrichment is not usually accompanied by warming as would be predicted by climate models because of 'the problem of securing long-term funding which is a bothersome limitation to a more general approach' (Woodward 2007). Recent results from models including the physiological impact of CO_2 on plants (more biomass, reduced g) suggest that rising CO_2 will increase the temperature driven water evaporation from the oceans resulting in an increased absolute water vapour content of the air. However, the decrease in evapotranspiration over land (because of g↓) would still lead to an overall decrease in relative humidity and to an increased evaporative demand according to current knowledge (Boucher et al. 2009). Plant surfaces should then heat up more because of stomatal closure which is similar to findings from field-based warming/CO_2 experiments (Musil et al. 2005) and adds to the complexity of expected responses difficult to trace and simulate in experiments.

It is exactly this complexity which necessitates a more global approach to setting up experimental systems to study the response of grapevines to the combined increase in temperature and CO_2. Few studies have investigated the response of grapevines to CO_2 either in small free air carbon dioxide enrichment systems (Bindi et al. 1995, Bindi et al. 2001a) or in open top chambers (Gonçalves et al. 2009), but these could only describe the impact of increasing CO_2 concentration in the absence of rising Tair. Nevertheless, the generally predicted increase in biomass was confirmed, yet the effects on water consumption remained unclear (Bindi et al. 1995, Bindi et al. 2001a). These experiments also showed that fruit sugar concentration should increase and acidity levels decrease under elevated CO_2

(Bindi et al. 2001b), but the response of other components contributing to flavour and aroma of grapes were heterogeneous and indicated a significant 'chamber effect', with plants grown outside responding differently than plants in open top chambers with or without elevated CO_2 (Gonçalves et al. 2009).

Data from single leaf experiments on grapevines (Figure 11) showed that the degree of stomatal closure induced by CO_2 matches the average closure response of about −20% in g (Figure 11(c)) analysed for different functional groups based on a literature survey and assuming C_a would rise from 380 to 560 µmol mol⁻¹ (Figure 11(c),(e)) (Ainsworth and Rogers 2007). The concomitant rise in WUE_{inst} could even be larger than this depending on LAVPD (Figure 11(c),(d)) and considering that respiration rate may also be suppressed by elevated CO_2 (Smart 2004). However, studies on individual leaves (or woody pieces in the case of Smart 2004) may not be representative of whole plant field experiments and the need to study the effects of elevated CO_2 and temperature in combination may be illustrated by the following example. Alonso et al. (2008) separated the effects of a doubling in CO_2 concentration and a 4°C increase in temperature on the photosynthetic apparatus of wheat in a field study with open top chambers over two successive years. Elevated CO_2 alone decreased the maximum rate of carboxylation of Rubisco (Vc_{max}) and so did an increase in temperature without changes in CO_2. However, an increase in temperature under elevated CO_2 increased Vc_{max} particularly at high measurement temperatures demonstrating the complex interactions between just these two environmental factors (Alonso et al. 2009). In an early experiment investigating the effects of elevated CO_2 (900 ppm) in combination with a heat therapy (35–40°C) of several weeks up to several months on grapevine functioning, Kriedemann et al. (1976) observed substantial changes in leaf anatomy with a much larger spongy mesophyll and an enrichment in the number of chloroplasts. Vine growth rate more than doubled and dry matter distribution was altered in favour of greater root growth. Photosynthetic capacity was enhanced, photorespiration depressed and high CO_2 mitigated the high temperature effect, similar to the more recent study on wheat mentioned above (Alonso et al. 2008).

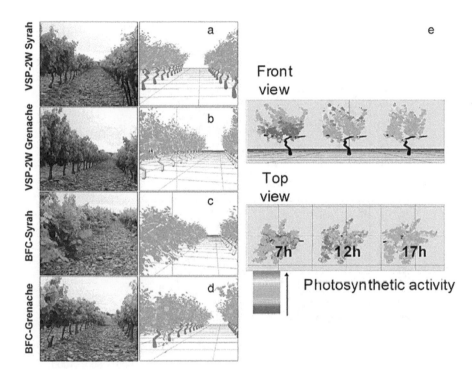

FIGURE 10: Comparison of photographs taken in a real vineyard (veraison) with the corresponding simulations for two-wire (VSP-2W) ((a), (b)) and one-wire (BFC) ((c), (d)) training systems for the varieties Syrah ((a), (c)) and Grenache ((b), (d)). (e) shows an example of for the coupling of a structural to a functional model on light interception and gas-exchange (e) (after Louarn et al. 2005, 2008, Annals of Botany, reproduced with permission of Oxford University Press).

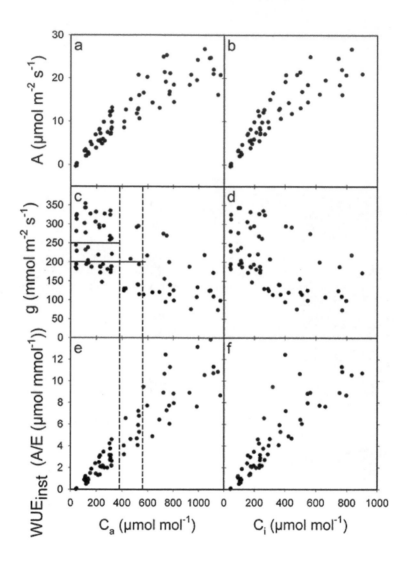

FIGURE 11: Response of A, ((a), (b)), g ((c), (d)) and instantaneous water use efficiency (WUEinst) ((e , (f)) to changes in C_a ((a), (c), (e)) and C_i ((b), (d), (f)). Data are from eight individual leaf CO_2 response curves from Zinfandel plants (4 years old) grown in the greenhouse (April–August) in 25 L pots at the University of California Davis, USA. Measurement temperature was 27.1 ± 0.5°C and leaf to air vapour pressure deficit (LAVPD) was maintained at 0.6–1.1 kPa. For experimental and measurement details see Schultz (2003b). Vertical dashed lines indicate current (380 ppm) and elevated (future) (560 ppm) C_a. The latter value is in reference to the survey on plant responses to CO_2 conducted by Ainsworth and Rogers (2007). Continuous horizontal lines (c) indicate g at current and elevated C_a.

REFERENCES

1.   Agati, G., Meyer, S., Matteini, P. and Cerovic, Z.G. (2007) Assessment of anthocyanins in grape (Vitis vinifera L.) berries using a noninvasive chlorophyll fluorescence method. Journal of Agriculture and Food Chemistry 55, 1053–1061.
2.   Ainsworth, E.A. and Rogers, A. (2007) The response of photosynthesis and stomatal conductance to rising [CO2]: mechanisms and environmental interactions. Plant, Cell and Environment 30, 258–270.
3.   Allen, M.T., Prusinkiewicz, P. and De Jong, T.M. (2005) Using L-Systems for modelling source sink interactions, architecture and physiology of growing trees. The L-Peach model. New Phytologist 166, 869–880.
4.   Alonso, A., Pérez, P. and Martinez-Carrasco, R. (2009) Growth in elevated CO2 enhances temperature response of photosynthesis in wheat. Physiologia Plantarum 135, 109–120.
5.   Alonso, A., Pérez, P., Morcuende, R. and Martinez-Carrasco, R. (2008) Future CO2 concentrations though not warmer temperatures enhance wheat photosynthesis temperature responses. Physiologia Plantarum 132, 102–112.
6.   Alsina, M.M., Herralde, F.D., Aranda, X., Savé, R. and Biel, C. (2007) Water relations and vulnerability to embolism are not related: experiments with eight grapevine cultivars. Vitis 46, 1–6.
7.   Aphalo, P.J. and Jarvis, P.G. (1993) An analysis of Ball's empirical model of stomatal conductance. Annals of Botany 72, 321–327.
8.   Atkin, O.K. and Macherel, D. (2009) The crucial role of plant mitochondria in orchestrating drought tolerance. Annals of Botany 103, 581–597.
9.   Atkin, O.K., Atkinson, J.L., Fisher, R.A., Campbells, C.E., Zaragoza-Castells, J., Pitchford, H.A., Woodward, F.I. and Hurry, V. (2008) Using temperature-dependent changes in leaf scaling relationships to quantitatively account for thermal acclimation of respiration in a coupled global climate-vegetation model. Global Change Biology 14, 2709–2726.
10. Bacon, M.A. (2004) Water use efficiency in plant biology. In: Water use efficiency in plant biology. Eds. M.A.Bacon (Blackwell Publishing: Oxford) pp. 1–22.
11. Badeck, F., Tcherkez, G., Nogués, S., Piel, C. and Ghashghaie, J. (2005) Post-photosynthetic fractionation of stable carbon isotopes between plant organs – a widespread phenomenon. Rapid Communication in Mass Spectrometry: RCM 19, 1381–1291.
12. Baldocchi, D.D. and Bowling, D.R. (2003) Modelling the discrimination of 13CO2 above and within a temperate broad-leaved forest canopy on hourly to seasonal time scales. Plant, Cell and Environment 26, 231–244.
13. Ball, J.T., Woodrow, I.E. and Berry, J.A. (1987) A model predicting stomatal conductance and its contribution to the control of photosynthesis under different environmental conditions. In: Progress in photosynthesis research. Ed. J.Biggins (Martinus Nijhoff: Dordrecht) pp. 221–224.
14. Bernacchi, C.J., Portis, A.R., Nakano, H., Von Caemmerer, S. and Long, S.P. (2002) Temperature response of mesophyll conductance. Implications for the determination of rubisco enzyme kinetics and for limitations to photosynthesis in vivo. Plant Physiology 130, 1992–1998.

15. Beyschlag, W. and Eckstein, J. (1998) Stomatal patchiness. In: Progress in botany. Eds. K.Behnke, K.Esser, J.W.Kadereit, U.Lüttge and M.Runge (Springer-Verlag: Berlin) pp. 293–298.
16. Bierhuizen, J.F. and Slatyer, R.O. (1965) Effect of atmospheric concentration of water vapour and CO_2 in determining transpiration-photosynthesis relationships of cotton leaves. Journal of Agricultural Meteorology 2, 259–270.
17. Bindi, M., Fibbi, L., Gozzini, B., Orlandini, S. and Miglietta, F. (1995) Experiments on the effects of increased temperature and/or elevated concentrations of carbon dioxide on crops. Mini Free Air Carbon dioxide Enrichment (FACE) experiments on grapevine. In: Climate change and agriculture in Europe: assessment of impacts and adaptations. Eds. P.A.Harrison, R.E.Butterfield and T.E.Downing (Research Report No. 9, Environmental Change Institute, University of Oxford: Oxford) pp. 125–137.
18. Bindi, M., Fibbi, L., Lanini, M. and Miglietta, F. (2001a) Free air CO_2 enrichment (FACE) of grapevine (Vitis vinifera L.): I. Development and testing of the system for CO_2 enrichment. European Journal of Agronomy 14, 135–143.
19. Bindi, M., Fibbi, L. and Miglietta, F. (2001b) Free air CO_2 enrichment (FACE) of grapevine (Vitis vinifera L.): II. Growth and quality of grape and wine in response to elevated CO_2 concentrations. European Journal of Agronomy 14, 145–155.
20. Bota, J., Flexas, J. and Medrano, H. (2001) Genetic variability of photosynthesis and water use in Balearic grapevine cultivars. Annals of Applied Biology 138, 353–361.
21. Boucher, O., Jones, A. and Betts, R.A. (2009) Climate response to the physiological impact of carbon dioxide on plants in the Met Office Unified Model HadCM3. Climate Dynamics 32, 237–249.
22. Bramley, R.G.V. and Hamilton, R.P. (2004) Understanding variability in wine grape production systems. I. Within vineyard variation in yield over several vintages. Australian Journal of Grape and Wine Research 10, 32–45.
23. Bravdo, B. and Naor, A. (1996) Effect of water regime on productivity and quality of fruit and wine. Acta Horticulturae 427, 15–26.
24. Bucci, S.J., Scholz, F.G., Goldstein, G., Meinzer, F.C., Hinojoso, J.A., Hoffman, W.A. and Franco, A.C. (2004) Processes preventing nocturnal equilibration between leaf and soil water potential in tropical savannah woody species. Tree Physiology 24, 1119–1127.
25. Caird, M.A., Richards, J.H. and Donovan, L.A. (2007) Night time stomatal conductance and transpiration in C3 and C4 plants. Plant Physiology 143, 4–10.
26. Carbonneau, A. and Cargnello, G. (2003) Architecture de la vigne et systèmes de conduites (Dunod Ed: Paris) p. 123.
27. Cerovic, Z.G., Ounis, A., Cartelat, A., Latouche, G., Goulas, Y., Meyer, S. and Moya, I. (2002) The use of chlorophyll fluorescence excitation spectra for the non-destructive in situ assessment of UV-absorbing compounds in leaves. Plant, Cell and Environment 25, 1663–1676.
28. Chaves, M., Santos, P.T., Souza, C.R., Ortuño, M.F., Rodrigues, M.L., Lopes, C.M., Maroco, J.P. and Pereira, J.S. (2007) Deficit irrigation in grapevine improves water-use efficiency while controlling vigour and production quality. Annals of Applied Biology 150, 237–252.

29. Chaves, M.M. and Oliveira, M.M. (2004) Mechanisms underlying plant resilience to water deficits: prospects for water-saving agriculture. Journal of Experimental Botany 55, 2365–2384

30. Chaves, M.M., Harley, P.C., Tenhunen, J.D. and Lange, O.L. (1987) Gas exchange studies in two Portuguese grapevine cultivars. Physiologia Plantarum 70, 639–647.

31. Choné, X., Van Leeuwen, C., Dubourdieu, D. and Gaudillères, J.-P. (2001) Stem water potential is a sensitive indicator of grapevine water status. Annals of Botany 87, 477–483.

32. Chouzouri, A. and Schultz, H.R. (2005) Hydraulic anatomy, cavitation susceptibility and gas-exchange of several grapevine cultivars of different geographical origin. Acta Horticulturae 689, 325–331.

33. Chu, C.R., Hsieh, C.-I., Wu, S.-A. and Phillips, N.G. (2009) Transient response of sap flow to wind speed. Journal of Experimental Botany 60, 249–255.

34. Cifre, J., Bota, J., Escalona, J.M., Medrano, H. and Flexas, J. (2005) Physiological tools for irrigation scheduling in grapevine (Vitis vinifera L.): an open gate to improve water-use efficiency? Agriculture, Ecosystems and Environment 106, 159–170.

35. Cochard, H., Venisse, J.-S., Barigah, T.S., Brunel, N., Herbette, S., Guillot, A., Tyree, M.T. and Sakr, S. (2007) Putative role of aquaporins in variable hydraulic conductance of leaves in response to light. Plant Physiology 143, 122–143.

36. Condon, A.G., Richards, R.A., Rebetzke, G.J. and Farquhar, G.D. (2004) Breeding for high water-use efficiency. Journal of Experimental Botany 55, 2447–2460.

37. Conejero, W., Alarcón, J.J., García-Orellana, Y., Nicolás, E. and Torrecillas, A. (2007) Evaluation of sap flow and trunk diameter sensors for irrigation scheduling in early maturing peach trees. Tree Physiology 27, 1753–1759.

38. Cowan, I.R. (1977) Stomatal behaviour and environment. Advances in Botanical Research 4, 117–228.

39. Domec, J.-C., Scholz, F.G., Bucci, S.J., Meinzer, F.C., Goldstein, G. and Villalobos-Vega, R. (2006) Diurnal and seasonal variation in root xylem embolism in neotropical savannah woody species: impact on stomatal control of plant water status. Plant, Cell and Environment 29, 26–35.

40. Donovan, L.A., Richards, J.H. and Linton, M.J. (2003) Magnitude and mechanisms of disequilibrium between predawn plant and soil water potentials. Ecology 84, 463–470.

41. Downey, M.O., Harvey, J.S. and Robinson, S.P. (2004) The effect of bunch shading on berry development and flavonoid accumulations in Shiraz grapes. Australian Journal of Grape and Wine Research 10, 55–73.

42. Downton, W.J.S., Loveys, B.R. and Grant, W.J.R. (1988) Non-uniform stomatal closure induced by water stress causes putative non-stomatal inhibition of photosynthesis. New Phytologist 110, 503–509.

43. Downton, W.J.S., Loveys, B.R. and Grant, W.J.R. (1990) Salinity effects on the stomatal behaviour of grapevine. New Phytologist 116, 499–503.

44. Dry, P.R., Loveys, B.R., Botting, D. and Düring, H. (1996) Effects of partial root zone drying on grapevine vigour, yield composition of fruit and use of water. Proceedings of 9th Australian Wine Industry Technical Conference, Adelaide, Australia (Winetitles: Adelaide) pp. 126–131.

45. Düring, H. (1987) Stomatal responses to alterations of soil and air humidity in grapevines. Vitis 26, 9–18.

46. Düring, H. (1992) Low air humidity causes non-uniform stomatal closure in heterobaric leaves of Vitis species. Vitis 31, 1–7.

47. Düring, H. and Loveys, B.R. (1982) Diurnal changes in water relations and abscisic acid in field-grown Vitis vinifera cvs. I. Leaf water potential components and leaf conductance under humid temperate and semiarid conditions. Vitis 21, 223–232.

48. Düring, H. and Loveys, B.R. (1996) Stomatal patchiness of field-grown Sultana leaves: diurnal changes and light effects. Vitis 35, 7–10.

49. Düring, H. and Scienza, A. (1980) Drought resistance of some Vitis species and cultivars. Proceedings of 3rd International Symposium on Grapevine Breeding (University of California: Davis, USA) pp. 179–190.

50. Eibach, R. and Alleweldt, G. (1984) Einfluss der Wasserversorgung auf Wachstum, Gaswechsel und Substanzproduktion traubentragender Reben. II. Der Gaswechsel. Vitis 23, 11–20.

51. Escalona, J.M., Flexas, J. and Medrano, H. (1999) Stomatal and non-stomatal limitations of photosynthesis under water stress in grapevines. Australian Journal of Plant Physiology 26, 421–433.

52. Farquhar, G.D. and Richards, R.A. (1984) Isotopic composition of plant carbon correlates with water use efficiency of wheat genotypes. Australian Journal of Plant Physiology 10, 205–226.

53. Farquhar, G.D., Caemmerer von, S. and Berry, J.A. (1980) A biochemical model of photosynthetic CO_2-assimilation in leaves of C3-species. Planta 149, 78–90.

54. Farquhar, G.D., Cernusak, L.A. and Barnes, B. (2007) Heavy water fractionation during transpiration. Plant Physiology 143, 11–18.

55. Farquhar, G.D., Ehleringer, J.R. and Hubick, K.T. (1989) Carbon isotope discrimination and photosynthesis. Annual Review of Plant Physiology and Plant Molecular Biology 40, 503–537.

56. Farquhar, G.D., O'Leary, M.H. and Berry, J.A. (1982) On the relationship between carbon isotope discrimination and intercellular carbon dioxide concentration in leaves. Australian Journal of Plant Physiology 9, 121–137.

57. Fereres, E. and Soriano, M.A. (2007) Deficit irrigation for reducing agricultural water use. Journal of Experimental Botany 58, 147–159.

58. Flexas, J., Bota, J., Cifre, J., Escalona, J.M., Galmés, J., Lefi, E., Martínez-Cañellas, S.F., Moreno, M.T., Ribas-Carbó, M., Riera, D., Sampol, B. and Medrano, H. (2004) Understanding down-regulation of photosynthesis under water stress: future prospects and searching for physiological tools for irrigation management. Annals of Applied Biology 144, 273–283.

59. Flexas, J., Bota, J., Escalona, J.M., Sampol, B. and Medrano, H. (2002) Effects of drought on photosynthesis in grapevines under field conditions: an evaluation of stomatal and mesophyll limitations. Functional Plant Biology 29, 461–471.

60. Flexas, J., Bota, J., Galmés, J., Medrano, H. and Ribas-Carbó, M. (2006) Keeping a positive carbon balance under adverse conditions: responses of photosynthesis and respiration to water stress. Physiologia Plantarum 127, 343–352.

61. Flexas, J., Diaz-Espejo, A., Galmés, J., Kaldenhoff, R., Medrano, H. and Ribas-Carbó, M. (2007) Rapid variations of mesophyll conductance in response to changes in CO2 concentrations around leaves. Plant, Cell and Environment 30, 1284–1298.

62. Flexas, J., Escalona, J.M. and Medrano, H. (1998) Down-regulation of photosynthesis by drought under field conditions in grapevine leaves. Australian Journal of Plant Physiology 25, 893–900.

63. Flexas, J., Galmes, J., Ribas-Carbó, M. and Medrano, H. (2005) The effects of water stress on plant respiration. In: Plant respiration. Eds. H.Lambers and M.Ribas-Carbó (Springer-Verlag: Berlin) pp. 85–94.

64. Flexas, J., Ribas-Carbó, M., Diaz-Espejo, A., Galmés, J. and Medrano, H. (2008) Mesophyll conductance to CO2: current knowledge and future prospects. Plant, Cell and Environment 31, 602–621.

65. Fouquet, R., Léon, C., Ollat, N. and Barrieu, F. (2008) Identification of grapevine aquaporins and expression analysis in developing berries. Plant Cell Reports 27, 1541–1550.

66. Fournier, C. and Andrieu, B. (1999) ADEL-maize: An L-system based model for the integration of growth processes from the organ to the canopy. Application to the regulation of growth by light availability. Agronomie 19, 313–325.

67. Franks, P.J., Drake, P.L. and Froend, R.H. (2007) Anisohydric but isohydrodynamic: seasonally constant plant water potential gradient explained by a stomatal control mechanism incorporating variable plant hydraulic conductance. Plant, Cell and Environment 30, 19–30.

68. Fuchs, M. (1990) Infrared measurement of canopy temperature and detection of plant water stress. Theoretical and Applied Climatology 42, 253–261.

69. Galmés, J., Pou, A., Alsina, M.M., Tomas, M., Medrano, H. and Flexas, J. (2007) Aquaporin expression in response to different water stress intensities and recovery in Richter-110 (Vitis sp.): relationship with ecophysiological status. Planta 226, 671–681.

70. Gates, D.M. (1964) Leaf temperature and transpiration. Agronomy Journal 56, 273–277.

71. Gaudillère, J.P., Van Leeuwen, C. and Ollat, N. (2002) Carbon isotope composition of sugars in grapevines, an integrated indicator of vineyard water status. Journal of Experimental Botany 369, 757–763.

72. Geissler, N., Hussin, S. and Koyro, H.-W. (2009) Elevated atmospheric CO2 concentration ameliorates effects of NaCl salinity on photosynthesis and leaf structure of Aster tripolium L. Journal of Experimental Botany 60, 137–151.

73. Gibberd, M.R., Walker, R.R., Blackmore, D.H. and Condon, A.G. (2001) Transpiration efficiency and carbon-isotope discrimination of grapevines grown under well-watered conditions in either glasshouse or vineyard. Australian Journal of Grape and Wine Research 7, 110–117.

74. Ginestar, C., Eastham, J., Gray, S. and Iland, P. (1998) Use of sap-flow sensors to schedule vineyard irrigation. I. Effects of post-veraison water deficits on water relations, vine growth, and yield of Shiraz grapevines. American Journal of Enology and Viticulture 49, 413–420.

75. Girona, J., Mata, M., Del Campo, J., Arbonés, A., Bartra, E. and Marsal, J. (2006) The use of midday leaf water potential for scheduling deficit irrigation in vineyards. Irrigation Science 24, 115–127.

76. Gladstone, E.A. and Dokoozlian, N.K. (2003) Influence of leaf area density and trellis/training system on the light microclimate within grapevine canopies. Vitis 42, 124–131.

77. Goldhamer, D.A. and Fereres, E. (2004) Irrigation scheduling of almond trees with trunk diameter sensors. Irrigation Science 23, 11–19.

78. Gonçalves, B., Falco, F., Moutinho-Pereira, H., Bacelar, E., Peixoto, F. and Correia, C. (2009) Effects of elevated CO_2 on grapevine (Vitis vinifera L.): volatile composition, phenolic content, and in vitro antioxidant activity of red wine. Journal of Agricultural and Food Chemistry 57, 265–273.

79. Goodwin, J. and Jerie, P. (1992) Regulated deficit irrigation: from concept to practice. Advances in vineyard irrigation. Australian and New Zealand Wine Industry Journal 7, 258–261.

80. Grant, O.M., Tronina, L., Jones, H.G. and Chaves, M.M. (2007) Exploring thermal imaging variables for the detection of stress responses in grapevine under different irrigation systems. Journal of Experimental Botany 58, 815–825.

81. Green, S.R., McNaughton, K.G. and Clothier, B.E. (1989) Observations of night-time water use in kiwi-fruit vines and apple trees. Agriculture and Forest Meteorology 48, 251–261.

82. Gruber, B.R. and Schultz, H.R. (2009) Comparison of physiological parameters for scheduling low-frequency irrigation on steep slopes in a cool climate grape (Vitis vinifera L.) growing region. Irrigation Science (in press).

83. Gómez del Campo, M., Baeza, P., Ruiz, C. and Lissarrague, J.R. (2004) Water-stress induced physiological changes in leaves of four container-grown grapevine cultivars (Vitis vinifera L.). Vitis 43, 99–105.

84. Hall, A., Lamb, D.W., Holzapfel, B. and Louis, J. (2002) Optical remote sensing applications in viticulture – a review. Australian Journal of Grape and Wine Research 8, 36–47.

85. Harley, P.C. and Tenhunen, J.D. (1991) Modeling the photosynthetic response of C3 leaves to environmental factors. In: Modeling crop photosynthesis: from biochemistry to canopy. Eds. K.J.Boote and R.S.Loomis (Special Publication of the Crop Science Society of America, Madison, WI) pp. 17–39.

86. Helliker, B.R. and Ehleringer, J.R. (2002a) Differential 18O enrichment of leaf cellulose in C3 versus C4 grasses. Functional Plant Biology 29, 435–442.

87. Helliker, B.R. and Ehleringer, J.R. (2002b) Grass blades as tree rings: environmentally induced changes in the oxygen isotope ratio of cellulose along the length of grass blades. New Phytologist 155, 417–424.

88. Hermann, A. and Voerkelius, S. (2008) Meteorological impact on oxygen isotope ratios of German wines. American Journal of Enology and Viticulture 59, 194–199.

89. IPCC (2008) Climate change and water. In: IPCC Tech. Paper VI (Eds. B.C.Bates, Z.W.Kundzewicz, S.Wu, J.P.Palutikof) (IPCC Secretariat, Geneva).

90. Idso, S.B. (1982) Non-water-stressed baselines: a key to measuring and interpreting plant water stress. Agriculture and Forest Meteorology 27, 59–70.

91. Idso, S.B., Jackson, R.D., Pinter, P.J., Reginato, R.H. and Hatfield, J.L. (1981) Normalising the stress degree-day parameter for environmental variability. Agriculture and Forest Meteorology 24, 45–55.
92. Jifon, J.L. and Syvertsen, J.P. (2003) Moderate shade can increase net gas exchange and reduce photo-inhibition in citrus leaves. Tree Physiology 23, 119–127.
93. Johanson, U., Karlsson, M., Johansson, I., Gustavsson, S., Sjovall, S., Fraysse, L., Weig, A.R. and Kjellbom, P. (2001) The complete set of genes encoding major intrinsic proteins in Arabidopsis provides a framework for a new nomenclature for major intrinsic proteins in plants. Plant Physiology 126, 1358–1369.
94. Johnson, L.F., Lobitz, B., Armstrong, R., Baldy, R., Weber, E., De Benedictis, J. and Bosch, D.F. (1996) Airborne imaging for vineyard canopy evaluation. California Agriculture 50, 14–18.
95. Johnson, L.F. (2001) Nitrogen influence on fresh-leaf NIR spectra. Remote Sensing of Environment 78, 314–320.
96. Johnson, L.F., Bosch, D.F., Williams, D.C. and Lobitz, B.M. (2001) Remote sensing of vineyard management zones: Implication for wine quality. Applied Engineering in Agriculture 17, 557–560.
97. Jones, G.V. (2007a) Climate change: observations, projections and general implications for viticulture and wine production. In: Work Paper No 7, Economics department Whitman College. Eds. E.Essick, P.Griffin, B.Keefer, S.Miller and K.Storchmann (Whitman College: Walla Walla, USA) pp. 1–15.
98. Jones, H.G. (2007b) Monitoring plant and soil water status: established and novel methods revisited and their relevance to studies of drought tolerance. Journal of Experimental Botany 58, 119–130.
99. Jones, G.V., Duchène, E., Tomasi, D., Yuste, J., Bratislavska, O., Schultz, H.R., Martinez, C., Boso, S., Langellier, F., Perruchot, C. and Guimberteau, G. (2005b) Changes in European winegrape phenology and relationships with climate. Proceedings XIV GESCO Symposium, Geisenheim, Germany, Vol. I (Forschungsanstalt: Geisenheim) pp. 55–61.
100. Jones, G.V., White, M.A., Cooper, O.R. and Storchmann, K. (2005a) Climate change and global wine quality. Climate Change 73, 319–343.
101. Jones, H.G. (1992) Plants and microclimate, 2nd edn (Cambridge University Press: Cambridge) p. 428.
102. Jones, H.G. (1999) Use of infrared thermometry for estimation of stomatal conductance as a possible aid to irrigation scheduling. Agriculture and Forest Meteorology 95, 139–149.
103. Jones, H.G., Stoll, M., Santos, T., Sousa, C. and Chaves, M.M. (2002) Use of infrared thermometry for monitoring stomatal closure in the field: application to grapevine. Journal of Experimental Botany 53, 2249–2260.
104. Katul, G.G., Ellsworth, D.S. and Lai, C.T. (2000) Modelling assimilation and intercellular CO_2 from measured conductance: a synthesis of approaches. Plant, Cell and Environment 23, 1313–1328.
105. Keller, M. (2005) Deficit irrigation and vine mineral nutrition. American Journal of Enology and Viticulture 56, 267–283.
106. Koundouras, S., Tsialtas, I.T., Zioziou, E. and Nikolaou, N. (2008) Rootstock effects on the adaptive strategies of grapevine (Vitis vinifera L. cv. Cabernet Sauvi-

gnon) under contrasting water status: leaf physiological and structural responses. Agriculture, Ecosystems and Environment 128, 86–96.

107. Kramer, P.J. and Boyer, J.S. (1995) Water relations of plants and soils (Academic Press: San Diego).

108. Kriedemann, P.E., Sward, R.J. and Downton, W.J.S. (1976) Vine response to carbon dioxide enrichment during heat therapy. Australian Journal of Plant Physiology 3, 605–618.

109. Laisk, A. (1983) Calculation of leaf photosynthetic parameters considering the statistical distribution of stomatal apertures. Journal of Experimental Botany 34, 1627–1635.

110. Lake, J.A. and Woodward, F.I. (2008) Response of stomatal numbers to CO_2 and humidity: control by transpiration rate and abscisic acid. New Phytologist 179, 591–628.

111. Lamb, D.W., Weedon, M.M. and Bramley, R.G.V. (2004) Using remote sensing to predict grape phenolics and colour at harvest in a Cabernet Sauvignon vineyard: Timing observations against vine phenology and optimising image resolution. Australian Journal of Grape and Wine Research 10, 45–54.

112. Lawlor, D.W. and Tezera, W. (2009) Review: causes of decreased photosynthetic rate and metabolic capacity in water-deficient leaf cells: a critical evaluation of mechanisms and integration of processes. Annals of Botany 103, 561–579.

113. Litton, C.M. and Giardina, C.P. (2008) Below-ground carbon flux and partitioning: global patterns and response to temperature. Functional Ecology 22, 947–945.

114. Lloyd, J., Grace, J., Miranda, A.C., Meir, P., Wong, S.C., Miranda, H.S., Wright, I.R., Gash, J.H.C. and McIntyre, J. (1995) A simple calibrated model of Amazon rainforest productivity based on leaf biochemical properties. Plant, Cell and Environment 18, 1127–1145.

115. Long, S.P., Ainsworth, E.A., Rogers, A. and Ort, D.R. (2004) Rising atmospheric carbon dioxide: plants FACE the future. Annual Review of Plant Biology 55, 591–628.

116. Lopes, C., Monteiro, A., Rückert, F.E., Gruber, B.R., Steinberg, B. and Schultz, H.R. (2004) Transpiration of grapevines and co-habitating cover crop and weed species in a vineyard. A 'snapshot' at diurnal trends. Vitis 43, 111–117.

117. Louarn, G., Lebon, E. and Lecoeur, J. (2005) «Top-vine», a topiary approach based architectural model to simulate vine canopy structure. Proceedings XIV GESCO Symposium, Geisenheim, Germany, Vol. II (Forschungsanstalt: Geisenheim) pp. 464–470.

118. Louarn, G., Lecoeur, J. and Lebon, E. (2008) A three-dimensional statistical reconstruction model of grapevine (Vitis vinifera L.) simulating canopy structure variability within and between cultivar/training system pairs. Annals of Botany 101, 1167–1184.

119. Loveys, B.R. (1984a) Abscisic acid transport and metabolism in grapevine (Vitis vinifera L.). New Phytologist 98, 575–582.

120. Loveys, B.R. (1984b) Diurnal changes in water relations and abscisic acid in field-grown Vitis vinifera cultivars: III. The influence of xylem-derived abscisic acid on leaf gas exchange. New Phytologist 98, 563–573.

121. Loveys, B.R., Stoll, M. and Davies, W.J. (2004) Physiological approaches to enhance water use efficiency in agriculture: exploiting plant signalling in novel irrigation practice. In: Water use efficiency in plant biology. Eds. M.A.Bacon (Blackwell Publishing: Oxford) pp. 113–141.

122. Lovisolo, C. and Schubert, A. (2006) Mercury hinders recovery of shoot hydraulic conductivity during grapevine rehydration: evidence from a whole-plant approach. New Phytologist 172, 469–478.

123. Lovisolo, C., Hartung, W. and Schubert, A. (2002) Whole-plant hydraulic conductance and root-to-shoot flow of abscisic acid are independently affected by water stress in grapevines. Functional Plant Biology 29, 1349–1356.

124. Lovisolo, C., Perrone, I. Hartung, W. and Schubert, A. (2008a) An abscisic acid-related reduced transpiration promotes gradual embolism repair when grapevines are rehydrated after drought. New Phytologist 180, 1–10.

125. Lovisolo, C., Tramantini, S., Flexas, J. and Schubert, A. (2008b) Mercurial inhibition of root hydraulic conductance in Vitis spp. rootstocks under water stress. Environmental and Experimental Botany 63, 178–182.

126. Luu, D.T. and Maurel, C. (2005) Aquaporins in a challenging environment: molecular gears for adjusting plant water status. Plant, Cell and Environment 28, 85–96.

127. Mabrouk, H. and Sinoquet, H. (1998) Indices of light microclimate and canopy structure of grapevines determined by 3D digitising and image analyses, and their relationship to grape quality. Australian Journal of Grape and Wine Research 4, 2–13.

128. Mabrouk, H., Carbonneau, A. and Sinoquet, H. (1997) Canopy structure and radiation regime in grapevine. I. Spatial and angular distribution of leaf area in two canopy systems. Vitis 36, 119–123.

129. Maherali, H., Moura, C.F., Caldeira, M.C., Willson, C.J. and Jackson, R.B. (2006) Functional coordination between leaf gas exchange and vulnerability to xylem cavitation in temperate forest trees. Plant, Cell and Environment 29, 571–583.

130. Martin, P., Zarco-Tejada, P.J., Gonzalez, M.R. and Berjon, A. (2007) Using hyperspectral remote sensing to map grape quality in 'Tempranillo' vineyards affected by iron deficiency chlorosis. Vitis 46, 7–14.

131. Masle, J., Gilmore, S.R. and Farquhar, G.D. (2005) The ERECTA gene regulates plant transpiration efficiency in Arabidopsis. Nature 436, 866–870.

132. Matthews, M.A., Anderson, M.M. and Schultz, H.R. (1987) Phenolic and growth responses to seasonal water deficits in Cabernet Franc. Vitis 26, 147–160.

133. Maurel, C., Verdoucq, L., Luu, D.T. and Santoni, V. (2008) Plant Aquaporins: membrane channels with multiple integrated functions. Annual Review of Plant Biology 59, 595–624.

134. McCarthy, M.G., Loveys, B.R., Dry, P.R. and Stoll, M. (2002) Water reports. FAO Publication number 22, Rome, pp. 79–87.

135. Medrano, H., Bota, H., Escalona, J.M., Ribas-Carbó, M. and Flexas, J. (2005) Variability of intrinsic water use efficiency in Mediterranean grapevines. Proceedings XIV GESCO Symposium, Geisenheim, Germany, Vol. II (Forschungsanstalt: Geisenheim) pp. 513–520.

136. Mo, X., Chen, J.M., Ju, W. and Black, T.A. (2008) Optimization of ecosystem model parameters through assimilating eddy covariance flux data with an ensemble Kalman filter. Ecological Modelling 217, 157–173.

137. Morgan, P.B., Mies, T.A., Bollero, G.A., Nelson, R.L. and Long, S.P. (2006) Seasonal-long elevation of ozone concentration to projected 2050 levels under fully open-air conditions substantially decreases the growth and production of soybean. New Phytologist 170, 333–343.

138. Morlat, R., Penavayre, M., Jacquet, A., Asselin, C. and Lemaitre, C. (1992) Influence des terroirs sur le fonctionnement hydrique et al photosynthèse de la vigne en millésime exceptionnellement sec (1990). Conséquence sur la maturation du raisin. International Journal of Vine and Wine Sciences 26, 197–218.

139. Mott, K.A. and Buckley, T.N. (1998) Stomatal heterogeneity. Journal of Experimental Botany 49, 407–417.

140. Mott, K.A. and Peak, D. (2007) Stomatal patchiness and task-performing networks. Annals of Botany 99, 219–226.

141. Musil, C.F., Schmiedel, V. and Midgley, C.F. (2005) Lethal effects of experimental warming approximating a future climate scenario on southern African quartz-field succulents: a pilot study. New Phytologist 165, 537–547.

142. Möller, M., Alchanatis, V., Cohen, Y., Meron, M., Tsipris, J., Naor, A., Ostrovsky, V., Sprintsin, M. and Cohen, S. (2007) Use of thermal and visible imagery for estimating crop water status of irrigated grapevine. Journal of Experimental Botany 58, 827–838.

143. Naor, A. (1998) Relations between leaf and stem water potentials and stomatal conductance in three field grown woody species. Journal of Horticultural Science and Biotechnology 73, 431–436.

144. Naor, A. and Bravdo, B. (2000) Irrigation and water relations interactions in grapevines. Acta Horticulturae 526, 109–113.

145. Naor, A., Bravdo, B. and Gelobter, J. (1994) Gas exchange and water relations in field-grown Sauvignon blanc grapevines. American Journal of Enology and Viticulture 45, 423–427.

146. Neil, S., Desikan, R. and Hancock, J. (2003) Nitric oxide signalling in plants. New Phytologist 159, 11–31.

147. Nilson, S.E. and Assmann, S.M. (2007) The control of transpiration. Insights from Arabidopsis. Plant Physiology 143, 19–27.

148. Olivo, N., Girona, J. and Marsal, J. (2009) Seasonal sensitivity of stem water potential to vapour pressure deficit in grapevine. Irrigation Science 27, 175–182.

149. Olmstead, M.A., Wample, R.L., Greene, S.L. and Tarara, J.M. (2001) Evaluation of potential cover crops for inland Pacific Northwest vineyards. American Journal of Enology and Viticulture 52, 292–303.

150. Ortuño, M.F., García-Orellana, Y., Conejero, W., Ruiz-Sánchez, M.C., Alarcón, J.J. and Torrecillas, A. (2006) Stem and leaf water potentials, gas exchange, sap flow, and trunk diameter fluctuations for detecting water stress in lemon trees. Trees 20, 1–8.

151. Osmond, C.B., Björkman, O. and Anderson, D.J. (1980) Physiological processes in plant ecology (Springer-Verlag: Berlin).

152. O'Leary, M.H. (1981) Carbon isotope fractionation in plants. Phytochemistry 20, 553–567.
153. Palliotti, A., Cartechini, A., Nasini, L., Silvestroni, O., Mattio, S. and Meri, D. (2004) Seasonal carbon balance of 'Sangiovese' grapevines in two different central Italy environments. Acta Horticulturae 652, 183–190.
154. Patakas, A. and Zotos, A. (2005) Possible role of nitric oxide production in drought stress signalling in grapevines. Proceedings XIV GESCO Symposium, Geisenheim, Germany, Vol. II, (Forschungsanstalt: Geisenheim) pp. 535–558.
155. Patakas, A., Noitsakis, B. and Chouzouri, A. (2005) Optimization of irrigation water use in grapevines using the relationship between transpiration and plant water status. Agriculture, Ecosystems and Environment 106, 253–259.
156. Peguy, C.P. (1970) Précis de Climatologie (Masson, Paris) p. 468.
157. Pellegrino, A. (2003) Elaboration d'un outil de diagnostic du stress hydrique utilisable sur la vigne en parcelle agricole par couplage d'un modèle de bilan hydrique et d'indicateurs de fonctionnement de la plante. Thèse: Diplôme de Doctorat, ENSA, Montpellier.
158. Pellegrino, A., Lebon, E., Simmoneau, T. and Wery, J. (2005) Towards a simple indicator of water stress in grapevine (Vitis vinifera L.) based on the differential sensitivities of vegetative growth components. Australian Journal of Grape and Wine Research 11, 306–315.
159. Poni, S., Bernizzoni, F., Civardi, S., Gatti, M., Porro, D. and Comin, F. (2009) Performance and water-use efficiency (single-leaf vs. whole-canopy) of well-watered and half-stressed split-root Lambrusco grapevines grown in Po Valley (Italy). Agriculture, Ecosystems and Environment 129, 97–106.
160. Pou, A., Flexas, J., Alsina, M., Bota, J., Carambula, C., Herralde, F., Galmés, J., Lovisolo, C., Jiminéz, M., Ribas-Carbó, M., Rusjan, D., Secchi, F., Tomàs, M., Zsófi, Z. and Medrano, H. (2008) Adjustments of water use efficiency by stomatal regulation during drought and recovery in the drought-adapted Vitis hybrid Richter-110 (V. berlandieri x V. rupestris). Physiologia Plantarum 134, 313–323.
161. Quan, L.-J., Zhang, B., Shi, W.-W. and Li, H.-Y. (2008) Hydrogen peroxide in plants: a versatile molecule of the reactive oxygen species network. Journal of International Plant Biology 50, 2–18.
162. Rascher, U. and Pieruschka, R. (2008) Spatio-temporal variations of photosynthesis: the potential of optical remote sensing to better understand and scale light use efficiency and stresses of plant ecosystems. Precision Agriculture 9, 355–366.
163. Reynolds, A.G. and Wardle, D.A. (1989) Influence of fruit microclimate on monoterpene levels of Gewürztraminer. American Journal of Enology and Viticulture 40, 149–154.
164. Rodrigues, M.L., Santos, T.P., Rodrigues, A.P., Souza, C.R., Lopes, C.M., Maroco, J.P., Pereira, J.S. and Chaves, M.M. (2008) Hydraulic and chemical signalling in the regulation of stomatal conductance and plant water use in field grapevines growing under deficit irrigation. Functional Plant Biology 35, 565–579.
165. Sade, N., Vinocur, B.J., Dibes, A., Shatil, A., Ronen, G., Nissan, H., Wallach, R., Karchi, H. and Moshelion, M. (2009) Improving plant stress tolerance and yield production: is the tonoplast aquaporin SITIP2;2 a key to isohydric to anisohydric conversion? New Phytologist 181, 651–661.

166. Sadras, V. (2009) Does partial root-zone drying improve irrigation water productivity in the field? A meta-analysis. Irrigation Science 27, 183–190.

167. Sang, J., Jiang, M., Lin, F., Xu, S., Zhang, A. and Tan, M. (2008) Nitric oxide reduces hydrogen peroxide accumulation involved in water stress-induced subcellular antioxidant defense in maize plants. Journal of International Plant Biology 50, 231–243.

168. Santos, T.P., Lopes, C.M., Rodrigues, M.L., De Souza, C.R., Ricardo-da Silva, J.M., Maroco, J.P. and Chaves, M.M. (2005) Effects of partial root-zone drying irrigation on cluster microclimate and fruit composition of field grown Castelão grapevines. Vitis 44, 117–125.

169. Schmid, J. (1997) Xylemflussmessungen an Reben. Geisenheimer Berichte, Band 33.

170. Schultz, H.R. (1995) Grape canopy structure, light microclimate and photosynthesis. I. A two-dimensional model of spatial distribution of surface area densities and leaf ages in two canopy systems. Vitis 34, 211–215.

171. Schultz, H.R. (1991) Seasonal and nocturnal changes in leaf age dependent dark respiration of grapevine (Vitis vinifera L.) sun and shade leaves: modelling the temperature effect. Viticultural and Enological Sciences 46, 129–140.

172. Schultz, H.R. (1996) Water relations and photosynthetic responses of two grapevine cultivars of different geographical origin during water stress. Acta Horticulturae 427, 251–266.

173. Schultz, H.R. (2003a) Differences in hydraulic architecture account for near-isohydric and anisohydric behaviour of two field-grown Vitis vinifera L. cultivars during drought. Plant, Cell and Environment 26, 1393–1405.

174. Schultz, H.R. (2003b) Extension of a Farquhar model for grapevines (cvs. Riesling and Zinfandel) for light environment, phenology and leaf age induced limitations of photosynthesis. Functional Plant Biology 30, 673–687.

175. Schultz, H.R. and Jones, G.V. (2008) Sozio-ökonomische Aspekte des Klimawandels: Gewinner und Verlierer. Veränderungen in der Landwirtschaft am Beispiel des Weinbaus. In: Warnsignal Klima. Eds. J.L.Lozán, H.Graßl, G.Jendritzky, L.Karbe, K.Reise (GEO Wissenschaftliche Abhandlungen: Hamburg) pp. 268–273.

176. Schultz, H.R., Lebon, E. and Rousseau, C. (1999) Suitability of the 'Ball, Woodrow, Berry'-model for the description of stomatal coupling to photosynthesis of different Vitis species and Vitis vinifera cultivars in different climatic regions at various levels of water deficit. Acta Horticulturae 493, 17–30.

177. Schultz, H.R., Pieri, P., Poni, S. and Lebon, E. (2001) Modelling water use and carbon assimilation of vineyards with different canopy structures and varietal strategies during water deficit. Proceedings of the 12th International Congress on Photosynthesis (CSIRO Publishing: Brisbane, Australia).

178. Schulze, E.-D. and Hall, A.E. (1982) Stomatal responses, water loss and CO2 assimilation rates of plants in contrasting environments. In: Encylopedia of plant physiology, physiological plant ecology II. Water relations and carbon assimilation, Vol. 12B. Eds. O.L.Lange, P.S.Nobel, C.B.Osmond, H.Ziegler (Springer: Berlin) pp. 181–230

179. Seelig, H.-D., Adams, W., Hoehn, A., Stodieck, L., Klaus, D. and Emery, W. (2008) Extraneous variables and their influence on reflectance-based measurements of leaf water content. Irrigation Science 26, 407–414.

180. Seibt, V., Rajabi, A., Griffith, H. and Berry, J.A. (2008) Carbon isotopes and water use efficiency: sense and sensitivity. Oecologia 155, 441–454.

181. Smart, D.R. (2004) Exposure to elevated carbon dioxide concentration in the dark lowers the respiration quotient of Vitis cane wood. Tree Physiology 24, 115–120.

182. Smart, R.E., Schaulis, N.J. and Lemon, R.E. (1982) The effect of Concord vineyard microclimate on yield. I. The effect of pruning, training und shoot positioning on radiation microclimate. American Journal of Enology and Viticulture 33, 99–108.

183. Snyder, K.A., Richards, H.H. and Donovan, L.H. (2003) Night time conductance in C3 and C4 species: do plants lose water at night? Journal of Experimental Botany 54, 861–865.

184. Soar, C.J. and Loveys, B.R. (2007) The effect of changing patterns in soil-moisture availability on grapevine root distribution, and viticultural implications for converting full-cover irrigations into a point source irrigation system. Australian Journal of Grape and Wine Research 13, 2–13.

185. Soar, C.J., Collins, M.J. and Sadras, V.O. (2009) Irrigated Shiraz vines (Vitris vinifera) up-regulate gas-exchange and maintain berry growth in response to short spells of high maximum temperatures in the field. Functional Plant Biology 36, 801–804.

186. Soar, C.J., Dry, P.R. and Loveys, B.R. (2006b) Scion photosynthesis and leaf gas exchange in Vitis vinifera L. cv. Shiraz: Mediation of rootstock effect via xylem sap ABA. Australian Journal of Grape and Wine Research 12, 82–96.

187. Soar, C.J., Speirs, J., Maffei, S.M., Penrose, A.B., McCarthy, M.G. and Loveys, B.R. (2006a) Grape vine varieties Shiraz and Grenache differ in their stomatal response to VPD: apparent links with ABA physiology and gene expression in leaf tissue. Australian Journal of Grape and Wine Research 12, 2–12.

188. Souza, C.R., Maroco, J.P., Dos Santos, T.P., Rodrigues, M.L., Lopes, C.M., Pereira, J.S. and Chaves, M.M. (2005b) Control of stomatal aperture and carbon uptake by deficit irrigation in two grapevine cultivars. Agriculture, Ecosystems and Environment 106, 261–274.

189. Souza, C.R., Maroco, J.P., Dos Santos, T.P., Rodrigues, M.L., Lopes, C.M., Pereira, J.S. and Chaves, M.M. (2005a) Impact of deficit irrigation on water use efficiency and carbon isotope composition (δ13C) of field-grown grapevines under Mediterranean climate. Journal of Experimental Botany 56, 2163–2172.

190. Stocker, O. (1956) Die Abhängigkeit der Transpiration von den Umweltfaktoren. In: Encyclopedia of plant physiology, Vol. 3. Eds. W.Ruhland (Springer-Verlag: Berlin) pp. 436–488.

191. Stoll, M. and Jones, H.G. (2007) Thermal imaging as a viable tool for monitoring plant stress. Journal International de Sciences de la Vigne et du Vin 41, 77–84.

192. Stoll, M., Loveys, B.R. and Dry, P.R. (2000) Hormonal changes induced by partial rootzone drying of irrigated grapevine. Journal of Experimental Botany 51, 1627–1634.

193. Tanner, C.B. (1963) Plant temperatures. Agronomy Journal 55, 210–211.

194. Tardaguila, J., Bertamini, M., Reniero, F. and Versini, G. (1997) Oxygen isotope composition of must-water in grapevine: effects of water deficit and rootstock. Australian Journal of Grape and Wine Research 3, 84–89.

195. Tenhunen, J.D., Sala Serra, A., Harley, P.C., Dougherty, R.L. and Reynolds, J.F. (1990) Factors influencing carbon fixation and water use by Mediterranean sclerophyll shrubs during summer drought. Oecologia 82, 381–393.

196. Tonietto, J. and Carbonneau, A. (2004) A multicriteria climatic classification system for grape-growing regions worldwide. Agricultural and Forest Meteorology 124, 81–97.

197. Turner, N.C., Schulze, E.-D. and Gollan, T. (1984) The response of stomata and leaf gas exchange to vapour pressure deficits and soil water content: I. Species comparisons at high soil water contents. Oecologia 63, 338–342.

198. Van Leeuwen, C., Gaudillère, J.-P. and Tregoat, O. (2001) L'évaluation du régime hydrique de la vigne à partir du rapport isotopique 13C/12C. L'intérêt de sa mesure sur les sucres du moût à maturité. Journal International de Sciences de Vigne et du Vin 35, 195–105.

199. Van Leeuwen, C. and Seguin, G. (1994) Incidences de l'alimentation en eau de la vigne, appreciée per l'etat hydrique du feuillage, sur le developpement de l'appareil végétatif et la maturation du raisin. Journal of Vine and Wine Sciences 28, 81–110.

200. Vandeleur, R.K., Mayo, G., Shelden, M.C., Gilliham, M., Kaiser, B.N. and Tyerman, S.D. (2009) The role of PIP aquaporins in water transport through roots: diurnal and drought stress responses reveal different strategies between isohydric and anisohydric cultivars of grapevine. Plant Physiology 149, 445–460.

201. Wang, Y.P., Baldochi, D., Leuning, R., Falge, E. and Vesal, T. (2007) Estimating parameters in a land-surface model by applying nonlinear inversion to eddy covariance flux measurements from eight FLUXNET sites. Global Change Biology 13, 652–670.

202. Webb, L.B., Whetton, P.H. and Barlow, E.W. (2008) Climate change and wine grape quality in Australia. Climate Research 36, 99–111.

203. Webb, L.B., Whetton, P.H. and Barlow, E.W.R. (2007) Modelled impact of future climate change on the phenology of gapevines in Australia. Australian Journal of Grape and Wine Research 13, 165–175.

204. Williams, L.E. and Araujo, F.J. (2002) Correlations among predawn leaf, midday leaf, and midday stem water potential and their correlations with other measures of soil and plant water status in Vitis vinifera. Journal of the American Society for Horticultural Science 127, 448–454.

205. Williams, L.E. and Matthews, M.A. (1990) Grapevine. In: Irrigation of Agricultural Crops-Agronomy (Monograph nor. 30, ASA-CSSA-SSSA), pp. 1019–1055.

206. Woodward, F.I. (2007) An inconvenient truth. New Phytologist 174, 470–473.

207. Zarco-Tejada, P.J., Berjon, A., Lopez-Lozano, R., Miller, J.R., Martin, P., Cachorro, V., Gonzalez, M.R. and De Frutos, A. (2005) Assessing vineyard condition with hyperspectral indices: leaf and canopy reflectance simulation in a row-structured discontinuous canopy. Remote Sensing of Environment 99, 271–287.

208. Zsófi, Z., Gál, L., Szilágy, Z., Szücs, E., Marschall, M., Nagy, Z. and Bálo, B. (2009) Use of stomatal conductance and pre-dawn water potential to classify terroir for the variety Kékfrankos. Australian Journal of Grape and Wine Research 15, 36–47.

CHAPTER 9

MANAGING GRAPEVINES TO OPTIMIZE FRUIT DEVELOPMENT IN A CHALLENGING ENVIRONMENT: A CLIMATE CHANGE PRIMER FOR VITICULTURISTS

M. KELLER

9.1 INTRODUCTION

The genus *Vitis* is home to the temperate climate zones of the northern hemisphere. Its members are woody shrubs or lianas that climb in trees by means of their leaf-opposed tendrils. Although this genus is thought to comprise about 60 extant species spread mostly throughout Asia and North America (Alleweldt and Possingham 1988, Wan et al. 2008), the Eurasian species *Vitis vinifera* L. gave rise to the great majority of modern grape cultivars that are being grown for the production of wine, table grapes, dried grapes (raisins), grape juice and brandy. While several American species have become important, and typically inter-specific, crossing partners in rootstock breeding programs, their direct genetic influence on

Keller M. Managing Grapevines to Optimise Fruit Development in a Challenging Environment: A Climate Change Primer for Viticulturists. Australian Journal of Grape and Wine Research **16**,S1 *(2010),* DOI: *10.1111/j.1755-0238.2009.00077.x. Reprinted with permission from the authors.*

wine, juice and other cultivars has been minor. Similarly, despite some interesting disease resistance traits, most Asian species have remained obscure (Wan et al. 2007, Li et al. 2008).

The evolutionary innovation of fleshy, sweet, and dark-skinned berries during the late Cretaceous period was probably an adaptation to the simultaneous evolution of birds and mammals (Hardie 2000). Birds are the principal seed dispersers of *V. vinifera* grapes, which, from a grower's viewpoint, makes them a major pest of cultivated grapevines. Moreover, the light-coloured skin of so-called white grapes has evolved from the dark-skinned 'default' version (Kobayashi et al. 2004, Walker et al. 2007). In spite of much wishful thinking on our part, the role of grape berries is only to enable the vine to spread its seeds and hence its genes, which the berries achieve by attracting seed dispersers using pigments and aroma volatiles and by 'paying' for transportation with energy-rich sugar and other nutritionally valuable components (Hardie 2000, Goff and Klee 2006). Charles Darwin noticed long ago that a fruit's 'beauty serves merely as a guide to birds and beasts in order that the fruit may be devoured and the manured seeds disseminated' (Darwin 1859). The high acidity and 'green' aroma of unripe berries prevents the seeds from being destroyed before they are mature, while tannins and other secondary metabolites may stave off microbes that destroy the fruit without scattering its seeds.

This paper provides an overview of published information about grapevine reproductive development and recent advances in research and practice concerning factors that influence yield formation and fruit composition at harvest. Because the impacts of nutrients and water deficit were recently reviewed elsewhere (Mpelasoka et al. 2003, Bell and Henschke 2005, Keller 2005), the focus of the present paper is mainly on two other major environmental determinants of yield and fruit quality: light and temperature. Novel discoveries from the last few years concerning water and nutrients will nevertheless be considered where appropriate. However, because of the overriding importance of temperature for grape and wine quality (Hofäcker et al. 1976, Downey et al. 2006), most of the discussion will be centred on a critical appraisal of possible consequences of global climate change and their implications for viticulture. Although climate change obviously also has implications for weed, pest and disease pressure in vineyards (e.g. Seem et al. 2000, Ainsworth and Rogers 2007), these issues will not be considered here because of space constraints and to keep the discussion focused. I will conclude by attempting to identify opportunities and priorities for future research on grapevine physiology and viticulture that aims to optimise fruit development in a challenging environment, particularly in response to the challenges and threats raised by a changing climate.

9.2 REPRODUCTIVE GROWTH: YIELD FORMATION

Grapevine reproductive development extends over 2 years: buds formed in the first growing season give rise to shoots that carry fruit in the following growing season. This process has been reviewed extensively, both from an anatomical and molecular genetic perspective (Pratt 1971, Srinivasan and Mullins 1981, Swanepoel and Archer 1988, Morrison 1991, Gerrath 1992, May 2004, Meneghetti et al. 2006, Carmona et al. 2008); therefore, only a brief overview is presented here. Reproductive growth begins with the formation of uncommitted primordia in the developing buds from spring through early summer. Uncommitted primordia differentiate into inflorescence, tendril, or even shoot primordia, before the buds enter dormancy. It seems that in most cultivars only the basal six to eight buds of a shoot are able to form inflorescence primordia, whereas Sultana produces fruitful buds over the entire length of its shoots (Sánchez and Dokoozlian 2005). Most *V. vinifera* cultivars normally initiate two, sometimes three or even four, inflorescence primordia, but environmental influences strongly modulate the actual number produced.

Although some researchers hold that a portion of the flower initials are formed before bud dormancy (Alleweldt 1966, Alleweldt and Ilter 1969, Agaoglu 1971), the commonly accepted view is that winter dormancy seasonally separates the process of inflorescence initiation and differentiation from that of flower initiation and differentiation (Srinivasan and Mullins 1981, Gerrath 1992, May 2004, Meneghetti et al. 2006, Carmona et al. 2008). This consensus view holds that inflorescence primordia produce branch meristems before the buds enter dormancy. When the buds begin to swell in early spring these branch meristems are reactivated to produce groups of three to four flower meristems. This flower initiation ostensibly ceases around the time of budburst and is followed by the development of floral organs (flower differentiation) while the inflorescences become visible on the shoot and then separate. Different genes or groups of genes identify each floral organ, but the mechanism by which these genes control floral organogenesis is not yet known (Díaz-Riquelme et al. 2009). Irrespective of how these processes are controlled, the degree of branching before and after dormancy and the extent of flower development after budburst determine the number of flowers formed on each inflorescence. This number is highly variable, even on the same shoot: the basal inflorescence usually forms the most flowers, and numbers decrease for higher inserted inflorescences (May 2004).

Soon after the flowers are fully formed and about 5–10 weeks after budburst, capfall (anthesis) marks the start of flowering. The anthers release the pollen onto the stigma (pollination), where the pollen grains germinate. Most *V. vinifera* cultivars are self-pollinated, and pollination often occurs shortly before capfall (Staudt 1999, Heazlewood and Wilson 2004). The pollen tubes grow down the style to fuse with the egg cells in the ovules (fertilisation), which leads to fruit set. Eggs that are not fertilised within 3 to 4 days after anthesis will degenerate, while fertilised ovules develop into seeds (Kassemeyer and Staudt 1981). Although pollination, rather than fertilisation, triggers initial ovary development, subsequent fruit formation, which involves differentiation of an exocarp (skin) and mesocarp (flesh or pulp), requires release of auxin by the seed(s) that stimulates gibberellin biosynthesis in the pericarp (Kassemeyer and Staudt 1983, Gillaspy et al. 1993, O'Neill 1997). Therefore, in most cultivars (except for parthenocarpic and stenospermocarpic cultivars that produce seedless berries or berries with seed traces, respectively), growth of grape berries requires pollination, fertilisation, and development of one or more seeds (Nitsch et al. 1960).

The percentage of flowers that set fruit is inversely related to the number of flowers per inflorescence and normally varies between 20 and 50%. Fruit set is highly variable among cultivars and is altered by environmental conditions and rootstock (Alleweldt and Hofäcker 1975, Keller et al. 2001). Environmental stress, ineffective leaf area, canopy shade, and overly vigorous shoot growth may diminish fruit set, and ovaries may be shed for up to 1 month after anthesis (Staudt and Kassrawi 1973, Kassemeyer and Staudt 1983, Candolfi-Vasconcelos and Koblet 1990). Differences in percentage fruit set combined with the extent of inflorescence branching and differences in bunch-stem cell elongation result in large differences in bunch size and architecture.

A period of cell division follows fertilisation, but while the cell number doubles about 17 times before anthesis, only about two more doublings are thought to occur after flowering (Coombe 1976). Cell division stops in the mesocarp within a month after anthesis but may continue in the skin for another 2 to 3 weeks, before seeded berries enter a lag phase of slow or no growth (Nakagawa and Nanjo 1965, Harris et al. 1968, Considine and Knox 1981). The duration of this lag phase varies by cultivar, suggesting that it is genetically determined, and is important in establishing the time of fruit maturity. Despite a roughly tenfold decrease in mesocarp cell turgor pressure during the lag phase (Thomas et al. 2008), cell expansion resumes after this phase, fuelled by phloem influx and sugar unloading (Keller et al. 2006), so that mesocarp cells may augment over 300-fold from anthesis to maturity (Coombe 1976). This second phase of cell expansion coincides with the beginning of ripening (veraison) and sugar accumulation in the berry. Seedless grape berries contrast with seeded berries in that they often do not display distinct growth phases; their size increases much more 'smoothly', and they usually remain smaller than seeded berries (Nitsch et al. 1960).

Cell division goes together with cell expansion, for growth essentially occurs via cell expansion. Consequently, the size of mature berries at harvest is a function of the number of cell divisions before and after flowering, the extent of expansion of these cells (Coombe 1976), and the extent of pre-harvest shrinkage. Because the early post-flowering rates of cell division and cell expansion are controlled by the seeds, final berry volume increases as the number of seeds per berry increases (Winkler and Williams 1935, Scienza et al. 1978, Gillaspy et al. 1993). The upper limit on mesocarp cell expansion is likely imposed by the elastic properties of the skin, whose extensibility declines during grape ripening (Matthews et al. 1987). Despite their potential importance for berry size, however, the changes in skin elasticity have received little research attention.

Recent research showed that cell compartmentation in the berry remains intact throughout most of the ripening phase. Loss of membrane integrity and senescence in most of the pericarp apparently do not begin until a berry has attained its maximal weight and may be associated with subsequent berry shrinkage, at least in some cultivars (Hardie et al. 1996, Krasnow et al. 2008, Tilbrook and Tyerman 2008).

9.3 YIELD: FROM POTENTIAL TO HARVEST

The individual steps of the process whereby the different components of yield cumulatively and interactively culminate in the harvestable crop yield are under genetic control. Yet, yield formation is subject to great spatial and temporal fluctuations. There is ample opportunity for environmental variables and management practices to alter each yield component of a given cultivar, and for the vines to respond to and partially compensate for any such alterations. Warm temperatures, high irradiance, and adequate water and nutrient supply are required for the formation of the maximum number of inflorescence primordia (Srinivasan and Mullins 1981, Meneghetti et al. 2006). Both cool (<20°C) and hot temperatures (>35°C) during the initiation phase favour tendril formation (Buttrose 1970, 1974), which may cause buds to be unfruitful. Low light, whether caused by clouds or high canopy density, interferes with inflorescence formation, and this effect appears to be mediated, at least to a considerable extent, by assimilate supply to the buds (Buttrose 1974, Keller and Koblet 1995, Dry 2000, Sánchez and Dokoozlian 2005). Support for this argument comes from the fact that severe loss of leaf area can also reduce inflorescence initiation and differentiation (Bennett et al. 2005, reviewed by Lebon et al. 2008). Both nutrient (especially nitrogen) and water deficit and excess can limit inflorescence initiation (reviewed by Keller 2005, Meneghetti et al. 2006). Perhaps this nutrient impact involves changes in canopy microclimate, although this does not explain why low nutrient availability also decreases fruitfulness. While mild water deficit may promote inflorescence initiation because of improved canopy microclimate, more severe stress reduces inflorescence numbers.

The environmental and developmental causes of the high variation in flower number per inflorescence are mostly unknown. It is thought that low temperatures before and during budburst increase flower numbers (Pouget

1981, Dunn and Martin 2000, Petrie and Clingeleffer 2005) and that rapid shoot growth in spring may compete with flower formation (Bowen and Kliewer 1990). Whether these two effects are related and whether they exert their influence via the supply of storage reserves remains to be investigated. Nonetheless, untimely loss of leaf area coupled with low carbohydrate reserve status has indeed been found to reduce the number of flowers per inflorescence in the following season (Bennett et al. 2005).

The next step in yield formation is always one of lowering the yield potential, because not all flowers formed on a vine will set fruit, even under the best of circumstances. Adverse environmental conditions, while not preventing pollination, diminish fertilisation by reducing pollen viability or germination rate, which leads to poor fruit set (Koblet 1966, Kliewer 1977, Staudt 1982). Thus, similar to their impact on inflorescence initiation, both cool (<15°C) and hot temperatures (>35°C) decrease fruit set. In addition, low light is also detrimental to fruit set, probably because it leads to carbon starvation in the inflorescences because of their low sink priority prior to fruit set (Keller and Koblet 1994, Lebon et al. 2008). Lack of carbon supply is also the likely reason early-season defoliation usually curtails fruit set (Coombe 1962). In extreme cases, unfavourable conditions may result in a syndrome named inflorescence necrosis or early bunch-stem necrosis that is associated with abscission of portions of or entire inflorescences (Jackson and Coombe 1988, Jackson 1991, Keller and Koblet 1994, Keller et al. 2001). Shoot tipping or hedging during flowering has been found to ameliorate adverse environmental effects on fruit set by temporarily eliminating strong sinks competing for carbon supply (Coombe 1962, Koblet 1969). Thus, conditions that maximise fruit set are much like those that maximise inflorescence formation, and berry number is most vulnerable to environmental stress at or just before flowering.

Grapevines may compensate for differences in berry number per vine resulting from variations in inflorescence formation and fruit set by changing the rate of berry growth. Consequently, final berry size tends to increase with decreasing berry number, although such compensatory berry growth ostensibly can be limited by imposing more severe water deficit after fruit set (Keller et al. 2008). Moreover, at that time both low (<15°C) and very high temperatures (>35°C) again slow down the rate of cell division, which may limit berry size (Kliewer 1977). Although heat stress also reduces cell expansion, heat seems to restrict berry size only if it occurs before the lag phase of growth but not during the post-veraison expansion phase (Hale and Buttrose 1974). Because sun-exposed berries are heated by the incoming radiation (Smart and Sinclair 1976, Spayd et al. 2002), this temperature effect implies that in warm climates early leaf removal may have the potential to limit berry size.

Although grape berries, especially during the early stages of development, are capable of photosynthetically fixing small amounts of CO_2 (Kriedemann 1968a, Palliotti and Cartechini 2001), they are dependent on sustained sugar influx for development, including cell division, cell expansion, and ripening. Whereas early-season limitations in carbon supply, whether imposed by environmental stress or by cultural practices (e.g. leaf removal or deficit irrigation), may restrict berry number and size, they usually do not impair ripening. In contrast, if limitations occur after the lag phase of berry growth, they evidently do impact fruit ripening. Although the influence of irrigation and nutrition strategies on yield formation was recently reviewed (Keller 2005), new information has since become available that is particularly relevant for late-season (pre-harvest) irrigation, especially with respect to the extended ripening ('hang time') that is currently in demand by many wineries. The practice of letting grapes hang on the vines beyond their physiological sugar maximum of about 24–25°Brix is often associated with yield losses to the growers in the order of 10–20% owing to water loss from the berries. Recent findings suggest that it may be possible to mitigate such losses by increasing the frequency, and perhaps the amount, of irrigation water supply after veraison (Keller et al. 2006). The strategy needs to be tested under field conditions to ensure it does not stimulate shoot regrowth and compromise grape composition and therefore wine quality.

It is clear, then, that vineyard management tools (e.g. canopy management, irrigation, nutrition) are available to manipulate individual yield components in order to optimise yield (Dry 2000, Keller 2005). The combination, timing and extent of such cultural practices need to be fine-tuned to maximise quality depending on the intended end-use of the grapes, adapted to match specific clone/cultivar/rootstock/site combinations, and modified depending on short-term variations in weather conditions and long-term changes in climate.

9.4 FRUIT COMPOSITION

Uniform grape composition is usually regarded as desirable for wine making as well as for table grape and raisin production. Composition changes continuously during berry development and ripening, and the associated metabolic pathways are under genetic control. As in the case of yield formation, however, environmental factors and cultural practices, and their interaction with the genotype of the cultivar, can alter the magnitude of these changes. Any factor altering vine growth and/or physiology directly or indirectly impacts fruit composition, which results in large quality variations from one year to another. Annual variations in climate, especially temperature (Hofäcker et al. 1976), are particularly important for wine production, because such variations, in addition to site, typically far outweigh any changes introduced by cultural practices (reviewed by Downey et al. 2006) and even those arising from differences in soil moisture (van Leeuwen et al. 2004, Pereira et al. 2006, Keller et al. 2008). Intriguingly, gene expression in grape berries has been reported to vary more among growing seasons than between pre- and post-veraison berries in the same season (Pilati et al. 2007).

The increase in the amount and concentration of sugars, amino acids and phenolic compounds (especially anthocyanins in red grapes) and the decrease in acidity (especially malic acid), which is associated with a rise in pH, during grape ripening are well known (reviewed by Ollat et al. 2002, Adams 2006). The viticultural literature also abounds with reports describing environmental effects and impacts of vineyard management practices on grape soluble solids, titratable acidity, pH and, often, colour. For the most part, the information provided in the extensive review by Jackson and Lombard (1993) still applies today. The accumulation and alterations in skin and seed tannins are less well understood, although it is now clear that biosynthesis occurs mostly, perhaps even exclusively, during the early stages of berry development, while the ripening phase is characterised by polymerisation reactions and other alterations to existing tannin units (reviewed by Adams 2006, Downey et al. 2006). Yet it is far from clear why tannin extractability during winemaking ranges from about 25 to 75%, why 20–50% of this tannin is derived from the seeds

(Cerpa-Calderón and Kennedy 2008), and why the tannin concentration in red wines of the same cultivar varies over two orders of magnitude (Harbertson et al. 2008). Moreover, though progress is being made, the influence of environmental factors and cultural practices on tannin accumulation, polymerisation and extractability remains largely unknown (reviewed by Downey et al. 2006). The gaps in our knowledge are considerably wider still with respect to flavour and aroma components. Nonetheless, real or purported (and supposedly beneficial) changes in both tannin composition and flavour volatiles during the later stages of ripening form the basis for the current 'hang-time' trend (Wilson et al. 1984, Fang and Qian 2006).

Adequate, but not excessive, exposure of grapes to sunlight has long been known to be beneficial for wine quality, especially red wine quality (Jackson and Lombard 1993). But it is often unclear if the advantageous effects of sun exposure arise from higher light or from higher temperature, which are both consequences of solar radiation. Depending on trellis and training system and vine vigour, light in the bunch zone can range from less than 1% of ambient (e.g. single-curtain 'sprawl' trellis) to about 10% (e.g. vertically shoot-positioned systems) and over 30% (e.g. double-curtain systems) (Dokoozlian and Kliewer 1995a,b). At the same time, sun-exposure has been found to elevate berry surface temperatures by as much as 17°C above ambient temperatures, while shaded berries are usually close to ambient (Smart and Sinclair 1976, Spayd et al. 2002, Tarara et al. 2008). It seems that severe canopy shade downregulates gene expression in the anthocyanin biosynthesis pathway (Jeong et al. 2004, Koyama and Goto-Yamamoto 2008), but at photon fluxes >100 $\mu mol/m^2/s$ on the berries temperature becomes the overriding variable in anthocyanin synthesis (Spayd et al. 2002, Downey et al. 2006, Tarara et al. 2008). Light exposure also promotes pre-veraison tannin formation in the skin and increases post-veraison tannin polymerisation, but ostensibly decreases tannin extractability (Cortell and Kennedy 2006, Koyama and Goto-Yamamoto 2008). Moreover, direct sunlight, possibly via elevated ultraviolet (UV) radiation, enhances both pre-veraison accumulation of carotenoids and their post-veraison degradation to aroma-active norisoprenoids (Razungles et al. 1998, Schultz 2000, Baumes et al. 2002, Düring and Davtyan 2002), but little is known about the influence of light on other aroma volatiles.

Pre-veraison malic acid biosynthesis is fastest at 20–25°C, whereas post-veraison malate degradation continues to accelerate up to a maximum of ~50°C (Lakso and Kliewer 1975). The optimum berry temperature for anthocyanin synthesis is around 30°C, but above 35°C anthocyanins stop accumulating (Spayd et al. 2002) or may even be degraded (Mori et al. 2007). In addition, high temperature apparently favours formation of malvidin-based anthocyanins over that of other pigments, which leads to shifts in the anthocyanin profile (Ortega-Regules et al. 2006, Tarara et al. 2008), an effect that has also been observed in response to low light (Keller and Hrazdina 1998). Similarly, in addition to stimulating anthocyanin pathway gene expression and enzyme activity, an effect that may be mediated by abscisic acid (ABA), water deficit may also induce shifts in the anthocyanin profile by favouring accumulation of malvidin- and petunidin-based anthocyanins (Jeong et al. 2004, Castellarin et al. 2007a,b) Lower temperatures, it seems, promote pre-veraison methoxypyrazine accumulation and slow down post-veraison degradation (Lacey et al. 1991, Roujou de Boubée et al. 2000). Norisoprenoids such as β-damascenone or β-ionone, which contribute to floral and fruity attributes in wine (Winterhalter and Schreier 1994), appear to be relatively insensitive to temperature, though they may be masked by methoxypyrazines under cool conditions. Nonetheless, the influence of temperature on most aroma and flavour compounds is not well understood, leaving plenty of work for future graduate students.

9.5 IMPENDING CHALLENGE: CLIMATE CHANGE

To state that grape and wine production is highly sensitive to climate variability is almost a platitude. After all, this sensitivity is the source of considerable variation in vintage quality which, in turn, forms a key basis for the existence of an entire 'industry' of wine judges, consumer magazines and related services. One competitive advantage of a particular wine region over another would be a comparatively lower climate variability making for consistency in vintage quality. It is important to recognise, however, that high variation around a particular average climatic index for a given region may be caused not only by wide fluctuations from year to year, but could also arise from a general trend of gradually more (or less) favourable climatic conditions. The analysis by Jones et al. (2005) showed that many of the world's wine regions are currently at or near their ideal climate for their respective grape cultivars and wine styles. Although the best wines in many regions have traditionally been produced in the warmest years (e.g. Jones and Davis 2000), the projected increases in average temperature and climate variability over the coming decades may threaten some regions' competitive advantage.

Grape growing regions are often classified into so-called 'Winkler regions' according to heat summation measured in cumulative growing degree days (GDD), a scheme originally proposed by Amerine and Winkler (1944). This method sums up the mean daily temperatures above a threshold typically set at 10°C over a 7-month 'standard' growing season (April–October in the northern hemisphere and October–April in the southern hemisphere). Winkler regions are grouped into five categories: I (<1389 GDD), II (1389–1667 GDD), III (1668–1944 GDD), IV (1945–2222 GDD), and V (>2222 GDD). Each 1°C increment in mean temperature adds 214 GDD to the standard growing season. Therefore, if one assumes an average increase from the present of 1.5°C by 2020, cumulative heat units would increase by 321 GDD. A 2.5°C increase by 2050 would add 535 GDD to the current heat units. This simple estimate shows that the projected rise in temperature associated with global climate change (IPCC 2007) will likely shift several of the world's growing regions into the next higher Winkler region by 2020, and that this shift will affect most regions by 2050. In addition, the predicted increase in spring and autumn temperatures will also lead to longer actual (as opposed to standard) growing seasons, which are determined by the frost-free period (days between the last spring and first autumn frosts). Longer frost-free seasons and higher summer temperatures combined with drier conditions are furthermore associated with an increase in the frequency of wildfires (Overpeck et al. 1990, Westerling et al. 2006, IPCC 2007).

On top of the rise in average temperatures and an associated increase in cumulative heat units and growing-season length, shifts will also occur in the number and extent of extreme weather events (IPCC 2007).

This conclusion follows directly from the statistical normal distribution of temperature data around a mean (cf. Cahill and Field 2008). As the mean shifts upward, so will the extremes (i.e. outliers) on either end of the curve. Therefore, although winter killing freezes and spring and autumn frosts will still occur in cool regions beyond 2050, they are likely to become less frequent and less severe. At the other end of the spectrum, however, summer heat waves will probably increase in both frequency and severity. An analysis for Victoria, Australia, shows that the probability of a run of five consecutive days over 35°C doubles for a 1°C warming and increases five-fold for a 3°C warming (Hennessy and Pittock 1995).

In addition to the predicted temperature rise, the atmospheric CO_2 concentration ($[CO_2]$), which is currently ~380 ppm, is projected to reach up to 600 ppm by the end of the 21st century, which is expected to accelerate the warming trend (IPCC 2007). It should be borne in mind, however, that these are conservative estimates, reached by both scientific and political consensus. The disturbing trend to date has been that each time models have been refined or supported by more accurate measurements, the projections for both temperature and $[CO_2]$ were corrected upward, never downward. Precipitation patterns are set to change, too. Rising temperatures will lead to greater proportions of winter precipitation to fall as rain instead of snow, and the average snowlines will move to higher elevations. Consequently, rivers whose flow depends on snow melt will experience a shift in peak flow rates from late spring/early summer to late winter/early spring, with diminished flow rates during summer (e.g. Mastin 2008). While such shifts in river flow are expected to increase the risk of flooding in early spring, there will be less water available during the critical irrigation season. Added to the earlier snow melt is the prediction that precipitation will increase less than evaporation, so that droughts will increase substantially in regions and seasons that are already relatively dry (Manabe et al. 2004). Thus, the reduced river flows in summer in dry regions are compounded by drier soils and higher evaporation from irrigated farmlands, which will greatly increase the demand, and therefore the competition, for water. Furthermore, the greater increase in evaporation than precipitation can also be expected to accelerate salinisation in dry (and drying) regions.

9.6 CONSEQUENCES: PHENOLOGY, GROWTH AND YIELD FORMATION

The annual succession of phenological stages of grapevines is commonly observed to be accelerated with a rise in temperature (Alleweldt et al. 1984, Jones and Davis 2000, Chuine et al. 2004, Duchêne and Schneider 2005, Wolfe et al. 2005, Webb et al. 2007). Such observations show a consistent trend towards earlier flowering, veraison and harvest. The timing of veraison may be of particular importance, because earlier veraison implies that the critical ripening period shifts towards the hotter part of the season. This has already been described for Alsace, France, where the period between budburst and harvest has become shorter, and ripening is occurring under increasingly warm conditions (Duchêne and Schneider 2005). Model calculations performed for Australian wine regions also project a forward shift in harvest date, which was arbitrarily defined as grapes reaching a soluble solids content of 20°Brix (Webb et al. 2007).

The optimum temperature for grapevine leaf photosynthesis is thought to be around 25–28°C (Kriedemann 1968b). Higher growth temperatures shift this temperature optimum upward (Schultz 2000), although the upper limit of this adaptation has not been established. In addition, a rise in $[CO_2]$ is thought to increase light-saturated photosynthesis while decreasing stomatal conductance (Schultz 2000, Düring 2003, Tognetti et al. 2005, Ainsworth and Rogers 2007). Models as well as empirical observations suggest that the simultaneous increase in temperature and $[CO_2]$ has synergistic effects on photosynthesis (Sage and Kubien 2007). All other plant processes affected by elevated $[CO_2]$ are thought to be consequences of these two basic responses (Long et al. 2004). This suggests that higher $[CO_2]$ will likely increase not only vine productivity, but also water-use efficiency. Therefore, grapes grown in arid regions are expected to benefit from rising $[CO_2]$ (Schultz 2000). In most agricultural systems, including vineyards in cool, humid regions, this gain would likely be offset by increased growth (and hence greater leaf area and yield) under rising $[CO_2]$ and temperature. But this may not be the case for wine grapes in dry regions, where shoot growth is controlled by regulated deficit irrigation (RDI), and crop load is often adjusted by bunch thinning (e.g. Dry et al. 2001, Keller et al. 2005, 2008). Assuming that the industry targets for shoot growth and crop load will not change in the future, the high [CO_2] scenario may thus result in savings of irrigation water—unless those savings are offset by higher evaporation rates. In fact, the expected increase in evapotranspiration and the associated decline of soil moisture (Manabe et al. 2004) might facilitate the implementation of RDI to control shoot growth. The same may not be true for table, raisin and juice grapes, as growers are likely to increase yields when the opportunity arises. The additional sinks on these vines may be able to utilise the extra carbon assimilated in a higher temperature and [CO_2] environment, enabling growers to ripen a larger crop. Nevertheless, Californian table grapes are currently grown near their temperature optimum (Lobell et al. 2006).

Because of the limitations on crop load often imposed by the makers of law (Old World) or wine (New World), wine grapes could either downregulate photosynthesis (via feedback inhibition because of sugar accumulation in the leaves) or export and store excess carbon as starch in the perennial structure of the vine. This might be beneficial for cold hardiness but could also stimulate vegetative growth and yield formation because of a cumulative effect over consecutive seasons arising from the higher availability of reserve carbohydrates for spring growth (Holzapfel et al. 2006). More vegetative growth would then have to be dealt with via more intensive canopy management practices, such as hedging. However, the generally low supply of nitrogen fertiliser in wine grape production may mitigate against this problem by imposing a sink limitation. Nitrogen, rather than carbohydrate, reserves have been found to determine both spring vegetative growth and fruiting of young *Vitis labruscana* grapevines (Cheng et al. 2004).

Although heat effects will certainly be important in a warmer world with higher [CO_2], the influence of climate change will be more important at the lower limits of temperature, because current and predicted temperature increases are higher at night, at higher latitudes, and in the winter (IPCC 2007). This will prolong the growing season and might decrease the risks of cold injury and the need for frost protection in cool regions. Whereas a longer growing season because of earlier spring warming (i.e. earlier budburst) and/or later autumn cooling (i.e. later leaf fall) may be expected to improve vine storage reserve status and cold hardiness, such positive outcomes will only occur if the warming trend does not compromise cold acclimation in autumn, cold hardiness in winter, and deacclimation in spring (cf. Schnabel and Wample 1987, Fennell 2004). A recent field trial, however, found that a 2.2°C higher running mean temperature during the cold acclimation phase in one year than in another was reflected in less cold-hardy grapevine buds and canes in the winter following the warmer autumn (Keller et al. 2008). Moreover, if budburst occurs earlier, then the risk for spring frost damage may not be reduced or may even increase, if the warming trend were associated with a higher frequency of cloudless nights in spring. The net effect of these differing trends might be close to zero, which would mean that we may not expect to see any change in winter injury because of a warming climate. At the other end of the spectrum, warmer winters may cause heightened problems with irregular budburst in warm regions because of reduced winter chilling, which is of particular concern to the table grape and raisin industries (Antcliff and May 1961, Kliewer and Soleimani 1972, Lavee and May 1997, Dokoozlian 1999, Webb et al. 2007).

Despite the apparent upper temperature limit of ~35°C for maximum yield formation discussed previously, higher wine grape yields have been predicted under warmer conditions in Australia (Webb et al. 2006). In contrast, Lobell et al. (2006) found that yields of Californian wine grapes are likely to change very little by the end of this century, whereas a declining trend was estimated for table grapes. However, this group modelled only the influence of temperature (projected to increase by 0.2–0.5°C per decade) and precipitation (projected to have a minor effect because of irrigation), ignoring the effect of CO_2. It is possible that the predicted increase in atmospheric $[CO_2]$ and its positive effect on yield (Bindi et al. 2001) might cancel out or reverse any negative trend because of temperature alone. On the other hand, a study modelling the combined effects of predicted changes in temperature, solar radiation and CO_2, found that the inter-annual variation of *V. vinifera* yield is likely to increase (Bindi et al. 1996).

9.7 CONSEQUENCES: FRUIT RIPENING AND QUALITY

In Bordeaux, France, a forward shift in the time of veraison has been correlated with elevated fruit sugar and lower acid concentrations, and generally better wine quality (Jones and Davis 2000). While accumulation of organic acids and tannins in grapes occurs mostly during the warmest part of the year (i.e. pre-veraison), the ripening-related accumulation of sugars, anthocyanins, and most flavour and aroma compounds, and degradation of malic acid and methoxypyrazines typically coincides with the gradual cooling trend towards the end of the growing season. Owing to their effects on fruit composition discussed previously, high temperatures tend to make the winemaking process more expensive, because low-acid grape juice requires addition of tartaric acid for processing to enhance microbial stability and mouthfeel. White wines such as Riesling, which are preferred with relatively high acidity and commonly do not undergo malolactic fermentation, may suffer from too low a concentration of malate. A titratable acidity of 6.5–8.5 g/L is considered optimal for the production of well-balanced wines (Conde et al. 2007). High temperatures may impact red wine quality, because (daytime) temperature, rather than light, appears to be the main driver of anthocyanin accumulation (Spayd et al. 2002, Tarara et al. 2008). Because the total amount and relative proportion of anthocyanins ultimately determine the colour potential of red wines, heat waves during the ripening phase tend to be detrimental to wine colour. Whereas warmer growing seasons may be associated with higher amino acid concentrations (Pereira et al. 2006), there is some preliminary evidence to suggest that the opposite may be true for vitamins necessary for yeast metabolism during fermentation (Hagen et al. 2008). Although the net impact of changing amino acid and vitamin concentrations is currently unknown, it may be speculated that high temperatures during grape ripening may be associated with a greater risk of incomplete fermentations. On the other hand, a rise in average growing-season temperatures should result in a lower incidence of wines with 'veggie', 'herbaceous' notes, because warmer temperatures depress methoxypyrazine accumulation and enhance their degradation (Lacey et al. 1991).

Because ripening grape berries are designed to minimise transpirational water loss (Radler 1965, Possingham et al. 1967, Blanke et al. 1999, Rogiers et al. 2004), they cannot take advantage of the evaporative cooling mechanism that protects leaves from overheating. Thus, while high temperatures tend to accelerate grape ripening, too much heat can inhibit or even denature berry proteins, and may lead to symptoms of sunburn. On balance, it seems reasonable to predict that the quality of red grapes (especially of late ripening cultivars such as Cabernet Sauvignon) in cooler regions stands to benefit from the projected warming trend of 1–2°C by 2020 and of 2–3°C by 2050, whereas the opposite may be true for warmer regions and for many white grapes (cf. Webb et al. 2008a,b). On the other hand, with ripening occurring at warmer temperatures, there may be an increased need for irrigation to prevent shrinkage of grape berries through water loss.

A heightened risk of wildfires (Overpeck et al. 1990) threatens not only established vineyards and winery facilities (and other human developments), but may also influence grape and wine quality. Exposure of grapes to smoke from wildfires may lead to smoke taint, imparting an off-odour to wine, which is already a concern in Australia (Kennison et al. 2008) and could also become a problem in other dry regions around the world.

9.8 CHALLENGES AND OPPORTUNITIES: MITIGATING CULTURAL PRACTICES

A recent report raised concerns that the projected increase in the frequency of hot (>35°C) summer days might compromise and eventually eliminate wine grape production in warm areas of the USA, with production partly shifting to cooler areas (White et al. 2006). This fear seems tenuous, given the stunning success of the Australian wine industry over the past 20 years. Yields of both red and white *V. vinifera* cultivars have increased there significantly during the last two decades of the 20th century and have since levelled off in both warm and hot regions (Dry and Coombe 2004), while the total vineyard area for wine grapes has almost tripled. Nonetheless, Webb et al. (2008b) reported significant negative correlations between grape prices and mean summer temperatures across Australian wine regions. For comparison, average prices for Cabernet Sauvignon grapes from California's Napa Valley exceeded $4100/ton in 2006, while those from the Central Valley sold for ~$260/ton; the latter region having a 2.7°C higher mean annual temperature than the former (Cahill and Field 2008). Although the potential decline in prices may be partly, but by no means completely, compensated by possible increases in yield, these findings are important because, as discussed previously, by the middle of the 21st century the projected warming trend will shift many warm regions closer to climatic conditions currently experienced in hot regions.

Although hot extremes and heat waves are set to become more frequent over the course of this century (IPCC 2007), the most imminent challenges facing the wine, table grape and raisin industries in arid and semi-arid regions are probably not heat waves per se, but increasing drought and salinity because of higher evaporation coupled with declining water availability (Schultz 2000, Stevens and Walker 2002). Rising salinisation of soils could pose a serious threat to grape growing, because most irrigated vineyards, especially deficit-irrigated vineyards, are at risk from salinisation owing to dissolved salts in irrigation water (in contrast to rain water). Salinity limits vine growth, photosynthesis, productivity, and fruit quality (Downton and Loveys 1978, Walker et al. 1981, Downton 1985, Shani et al. 1993, Cramer et al. 2007). Mitigation practices include abundant watering at the end of each season to leach salts down the soil profile (provided fresh water is available), application of straw or other mulch to limit evaporation, and less soil tilling to conserve soil structure. In addition, some rootstocks derived from American Vitis species (e.g. Ramsey, 1103 Paulsen, Ruggeri 140, 101–14) are relatively tolerant of saline conditions (Downton 1985, Stevens and Walker 2002). However, this tolerance may decrease with prolonged salt exposure.

Sun-exposed grape berries are often subject to sunburn and subsequent shrivelling as a consequence of overheating and excess UV and/or visible light. The projected increase in the frequency of hot summer days will undoubtedly exacerbate this problem, especially on the afternoon side of canopies (Spayd et al. 2002). This may require adaptations in row direction (e.g. away from the prevailing north-south orientation) and alterations in trellis design and training systems (e.g. away from the relatively common current practice of manually positioning shoots vertically upward toward 'sprawl' systems without shoot positioning). Sprawl systems are cheaper to construct: often, only one wire is required to support the permanent cordon, sometimes with the addition of one pair of 'foliage' wires to prevent excessive wind damage, which contrasts with the multiple wires necessary for vertical shoot-positioning systems. In addition, changes in cultural practices may include less shoot positioning and less leaf removal in the fruit zone, which would also reduce labour costs. Other practices may include the use of cover crops or resident vegetation to improve the canopy microclimate through their cooling effect (Nazrala 2007), or installation of undervine or overhead sprinkler systems for evaporative canopy cooling. Both of these approaches would also tend to reduce soil temperature and limit daily thermal amplitudes in the root zone (Pradel and Pieri 2000). However, such practices will not only increase overall vineyard water use but also make grape production more expensive (Tesic et al. 2007, Celette et al. 2009). Maintaining a green cover crop throughout the growing season in dry regions typically requires installation of additional irrigation hardware, such as micro-sprinklers. Moreover, cover crops compete with grapevines for water and nutrients, especially in warm/dry regions, so that vineyard fertiliser requirements may increase if vine productivity is to be maintained (Keller et al. 2001, Keller 2005, Tesic et al. 2007, Celette et al. 2009). Such mitigating practices notwithstanding, excessive sunburn might lead to susceptible cultivars becoming unsuitable for planting in warmer regions, especially those that also experience high solar radiation during the growing season.

A relatively simple strategy for wine grape growers to delay fruit maturation such that it occurs during the cooler end of the season would be to markedly increase the crop load carried by the vines. In the Napa and Sonoma Valleys of California, increasing yields have been accompanied by better wine quality because of an asymmetric warming trend (at night and in spring) after 1950 (Nemani et al. 2001). However, while this may be an attractive option for growers, it is unpopular among winemakers and is often forbidden by law in Europe. Moreover, larger crops would tend to offset gains in irrigation water savings arising from better water-use efficiency.

In many areas, the consequences for wine grape production of the projected decline in irrigation water availability may be relatively minor owing to their already low water use met by drip irrigation, usually combined with deficit irrigation strategies (Dry et al. 2001, Kriedemann and Goodwin 2003, Keller 2005). In the early 2000s, ultra-premium-quality Cabernet Sauvignon grapes were grown in eastern Washington, USA, with an annual water supply from both rainfall and drip irrigation of as little as 308 mm (Keller et al. 2008). This contrasts with table, raisin and juice grapes, whose larger canopies and heavier crops require substantially more water. For example, well-watered Concord grapes may use as much as three times more water than deficit-irrigated red wine grapes (Tarara and Ferguson 2006). Moreover, many non-wine-grape growers still supply water by flood, furrow or overhead-sprinkler irrigation, methods that are inherently far less water-efficient than is drip irrigation. For the most part, these vineyards will have to be converted to drip irrigation to conserve water. Although this will put an additional short-term financial burden on growers, there may be savings in the longer term, because labour costs for operation and maintenance tend to be lower with drip irrigation.

Because grape cultivars differ in their suitability for and adaptability to different climates, shifts in the cultivar profile of different regions, and possibly the emergence of hitherto unsuitable lesser-known or even novel cultivars, can be expected over the coming decades. A shift of grape production to cooler regions of the world, i.e. towards higher latitudes and altitudes, is another likely scenario as a result of global warming (cf. Schultz 2000). However, such shifts imply that some vineyards located in the warmest and/or driest regions may be abandoned, which has implications for the quality of life in rural areas. Moreover, vineyard development in novel areas is dependent on the availability of affordable land, irrigation water and labour force. It will also require substantial investments in infrastructure and vineyard establishment. With the typical life of a vineyard exceeding 30 years, decisions on cultivars, clones, rootstocks, and vineyard sites will have to be made on a long-term basis. Moreover, vineyards that are planted now will experience an essentially new climate 20 years from now (Cahill and Field 2008), making such decisions challenging. An additional issue that has to be taken into account is harvest logistics: because grapes ripen more rapidly in warmer climates, the 'harvest window' tends to be more compressed, so that grape intake to accommodate 'optimum' maturity for different cultivars may pose scheduling, labour and capacity problems for growers and wineries alike.

9.9 OPPORTUNITIES AND PRIORITIES: FUTURE RESEARCH

From the above-mentioned overview of published information, one may conclude that grapevine reproductive development has an optimum temperature range from about 20 to 30°C, with temperatures below 15°C and above 35°C leading to marked reductions in yield formation and fruit ripening. But this conclusion has an important caveat: it is not clear whether this temperature range applies to ambient or to tissue temperatures. Most studies attempting to uncover temperature effects were conducted indoors, often in growth chambers, where tissue temperatures typically equal 'room' temperatures. In contrast, plant tissues exposed to sunlight normally are heated above ambient by solar radiation but fall below ambient at night. One elegant study conducted with field-grown vines avoided this pitfall by heating the measured berry-skin temperature of shaded bunches to the berry-skin temperature of sun-exposed bunches and cooling exposed bunches to the temperature of shaded bunches, thereby also separating the potential effects of temperature from those of light (Spayd et al. 2002, Tarara et al. 2008). Without trying to diminish the value of growth-chamber studies, it is probably fair to ask for more such innovative experiments that manipulate temperature and/or light in the field.

The expected increase in climate variation (IPCC 2007) flies in the face of growers' attempts to minimise spatial and annual variation in grape yield and quality. This is a concern, for a recent analysis with Cabernet Sauvignon wines from California's Napa Valley found that wine prices were closely related to seasonal weather between 1970 and 2004 (Ramirez 2009). Studies aimed at understanding the consequences of climatic change and variability are crucial for the many regional wine industries to remain competitive. The only free air CO_2 enrichment study conducted with grapevines thus far (Bindi et al. 2001, 2005, Tognetti et al. 2005) found increases of 40–50% in both vegetative and reproductive biomass with little change in fruit and wine composition. The authors concluded that rising atmospheric [CO_2] may strongly stimulate vine growth and productivity while not affecting fruit and wine quality. One might add that 'no effect' also implies that there may not be any beneficial effects on wine quality. Yet it is puzzling that to date not a single study has investigated the interactive effects on grapevines of the predicted simultaneous rise in temperature and atmospheric [CO_2]. In spite of the obvious importance for the global wine industry, we do not know how rising [CO_2] influences the widely studied effects (see previous discussion) of temperature variation and water supply on vine growth, phenology, yield formation, fruit ripening and composition and, ultimately, wine quality. Such studies, conducted over long periods (multiple years), are critical to enable development of future mitigation strategies and to test cultivar suitability in a changing climate. Growers will require knowledge to choose from among alternative options to prepare for warmer growing seasons with less water and, in some areas, increasingly saline soils.

One option is the choice of better-adapted planting material, but this requires a coordinated approach to evaluating alternative cultivars, clones, and rootstocks in a range of climates. With the roughly 500 million bases of the *V. vinifera* genome now sequenced (Jaillon et al. 2007, Velasco et al. 2007) and progress in genomics adding to our understanding of the function of important genes (e.g. Terrier et al. 2005, Cramer et al. 2007, Deluc et al. 2007, Pilati et al. 2007), the new tools of the '-omics age' (functional genomics, proteomics, metabolomics, etc.) will need to be put to use to investigate in more detail the developmental and environmental regulation of yield formation, fruit development and ripening. They will also need to be integrated with more general grape physiology and viticulture research. And perhaps it is time to begin developing genetically modified cultivars that will be able not only to cope with warmer temperature, higher $[CO_2]$ and less water of higher salinity, but will also produce high-quality fruit under such conditions. Alas, we continue to have a poor understanding of the concept of fruit and wine quality. Professional judges do not agree or do not consistently recognise wine quality (Hodgson 2008), and there is no consensus on what constitutes quality-relevant or quality-impact compounds in grapes. The identification and definition of such key components is critical for better vineyard management and harvest decisions to produce grapes according to end-use specifications. Clear specifications will enhance the ability to differentiate wines and other nutritionally valuable grape-related products according to consumer demand.

Soil management will have to take its place alongside canopy management as a key component of the sustainable vineyard management 'toolbox' (Keller 2005). This calls for research into the integration of appropriate and refined (deficit) irrigation techniques with vine nutrition, salinity management and vineyard floor management to optimise vine productivity, maximise fruit quality and ensure long-term soil fertility, perhaps in conjunction with enhanced carbon sequestration (e.g. Morlat and Chaussod

2008). Crop load and canopy management should be fine-tuned according to the desired end-use of the grapes from a particular vineyard block. This will not only require more quantitative assessments of interactions among treatment combinations, but also integration with precision viticulture approaches, and adaptations of trellis designs to facilitate mechanisation. Quantitative data will need to be incorporated into models, and hard- and software, including (remote) sensor technology, will need to be developed for mechanisation and, ultimately, automation of cultural practices and vineyard sampling that are fully integrated with real-time decision-management support systems and precision viticulture technology.

Interdisciplinary research conducted by teams of various combinations of molecular biologists, physiologists, viticulturists, oenologists, sensory scientists, chemists, physicists, mathematicians, computer scientists and economists will have to tackle these issues. In addition to the undisputed need for extension of applied research results to industry, there is also a requirement for fundamental research that can be used as a basis to develop practical outcomes. It should be clear that a one-sided focus on applied, practical research that promises short-term returns on investment will eventually deplete the novel ideas that give rise to unforeseen applicable questions.

REFERENCES

1.   Adams, D.O. (2006) Phenolics and ripening in grape berries. American Journal of Enology and Viticulture 57, 249–256.
2.   Agaoglu, Y.S. (1971) A study on the differentiation and the development of floral parts in grapes (Vitis vinifera L. var.). Vitis 10, 20–26.
3.   Ainsworth, E.A. and Rogers, A. (2007) The response of photosynthesis and stomatal conductance to rising [CO2]: mechanisms and environmental interactions. Plant, Cell and Environment 30, 258–270.
4.   Alleweldt, G. (1966) Die Differenzierung der Blütenorgane der Rebe. Wein-Wissenschaft 21, 393–402.
5.   Alleweldt, G. and Hofäcker, W. (1975) Einfluss von Umweltfaktoren auf Austrieb, Blüte, Fruchtbarkeit und Triebwachstum bei der Rebe. Vitis 14, 103–115.
6.   Alleweldt, G. and Ilter, E. (1969) Untersuchungen über die Beziehungen zwischen Blütenbildung und Triebwachstum bei Reben. Vitis 8, 286–313.
7.   Alleweldt, G. and Possingham, J.V. (1988) Progress in grapevine breeding. Theoretical and Applied Genetics 75, 669–673.

8. Alleweldt, G., Düring, H. and Jung, K.H. (1984) Zum Einfluss des Klimas auf Beerenentwicklung, Ertrag und Qualität bei Reben: Ergebnisse einer siebenjährigen Faktorenanalyse. Vitis 23, 127–142.

9. Amerine, M.A. and Winkler, A.J. (1944) Composition and quality of musts and wines of California grapes. Hilgardia 15, 493–675.

10. Antcliff, A.J. and May, P. (1961) Dormancy and bud burst in Sultana vines. Vitis 3, 1–14.

11. Baumes, R., Wirth, J., Bureau, S., Gunata, Y. and Razungles, A. (2002) Biogeneration of C13-norisoprenoid compounds: experiments supportive for an apo-carotenoid pathway in grapevines. Analytica Chimica Acta 458, 3–14.

12. Bell, S.J. and Henschke, P.A. (2005) Implications of nitrogen nutrition for grapes, fermentation and wine. Australian Journal of Grape and Wine Research 11, 242–295.

13. Bennett, J., Jarvis, P., Creasy, G.L. and Trought, M.C.T. (2005) Influence of defoliation on overwintering carbohydrate reserves, return bloom, and yield of mature Chardonnay grapevines. American Journal of Enology and Viticulture 56, 386–393.

14. Bindi, M., Fibbi, L., Gozzini, B., Orlandini, S. and Miglietta, F. (1996) Modelling the impact of future climate scenarios on yield and yield variability of grapevine. Climate Research 7, 213–224.

15. Bindi, M., Fibbi, L. and Miglietta, F. (2001) Free Air CO2 Enrichment (FACE) of grapevine (Vitis vinifera L.): II. Growth and quality of grape and wine in response to elevated CO2 concentrations. Europen Journal of Agronomy 14, 145–155.

16. Bindi, M., Raschi, A., Lanini, M., Maglietta, F. and Tognetti, R. (2005) Physiological and yield response of grapevine (Vitis vinifera L.) exposed to elevated CO2 concentrations in a Free Air CO2 Enrichment (FACE). Journal of Crop Improvement 13, 345–359.

17. Blanke, M.M., Pring, R.J. and Baker, E.A. (1999) Structure and elemental composition of grape berry stomata. Journal of Plant Physiology 154, 477–481.

18. Bowen, P.A. and Kliewer, W.M. (1990) Relationships between the yield and vegetative characteristics of individual shoots of 'Cabernet Sauvignon' grapevines. Journal of the American Society for Horticultural Science 115, 534–539.

19. Buttrose, M.S. (1970) Fruitfulness in grape-vines: the response of different cultivars to light, temperature and daylength. Vitis 9, 121–125.

20. Buttrose, M.S. (1974) Climatic factors and fruitfulness in grapevines. Horticultural Abstracts 44, 319–325.

21. Cahill, K.N. and Field, C.B. (2008) Future of the wine industry: climate change science. Practical Winery and Vineyard 29, 16, 18, 20–22, 24–26, 28–30, 32–33.

22. Candolfi-Vasconcelos, M.C. and Koblet, W. (1990) Yield, fruit quality, bud fertility and starch reserves of the wood as a function of leaf removal in Vitis vinifera – Evidence of compensation and stress recovering. Vitis 29, 199–221.

23. Carmona, M.J., Chaïb, J., Martínez-Zapater, J.M. and Thomas, M.R. (2008) A molecular genetic perspective of reproductive development in grapevine. Journal of Experimental Botany 59, 2579–2596.

24. Castellarin, S.D., Matthews, M.A., Di Gaspero, G. and Gambetta, G.A. (2007a) Water deficits accelerate ripening and induce changes in gene expression regulating flavonoid biosynthesis in grape berries. Planta 227, 101–112.

25. Castellarin, S.D., Pfeiffer, A., Sivilotti, P., Degan, M., Peterlunger, E. and Di Gaspero, G. (2007b) Transcriptional regulation of anthocyanin biosynthesis in ripening fruits of grapevine under seasonal water deficit. Plant, Cell and Environment 30, 1381–1399.

26. Celette, F., Findeling, A. and Gary, C. (2009) Competition for nitrogen in an unfertilised intercropping system: the case of an association of grapevine and grass cover in a Mediterranean climate. Europen Journal of Agronomy 30, 41–51.

27. Cerpa-Calderón, F.K. and Kennedy, J.A. (2008) Berry integrity and extraction of skin and seed proanthocyanidins during red wine fermentation. Journal of Agricultural and Food Chemistry 56, 9006–9014.

28. Cheng, L., Xia, G. and Bates, T. (2004) Growth and fruiting of young 'Concord' grapevines in relation to reserve nitrogen and carbohydrates. Journal of the American Society for Horticultural Science 129, 660–666.

29. Chuine, I., Yiou, P., Viovy, N., Seguin, B., Daux, V. and Le Roy Ladurie, E. (2004) Grape ripening as a past climate indicator. Nature 432, 289–290.

30. Conde, C., Silva, P., Fontes, N., Dias, A.C.P., Tavares, R.M., Sousa, M.J., Agasse, A., Delrot, S. and Gerós, H. (2007) Biochemical changes throughout grape berry development and fruit and wine quality. Food (Global Science Books) 1, 1–22.

31. Considine, J.A. and Knox, R.B. (1981) Tissue origins, cell lineages and patterns of cell division in the developing dermal system of the fruit of Vitis vinifera L. Planta 151, 403–412.

32. Coombe, B.G. (1962) The effect of removing leaves, flowers and shoot tips on fruit-set in Vitis vinifera L. Journal of Horticultural Science 37, 1–15.

33. Coombe, B.G. (1976) The development of fleshy fruits. Annual Review of Plant Physiology 27, 507–528.

34. Cortell, J.M. and Kennedy, J.A. (2006) Effect of shading on accumulation of flavonoid compounds in (Vitis vinifera L.) Pinot noir fruit and extraction in a model system. Journal of Agricultural and Food Chemistry 54, 8510–8520.

35. Cramer, G.R., Ergül, A., Grimplet, J., Tillett, R.L., Tattersall, E.A.R., Bohlman, M.C., Vincent, D., Sonderegger, J., Evans, J., Osborne, C., Quilici, D., Schlauch, K.A., Schooley, D.A. and Cushman, J.C. (2007) Water and salinity stress in grapevines: early and late changes in transcript and metabolite profiles. Functional and Integrative Genomics 7, 111–134.

36. Darwin, C. (1859) On the origin of species by means of natural selection, or the preservation of favoured races in the struggle for life (John Murray: London).

37. Deluc, L.G., Grimplet, J., Wheatley, M.D., Tillett, R.L., Quilici, D.R., Osborne, C., Schooley, D.A., Schlauch, K.A., Cushman, J.C. and Cramer, G.R. (2007) Transcriptomic and metabolite analysis of Cabernet Sauvignon grape berry development. BMC Genomics 8, 429, doi:10.1186/1471-2164-8-429.

38. Dokoozlian, N.K. (1999) Chilling temperature and duration interact on the budbreak of Perlette grapevine cuttings. HortScience 34, 1054–1056.

39. Dokoozlian, N.K. and Kliewer, W.M. (1995a) The light environment within grapevine canopies. I. Description and seasonal changes during fruit development. American Journal of Enology and Viticulture 46, 209–218.

40. Dokoozlian, N.K. and Kliewer, W.M. (1995b) The light environment within grapevine canopies. II. Influence of leaf area density on fruit zone light environment and

some canopy assessment parameters. American Journal of Enology and Viticulture 46, 219–226.

41. Downey, M.O., Dokoozlian, N.K. and Krstic, M.P. (2006) Cultural practice and environmental impacts on the flavonoid composition of grapes and wine: a review of recent research. American Journal of Enology and Viticulture 57, 257–268.

42. Downton, W.J.S. (1985) Growth and mineral composition of the Sultana grapevine as influenced by salinity and rootstock. Australian Journal of Agricultural Research 36, 425–434.

43. Downton, W.J.S. and Loveys, B.R. (1978) Compositional changes during grape berry development in relation to abscisic acid and salinity. Australian Journal of Plant Physiology 5, 415–423.

44. Dry, P.D. (2000) Canopy management for fruitfulness. Australian Journal of Grape and Wine Research 6, 109–115.

45. Dry, P.R. and Coombe, B.G. (2004) Viticulture. Volume 1 – Resources, 2nd edn. (Winetitles: Adelaide).

46. Dry, P.R., Loveys, B.R., McCarthy, M.G. and Stoll, M. (2001) Strategic irrigation management in Australian vineyards. Journal International des Sciences de la Vigne et du Vin 35, 45–61.

47. Duchêne, E. and Schneider, C. (2005) Grapevine and climatic changes: a glance at the situation in Alsace. Agronomy for Sustainable Development 25, 93–99.

48. Dunn, G.M. and Martin, S.R. (2000) Do temperature conditions at budburst affect flower number in Vitis vinifera L. cv. Cabernet Sauvignon? Australian Journal of Grape and Wine Research 6, 116–124.

49. Düring, H. (2003) Stomatal and mesophyll conductances control CO_2 transfer to chloroplasts in leaves of grapevine (Vitis vinifera L.). Vitis 42, 65–68.

50. Düring, H. and Davtyan, A. (2002) Developmental changes of primary processes of photosynthesis in sun- and shade-adapted berries of two grapevine cultivars. Vitis 41, 63–67.

51. Diaz-Riquelme, J., Lijavetzky, D., Martínez-Zapater, J.M. and Carmona, M.J. (2009) Genome-wide analysis of MIKCC-type MADS box genes in grapevine. Plant Physiology 149, 354–369.

52. Fang, Y. and Qian, M.C. (2006) Quantification of selected aroma-active compounds in Pinot noir wines from different grape maturities. Journal of Agricultural and Food Chemistry 54, 8567–8573.

53. Fennell, A. (2004) Freezing tolerance and injury in grapevines. In: Adaptations and responses of woody plants to environmental stresses. Ed. R.Arora (Food Products Press, imprint of Haworth Press: Binghamton, New York) pp. 201–235.

54. Gerrath, J.M. (1992) Developmental morphology and anatomy of grape flowers. Horticultural Reviews 13, 315–337.

55. Gillaspy, G., Ben-David, H. and Gruissem, W. (1993) Fruits: a developmental perspective. The Plant Cell 5, 1439–1451.

56. Goff, S.A. and Klee, H.J. (2006) Plant volatile compounds: sensory cues for health and nutritional value? Science 311, 815–819.

57. Hagen, K.M., Keller, M. and Edwards, C.G. (2008) Survey of biotin, pantothenic acid, and assimilable nitrogen in wine grapes from the Pacific Northwest. American Journal of Enology and Viticulture 59, 432–436.

58. Hale, C.R. and Buttrose, M.S. (1974) Effect of temperature on ontogeny of berries of Vitis vinifera L. cv. Cabernet Sauvignon. Journal of the American Society for Horticultural Science 99, 390–394.

59. Harbertson, J.F., Hodgins, R.E., Thurston, L.N., Schaffer, L.J., Reid, M.S., Landon, J.L., Ross, C.F. and Adams, D.O. (2008) Variability of tannin concentration in red wines. American Journal of Enology and Viticulture 59, 210–214.

60. Hardie, W.J. (2000) Grapevine biology and adaptation to viticulture. Australian Journal of Grape and Wine Research 6, 74–81.

61. Hardie, W.J., Aggenbach, S.J. and Jaudzems, V.G. (1996) The plastids of the grape pericarp and their significance in isoprenoid synthesis. Australian Journal of Grape and Wine Research 2, 144–154.

62. Harris, J.M., Kriedemann, P.E. and Possingham, J.V. (1968) Anatomical aspects of grape berry development. Vitis 7, 106–119.

63. Heazlewood, J.E. and Wilson, S. (2004) Anthesis, pollination and fruitset in Pinot Noir. Vitis 43, 65–68.

64. Hennessy, K.J. and Pittock, A.B. (1995) The potential impact of greenhouse warming on threshold temperature events in Victoria. International Journal of Climatology 15, 591–692.

65. Hodgson, R.T. (2008) An examination of judge reliability at a major U.S. wine competition. Journal of Wine Economics 3, 105–113.

66. Hofäcker, W., Alleweldt, G. and Khader, S. (1976) Einfluss der Umweltfaktoren auf Beerenwachstum und Mostqualität bei der Rebe. Vitis 15, 96–112.

67. Holzapfel, B.P., Smith, J.P., Mandel, R.M. and Keller, M. (2006) Manipulating the postharvest period and its impact on vine productivity of Semillon grapevines. American Journal of Enology and Viticulture 57, 148–157.

68. IPCC (2007) Climate Change 2007: The Physical Science Basis. Summary for Policymakers. http://www.ipcc.ch [accessed 27/07/09.

69. Jackson, D.I. (1991) Environmental and hormonal effects on development of early bunch stem necrosis. American Journal of Enology and Viticulture 42, 290–294.

70. Jackson, D.I. and Coombe, B.G. (1988) Early bunchstem necrosis in grapes – A cause of poor fruit set. Vitis 27, 57–61.

71. Jackson, D.I. and Lombard, P.B. (1993) Environmental and management practices affecting grape composition and wine quality – A review. American Journal of Enology and Viticulture 44, 409–430.

72. Jaillon, O., Aury, J.M., Noel, B., Policriti, A., Clepet, C., Casagrande, A., Choisne, N., Aubourg, S., Vitulo, N., Jubin, C., Vezzi, A., Legeai, F., Hugueney, P., Dasilva, C., Horner, D., Mica, E., Jublot, D., Poulain, J., Bruyère, C., Billault, A., Segurens, B., Gouyvenoux, M., Ugarte, E., Cattonaro, F., Anthouard, V., Vico, V., Del Fabbro, C., Alaux, M., Di Gaspero, G., Dumas, V., Felice, N., Paillard, S., Juman, I., Moroldo, M., Scalabrin, S., Canaguier, A., Le Clainche, I., Malacrida, G., Durand, E., Pesole, G., Laucou, V., Chatelet, P., Merdinoglu, D., Delledonne, M., Pezzotti, M., Lecharny, A., Scarpelli, C., Artiguenave, F., Pè, M.E., Valle, G., Morgante, M., Caboche, M., Adam-Blondon, A.F., Weissenbach, J., Quétier, F. and Wincker, P. for The French-Italian Public Consortium for Grapevine Genome Characterization (2007) The grapevine genome sequence suggests ancestral hexaploidization in major angiosperm phyla. Nature 449, 463–468.

73. Jeong, S.T., Goto-Yamamoto, N., Kobayashi, S. and Esaka, M. (2004) Effects of plant hormones and shading on the accumulation of anthocyanins and the expression of anthocyanin biosynthetic genes in grape berry skins. Plant Science 167, 247–252.

74. Jones, G.V. and Davis, R.E. (2000) Climate influences on grapevine phenology, grape composition, and wine production and quality for Bordeaux, France. American Journal of Enology and Viticulture 51, 249–261.

75. Jones, G.V., White, M.A., Cooper, O.R. and Storchmann, K. (2005) Climate change and global wine quality. Climatic Change 73, 319–343.

76. Kassemeyer, H.H. and Staudt, G. (1981) Über die Entwicklung des Embryosacks und die Befruchtung der Reben. Vitis 20, 202–210.

77. Kassemeyer, H.H. and Staudt, G. (1983) Über das Wachstum von Endosperm, Embryo und Samenanlagen von Vitis vinifera. Vitis 22, 109–119.

78. Keller, M. (2005) Deficit irrigation and vine mineral nutrition. American Journal of Enology and Viticulture 56, 267–283.

79. Keller, M. and Hrazdina, G. (1998) Interaction of nitrogen availability during flowering and light intensity during veraison: II. Effects on anthocyanin and phenolic development during grape ripening. American Journal of Enology and Viticulture 49, 341–349.

80. Keller, M. and Koblet, W. (1994) Is carbon starvation rather than excessive nitrogen supply the cause of inflorescence necrosis in Vitis vinifera L.? Vitis 33, 81–86.

81. Keller, M. and Koblet, W. (1995) Dry matter and leaf area partitioning, bud fertility and second-season growth of Vitis vinifera L.: responses to nitrogen supply and limiting irradiance. Vitis 34, 77–83.

82. Keller, M., Kummer, M. and Vasconcelos, M.C. (2001) Reproductive growth of grapevines in response to nitrogen supply and rootstock. Australian Journal of Grape and Wine Research 7, 12–18.

83. Keller, M., Mills, L.J., Wample, R.L. and Spayd, S.E. (2005) Bunch thinning effects on three deficit-irrigated Vitis vinifera cultivars. American Journal of Enology and Viticulture 56, 91–103.

84. Keller, M., Smith, J.P. and Bondada, B.R. (2006) Ripening grape berries remain hydraulically connected to the shoot. Journal of Experimental Botany 57, 2577–2587.

85. Keller, M., Smithyman, R.P. and Mills, L.J. (2008) Interactive effects of deficit irrigation and crop load on Cabernet Sauvignon in an arid climate. American Journal of Enology and Viticulture 59, 221–234.

86. Kennison, K.R., Gibberd, M.R., Pollnitz, A.P. and Wilkinson, K.L. (2008) Smoke-derived taint in wine: the release of smoke-derived volatile phenols during fermentation of Merlot juice following grapevine exposure to smoke. Journal of Agricultural and Food Chemistry 56, 7379–7383.

87. Kliewer, W.M. (1977) Effect of high temperatures during the bloom-set period on fruit-set, ovule fertility, and berry growth of several grape cultivars. American Journal of Enology and Viticulture 28, 215–222.

88. Kliewer, W.M. and Soleimani, A. (1972) Effect of chilling on budbreak in 'Thompson Seedless' and 'Carignane' grapevines. American Journal of Enology and Viticulture 23, 31–34.

89. Kobayashi, S., Goto-Yamamoto, N. and Hirochika, H. (2004) Retrotransposon-induced mutations in grape skin color. Science 304, 982.

90. Koblet, W. (1966) Fruchtansatz bei Reben in Abhängigkeit von Triebbehandlungen und Klimafaktoren. Wein-Wissenschaft 21, 297–321 and 346–379.

91. Koblet, W. (1969) Wanderung von Assimilaten in Rebtrieben und Einfluss der Blattfläche auf Ertrag und Qualität der Trauben. Wein-Wissenschaft 24, 277–319.

92. Koyama, K. and Goto-Yamamoto, N. (2008) Bunch shading during different developmental stages affects the phenolic biosynthesis in berry skins of 'Cabernet Sauvignon' grapes. Journal of the American Society for Horticultural Science 133, 743–753.

93. Krasnow, M., Matthews, M. and Shackel, K. (2008) Evidence for substantial maintenance of membrane integrity and cell viability in normally developing grape (Vitis vinifera L.) berries throughout development. Journal of Experimental Botany 59, 849–859.

94. Kriedemann, P.E. (1968b) Photosynthesis in vine leaves as a function of light intensity, temperature, and leaf age. Vitis 7, 213–220.

95. Kriedemann, P.E. (1968a) Observations on gas exchange in the developing sultana berry. Australian Journal of Biological Sciences 21, 907–916.

96. Kriedemann, P.E. and Goodwin, I. (2003) Regulated deficit irrigation and partial rootzone drying. An overview of principles and applications (Land and Water Australia: Canberra).

97. Lacey, M.J., Allen, M.S., Harris, R.L.N. and Brown, W.V. (1991) Methoxypyrazines in Sauvignon blanc grapes and wines. American Journal of Enology and Viticulture 42, 103–108.

98. Lakso, A.N. and Kliewer, W.M. (1975) The influence of temperature on malic acid metabolism in grape berries. I. Enzyme responses. Plant Physiology 56, 370–372.

99. Lavee, S. and May, P. (1997) Dormancy of grapevine buds – Facts and speculation. Australian Journal of Grape and Wine Research 3, 31–46.

100. Lebon, G., Wojnarowiez, G., Holzapfel, B., Fontaine, F., Vaillant-Gaveau, N. and Clément, C. (2008) Sugars and flowering in the grapevine (Vitis vinifera L.). Journal of Experimental Botany 59, 2565–2578.

101. Li, D., Wan, Y., Wang, Y. and He, P. (2008) Relatedness of resistance to anthracnose and to white rot in Chinese wild grapes. Vitis 47, 213–215.

102. Lobell, D.B., Field, C.B., Nicholas Cahill, K. and Bonfils, C. (2006) Impacts of future climate change on California perennial crop yields: model projections with climate and crop uncertainties. Agricultural and Forest Meteorology 141, 208–218.

103. Long, S.P., Ainsworth, E.A., Rogers, A. and Ort, D.R. (2004) Rising atmospheric carbon dioxide: plants FACE the future. Annual Review of Plant Biology 55, 591–628.

104. Manabe, S., Wetherald, R.T., Milly, P.C.D., Delworth, T.L. and Stouffer, R.J. (2004) Century-scale change in water availability: CO2-quadrupling experiment. Climate Change 64, 59–76.

105. Mastin, M.C. (2008) Effects of potential future warming on runoff in the Yakima River Basin, Washington. U.S. Geological Survey Scientific Investigations Report 2008-5124. http://pubs.usgs.gov[accessed 27/07/09.

106. Matthews, M.A., Cheng, G. and Weinbaum, S.A. (1987) Changes in water potential and dermal extensibility during grape berry development. Journal of the American Society for Horticultural Science 112, 314–319.

107. May, P. (2004) Flowering and fruitset in grapevines (Lythrum Press: Adelaide).

108. Meneghetti, S., Gardiman, M. and Calò, A. (2006) Flower biology of grapevine. A review. Advances in Horticultural Science 20, 317–325.

109. Mori, K., Goto-Yamamoto, N., Kitayama, M. and Hashizume, K. (2007) Loss of anthocyanins in red-wine grape under high temperature. Journal of Experimental Botany 58, 1935–1945.
110. Morlat, R. and Chaussod, R. (2008) Long-term additions of organic amendments in a Loire Valley vineyard. I. Effects on properties of a calcareous sandy soil. American Journal of Enology and Viticulture 59, 353–363.
111. Morrison, J.C. (1991) Bud development in Vitis vinifera L. Botanical Gazette 152, 304–315.
112. Mpelasoka, B.S., Schachtmann, D.P., Treeby, M.T. and Thomas, M.R. (2003) A review of potassium nutrition in grapevines with special emphasis on berry accumulation. Australian Journal of Grape and Wine Research 9, 154–168.
113. Nakagawa, S. and Nanjo, Y. (1965) A morphological study of Delaware grape berries. Journal of the Japanese Society for Horticultural Science 34, 85–95.
114. Nazrala, J.J.B. (2007) Microclima de la canopia de la vid: influencia del manejo del suelo y coberturas vegetales. Revista de la Facultad de Ciencias Agrarias Universidad Nacional de Cuyo 39, 1–13.
115. Nemani, R.R., White, M.A., Cayan, D.R., Jones, G.V., Running, S.W. and Coughlan, J.C. (2001) Asymmetric climatic warming improves California vintages. Climate Research 19, 25–34.
116. Nitsch, J.P., Pratt, C., Nitsch, C. and Shaulis, N.J. (1960) Natural growth substances in Concord and Concord Seedless grapes in relation to berry development. American Journal of Botany 47, 566–576.
117. Ollat, N., Diakou-Verdin, P., Carde, J.P., Barrieu, F., Gaudillère, J.P. and Moing, A. (2002) Grape berry development: a review. Journal International des Sciences de la Vigne et du Vin 36, 109–131.
118. Ortega-Regules, A., Romero-Cascales, I., López-Roca, J.M., Ros-García, J.M. and Gómez-Plaza, E. (2006) Anthocyanin fingerprint of grapes: environmental and genetic variations. Journal of the Science of Food and Agriculture 86, 1460–1467.
119. Overpeck, J.T., Rind, D. and Goldberg, R. (1990) Climate-induced changes in forest disturbance and vegetation. Nature 343, 51–53.
120. O'Neill, S.D.O. (1997) Pollination regulation of flower development. Annual Review of Plant Physiology and Plant Molecular Biology 48, 547–574.
121. Palliotti, A. and Cartechini, A. (2001) Developmental changes in gas exchange activity in flowers, berries, and tendrils of field-grown Cabernet Sauvignon. American Journal of Enology and Viticulture 52, 317–323.
122. Pereira, G.E., Gaudillere, J.P., Van Leeuwen, C., Hilbert, G., Maucourt, M., Deborde, C., Moing, A. and Rolin, D. (2006) 1H NMR metabolite fingerprints of grape berry: comparison of vintage and soil effects in Bordeaux grapevine growing areas. Analytica Chimica Acta 563, 346–352.
123. Petrie, P.R. and Clingeleffer, P.R. (2005) Effects of temperature and light (before and after budburst) on inflorescence morphology and flower number of Chardonnay grapevines (Vitis vinifera L.). Australian Journal of Grape and Wine Research 11, 59–65.
124. Pilati, S., Perazzolli, M., Malossini, A., Cestaro, A., Demattè, L., Fontana, P., Dal Ri, A., Viola, R., Velasco, R. and Moser, C. (2007) Genome-wide transcriptional analysis of grapevine berry ripening reveals a set of genes similarly modulated during

three seasons and the occurrence of an oxidative burst at veraison. BMC Genomics 8, 428, doi:10.1186/1471-2164-8-428.

125. Possingham, J.V., Chambers, T.C., Radler, F. and Grncarevic, M. (1967) Cuticular transpiration and wax structure and composition of leaves and fruit of Vitis vinifera. Australian Journal of Biological Sciences 20, 1149–1153.

126. Pouget, R. (1981) Action de la température sur la différenciation des inflorescences et des fleurs durant les phases de pré-débourrement et de post-débourrement des bourgeons latents de la vigne. Connaissance Vigne et Vin 15, 65–79.

127. Pradel, E. and Pieri, P. (2000) Influence of a grass layer on vineyard soil tempera-ture. Australian Journal of Grape and Wine Research 6, 59–67.

128. Pratt, C. (1971) Reproductive anatomy in cultivated grapes – A review. American Journal of Enology and Viticulture 22, 92–109.

129. Radler, F. (1965) Reduction of loss of moisture by the cuticle wax components of grapes. Nature 207, 1002–1003.

130. Ramirez, C. (2009) Wine quality, wine prices, and the weather: is Napa 'different'? Journal of Wine Economics 3, 114–131.

131. Razungles, A.J., Baumes, R.L., Dufour, C., Sznaper, C.N. and Bayonove, C.L. (1998) Effect of sun exposure on carotenoids and C-13-norisoprenoid glycosides in Syrah berries (Vitis vinifera L.). Sciences des Aliments 18, 361–373.

132. Rogiers, S.Y., Hatfield, J.M., Jaudzems, V.G., White, R.G. and Keller, M. (2004) Grape berry cv. Shiraz epicuticular wax and transpiration during ripening and preharvest weight loss. American Journal of Enology and Viticulture 55, 121–127.

133. Roujou de Boubée, D., Van Leeuwen, C. and Dubourdieu, D. (2000) Organoleptic impact of 2-methoxy-3-isobutylpyrazine on red Bordeaux and Loire wines. Effect of environmental conditions on concentrations in grapes during ripening. Journal of Agricultural and Food Chemistry 48, 4830–4834.

134. Sage, R.F. and Kubien, D.S. (2007) The temperature response of C3 and C4 photo-synthesis. Plant, Cell and Environment 30, 1086–1106.

135. Schnabel, B.J. and Wample, R.L. (1987) Dormancy and cold hardiness in Vitis vinif-era L. cv. White Riesling as influenced by photoperiod and temperature. American Journal of Enology and Viticulture 38, 265–272.

136. Schultz, H.R. (2000) Climate change and viticulture: a European perspective on cli-matology, carbon dioxide and UV-B effects. Australian Journal of Grape and Wine Research 6, 2–12.

137. Scienza, A., Miravalle, R., Visai, C. and Fregoni, M. (1978) Relationships between seed number, gibberellin and abscisic acid levels and ripening in Cabernet Sauvi-gnon grape berries. Vitis 17, 361–368.

138. Seem, R.C., Magarey, R.D., Zack, J.W. and Russo, J.M. (2000) Estimating disease risk at the whole plant level with General Circulation Models. Environmental Pol-lution 108, 389–395.

139. Shani, U., Waisel, Y., Eshel, A., Xue, S. and Ziv, G. (1993) Responses to salinity of grapevine plants with split root systems. New Phytologist 124, 695–701.

140. Smart, R.E. and Sinclair, T.R. (1976) Solar heating of grape berries and other spheri-cal fruits. Agricultural Meteorology 17, 241–259.

141. Spayd, S.E., Tarara, J.M., Mee, D.L. and Ferguson, J.C. (2002) Separation of sunlight and temperature effects on the composition of Vitis vinifera cv. Merlot berries. American Journal of Enology and Viticulture 53, 171–182.

142. Srinivasan, C. and Mullins, M.G. (1981) Physiology of flowering in the grapevine – a review. American Journal of Enology and Viticulture 32, 47–63.

143. Staudt, G. (1982) Pollenkeimung und Pollenschlauchwachstum in vivo bei Vitis und die Abhängigkeit von der Temperatur. Vitis 21, 205–216.

144. Staudt, G. (1999) Opening of flowers and time of anthesis in grapevines, Vitis vinifera L. Vitis 38, 15–20.

145. Staudt, G. and Kassrawi, M. (1973) Untersuchungen über das Rieseln di- und tetraploider Reben. Vitis 12, 1–15.

146. Stevens, R.M. and Walker, R.R. (2002) Response of grapevines to irrigation-induced saline–sodic soil conditions. Australian Journal of Experimental Agriculture 42, 323–331.

147. Swanepoel, J.J. and Archer, E. (1988) The ontogeny and development of Vitis vinifera L. cv. Chenin blanc inflorescence in relation to phenological stages. Vitis 27, 133–141.

148. Sánchez, L.A. and Dokoozlian, N.K. (2005) Bud microclimate and fruitfulness in Vitis vinifera L. American Journal of Enology and Viticulture 56, 319–329.

149. Tarara, J.M. and Ferguson, J.C. (2006) Two algorithms for variable power control of heat-balance sap flow gauges under high flow rates. Agronomy Journal 98, 830–838.

150. Tarara, J.M., Lee, J., Spayd, S.E. and Scagel, C.F. (2008) Berry temperature and solar radiation alter acylation, proportion, and concentration of anthocyanin in Merlot grapes. American Journal of Enology and Viticulture 59, 235–247.

151. Terrier, N., Glissant, D., Grimplet, J., Barrieu, F., Abbal, P., Couture, C., Ageorges, A., Atanassova, R., Léon, C., Renaudin, J.P., Dédaldéchamp, F., Romieu, C., Delrot, S. and Hamdi, S. (2005) Isogene specific oligo arrays reveal multifaceted changes in gene expression during grape berry (Vitis vinifera L.) development. Planta 222, 832 847.

152. Tesic, D., Keller, M. and Hutton, R.J. (2007) Influence of vineyard floor management practices on grapevine vegetative growth, yield, and fruit composition. American Journal of Enology and Viticulture 58, 1–11.

153. Thomas, T.R., Shackel, K.A. and Matthews, M.A. (2008) Mesocarp cell turgor in Vitis vinifera L. berries throughout development and its relation to firmness, growth, and the onset of ripening. Planta 228, 1067–1076.

154. Tilbrook, J. and Tyerman, S.D. (2008) Cell death in grape berries: varietal differences linked to xylem pressure and berry weight loss. Functional Plant Biology 35, 173–184.

155. Tognetti, R., Raschi, A., Longobucco, A., Lanini, M. and Bindi, M. (2005) Hydraulic properties and water relations of Vitis vinifera L. exposed to elevated CO2 concentrations in a free air CO2 enrichment (FACE). Phyton 45, 243–256.

156. Velasco, R., Zharkikh, A., Troggio, M., Cartwright, D.A., Cestaro, A., Pruss, D., Pindo, M., FitzGerald, L.M., Vezzulli, S., Reid, J., Malacarne, G., Iliev, D., Coppola, G., Wardell, B., Micheletti, D., Macalma, T., Facci, M., Mitchell, J.T., Perazzolli, M., Eldredge, G., Gatto, G., Oyzerski, R., Moretto, M., Gutin, N., Stefanini, M., Chen, Y., Segala, C., Davenport, C., Demattè, L., Mraz, A., Battilana, J., Stormo, K., Costa, F., Tao, Q., Si-Ammour, A., Harkins, T., Lackey, A., Perbost, C., Taillon, B., Stella, A., Solovyev, V., Fawcett, J.A., Sterck, L., Vandepoele, K., Grando, S.M., Toppo,

S., Moser, C., Lanchbury, J., Bogden, R., Skolnick, M., Sgaramella, V., Bhatnagar, S.K., Fontana, P., Gutin, A., Van de Peer, Y., Salamini, F. and Viola, R. (2007) A high quality draft consensus sequence of the genome of a heterozygous grapevine variety. PLoS ONE 2, e1326. doi:10.1371/journal.pone.0001326.

157. Walker, A.R., Lee, E., Bogs, J., McDavid, D.A.J., Thomas, M.R. and Robinson, S.P. (2007) White grapes arose through the mutation of two similar and adjacent regulatory genes. The Plant Journal 49, 772–785.

158. Walker, R.R., Törökfalvy, E., Scott, N.S. and Kriedemann, P.E. (1981) An analysis of photosynthetic response to salt treatment in Vitis vinifera. Australian Journal of Plant Physiology 8, 359–374.

159. Wan, Y., Schwaninger, H., He, P. and Wang, Y. (2007) Comparison of resistance to powdery mildew and downy mildew in Chinese wild grapes. Vitis 46, 132–136.

160. Wan, Y., Wang, Y., Li, D. and He, P. (2008) Evaluation of agronomic traits in Chinese wild grapes and screening superior accessions for use in a breeding program. Vitis 47, 153–158.

161. Webb, L., Whetton, P. and Barlow, E.W.R. (2008a) Climate change and wine grape quality in Australia. Climate Research 36, 99–111.

162. Webb, L., Whetton, P. and Barlow, E.W.R. (2008b) Modeling the relationship between climate, winegrape price, and winegrape quality in Australia. Climate Research 36, 89–98.

163. Webb, L.B., Whetton, P.H. and Barlow, E.W.R. (2006) Potential impacts of projected greenhouse gas-induced climate change on Australian viticulture. Australian and New Zealand Wine Industry Journal 21, 16–20.

164. Webb, L.B., Whetton, P.H. and Barlow, E.W.R. (2007) Modelled impact of future climate change on the phenology of winegrapes in Australia. Australian Journal of Grape and Wine Research 13, 165–175.

165. Westerling, A.L., Hidalgo, H.G., Cayan, D.R. and Swetnam, T.W. (2006) Warming and earlier spring increase western US forest wildfire activity. Science 313, 940–943.

166. White, M.A., Diffenbaugh, N.S., Jones, G.V., Pal, J.S. and Giorgi, F. (2006) Extreme heat reduces and shifts United States premium wine production in the 21st century. Proceedings of the National Academy of Sciences of the USA 103, 11217–11222.

167. Wilson, B., Strauss, C.R. and Williams, P.J. (1984) Changes in free and glycosidically bound monoterpenes in developing Muscat grapes. Journal of Agricultural and Food Chemistry 32, 919–924.

168. Winkler, A.J. and Williams, W.O. (1935) Effect of seed development on the growth of grapes. Proceedings of the American Society for Horticultural Science 33, 430–434.

169. Winterhalter, P. and Schreier, P. (1994) C13-Norisoprenoid glycosides in plant tissues: an overview on their occurrence, composition and role as flavour precursors. Flavour and Fragrance Journal 9, 281–287.

170. Wolfe, D.W., Schwartz, M.D., Lakso, A.N., Otsuki, Y., Pool, R.M. and Shaulis, N.J. (2005) Climate change and shifts in spring phenology of three horticultural woody perennials in northeastern USA. International Journal of Biometeorology 49, 303–309.

171. Van Leeuwen, C., Friant, P., Choné, X., Tregoat, O., Koundouras, S. and Dubourdieu, D. (2004) Influence of climate, soil, and cultivar on terroir. American Journal of Enology and Viticulture 55, 207–217.

CHAPTER 10

MOLECULAR STRATEGIES TO ENHANCE THE GENETIC RESISTANCE OF GRAPEVINES TO POWDERY MILDEW

I. B. DRY, A. FEECHAN, C. ANDERSON, A. M. JERMAKOW, A. BOUQUET, A.-F. ADAM-BLONDON, AND M. R. THOMAS

10.1 CONTROL OF FUNGAL DISEASES IS A MAJOR PROBLEM FOR VITICULTURE WORLDWIDE

The world wine industry is based on cultivation of the grape species, *Vitis vinifera*. However, this species is highly susceptible to a number of pathogens that can cause economically devastating diseases, including powdery mildew, caused by *Erysiphe necator* (syn. *Uncinula necator*), downy mildew (*Plasmopora viticola*) and Botrytis bunch rot (*Botrytis cinerea*). From a global perspective, powdery mildew is the most important of these three pathogens because it does not require specific humidity and temperature conditions for infection, as is the case with downy mildew and Botrytis infection. Consequently, although the severity of powdery mildew infection can vary from season to season, it is a constant threat to grapegrowers.

Dry IB, Feechan A, Anderson C, Jermakow AM, Bouquet A, Adam-Blondon A-F and Thomas MR. Molecular Strategies to Enhance the Genetic Resistance of Grapevines to Powdery Mildew. Australian Journal of Grape and Wine Research *16,S1 (2010), DOI: 10.1111/j.1755-0238.2009.00076.x. Reprinted with permission from the authors.*

Powdery mildew infects all green grapevine tissues including the leaves, stems and berries. Severe infection can debilitate vines, reduce net photosynthesis, retard ripening and degrade wine quality (Gadoury et al. 2001, Calonnec et al. 2004, Stummer et al. 2005). Even minor berry infections have been shown to be associated with elevated populations of spoilage microorganisms, increased infestation by insects and increased Botrytis bunch rot at harvest (Gadoury et al. 2007).

Modern grapevine cultivation relies heavily upon the use of chemical fungicides such as sulphur and sterol biosynthesis inhibitors to control this pathogen and most grapegrowers would apply between 6–10 fungicide sprays per season. In France alone, the cost of fungicides for powdery mildew control is around 75 million Euros per year (A. Bouquet, pers. comm., 2008) and this does not take into account the increasing fuel costs associated with the application of these fungicides. Furthermore, fungal strains have been reported that have evolved resistance to a number of the commonly used fungicides (Erickson and Wilcox 1997, Savocchia et al. 2004, Baudoin et al. 2008). There is also now increasing pressure by legislators to restrict the use of certain agrochemicals because they consider them to be detrimental to the environment and may pose a risk to human health. Indeed, the European Commission is currently considering a proposal to ban the use of a large number of agrochemicals in Europe by 2013 which are routinely used for the control of a number of fungal diseases in grapes including powdery mildew in grapes (Pesticides Safety Directorate 2008). Thus, the introduction of effective genetic resistance into winegrape cultivars to reduce the dependence of viticulture on chemical inputs would be of significant economic and environmental benefit.

Powdery mildews are obligate biotrophic pathogens, meaning that they require a living plant host for their growth and reproduction. Upon landing on the plant surface, powdery mildew spores (conidia) germinate to produce chains of elongated hyphal cells that grow across the surface of the plant. In order to obtain nutrients from the plant, the pathogen periodically produces a specialised structure called an appressorium that generates sufficient pressure to allow the fungus to insert a penetration peg through the cuticle and cell wall of a host epidermal cell. Once the cell wall has been breached, the pathogen invaginates the host cell membrane forming a multi-lobed structure called a 'haustorium' via which nutrients are obtained to allow the pathogen to reproduce and complete its life-cycle (Heintz and Blaich 1990, Glawe 2008).

The highly susceptible nature of *V. vinifera* to powdery mildew suggests that it lacks genetic mechanisms to protect itself from this pathogen. This is not surprising when one considers that *V. vinifera* is indigenous to Eurasia (This et al. 2006) whereas *E. necator* is endemic to North America and was only introduced into Europe in the 1840s, most probably on ornamental vines imported for use in European gardens. In evolutionary terms, therefore, *V. vinifera* has only been exposed to the *E. necator* pathogen for an extremely short period of time. In contrast, many grapevine species that are endemic to the USA display varying levels of resistance to powdery mildew (Pearson and Gadoury 1992) but these species lack the superior berry quality of the *V. vinifera* cultivars. Consequently, when attempts have been made to introduce powdery mildew resistance by conventional hybridisation between *V. vinifera* and North American *Vitis* sp., the resulting resistant hybrids have been generally unacceptable to growers and winemakers. The adoption of these 'French-American' hybrids has therefore been very limited throughout most viticultural regions of the world apart from eastern and mid-western USA (Pollefeys and Bousquet 2003). New molecular approaches are therefore required to introduce powdery mildew resistance into winegrape cultivars while maintaining wine quality.

10.2 CURRENT MODEL OF PLANT RESISTANCE TO PATHOGENS

Bent and Mackey (2007) recently published an excellent review in which they proposed a revised four-part model to describe our current understanding of disease resistance mechanisms in plants (Figure 1). One of the key features of this model is that it helps describe the link, in both functional and evolutionary terms, between what we now understand are the two major mechanisms plants use in defence against pathogen attack.

The first of these is referred to as non-host resistance or basal immunity. This describes resistance which is effective across an entire plant species against all known isolates of a pathogen species. In cellular or molecular terms this encompasses both the preformed physical and chemical barriers that plant tissues have to pathogen invasion and the inducible immune response that is triggered by the recognition of a microbe-associated molecular pattern (MAMP) by an extracellular receptor-like kinase (RLK). MAMPs can be considered as defence elicitors that are evolutionarily stable, forming a core component of the microorganism that generally cannot be sacrificed or altered significantly without seriously impairing viability (Bent and Mackey 2007). In the case of powdery mildew, this is likely to include the chitin which makes up fungal cell walls (Bittel and Robatzek 2007). The recognition of a MAMP by the RLK triggers a basal immune response (also called MAMP-triggered immunity) which comprises a variety of defence responses including cytoskeleton rearrangements, callose deposition and the induction of antimicrobial compounds. Plant pathologists have long been aware of the existence of non-host resistance, but until recently much of the information we possessed was descriptive in nature because it was not amenable to classical genetic analysis. However, large-scale forward genetic screens on Arabidopsis have now identified a number of key genes designated *PENETRATION1* (*PEN1*), *PEN2* and *PEN3* which play major roles in the basal immune response against powdery mildews (Collins et al. 2003, Lipka et al. 2005, Stein et al. 2006). Of particular interest is *PEN1* which encodes a syntaxin involved in membrane fusion events at the plasma membrane. This protein has been shown to redistribute within cells, under powdery mildew attack, to focus beneath the point of attempted penetration (Assaad et al. 2004). Furthermore, *PEN1* redistribution appears to be a unique response of plant cells to attempted penetration by powdery mildews and not by other fungal pathogens (Meyer et al. 2009). This suggests that *PEN1* may have a role in trafficking of secretory vesicles to the plasma membrane containing cargo that is required for penetration resistance against powdery mildews.

In stage two of the model (Figure 1), certain species of the pathogen become 'adapted' to the host through the evolution of effector molecules that actively suppress the basal immune response and manipulate plant cell functions to facilitate infection. In the case of grapevines, *E. necator* appears to be the only powdery mildew species which has become adapted on this plant host. While there is a wealth of knowledge relating to the identity of the multiple effectors released by bacteria and oomycetes into plant cells during infection (Bent and Mackey 2007), to date there is only one report of two effectors identified from the cereal powdery mildew *Blumeria graminis f. sp. hordei* (Ridout et al. 2006).

Stage three of the model (Figure 1) describes the initial stages of the plant-pathogen 'arms race', whereby plant hosts which are susceptible to adapted pathogens, evolve specific resistance (R) genes which enable the plant to detect the defence-suppressing effectors. This interaction initiates effector-triggered immune responses, thereby restoring resistance to the pathogen. Commonly, the result of defence activation involving R genes is the localised death of the infected cells, termed a hypersensitive response, which prevents the pathogen from obtaining nutrients and completing its life cycle (Mur et al. 2008). The vast majority of R genes encode proteins that contain leucine-rich repeats (LRRs), a central nucleotide-binding site (NBS) domain and a variable domain at the N-terminus, consisting of either a coiled-coil (CC) domain or a domain that has homology to the Toll/Interleukin-1 Receptor (TIR). With some exceptions, LRRs are mainly involved in recognition, the amino-terminal domain determines signalling specificity and the NBS domain appears to function as a molecular switch (Takken et al. 2006).

The final stage of the model describes a further evolution of the pathogen to avoid detection by the R protein by modifying or eliminating the effector(s) that triggers those defences (Figure 1). Clearly, any such changes in the pathogen effector must still be compatible with its role as a virulence factor or the changes will not be maintained. This effectively returns the plant-pathogen interaction back to Stage 2, except that the pathogen has had to alter or lose an effector protein and resistance will only be re-established through the introduction of another R gene that recognises the modified effector.

The major message from this model is that plants have evolved two overlapping defence pathways to detect and restrict the growth of invading pathogens. In this short review, we will focus on the different molecular strategies that we are currently employing in our laboratory to develop resistance to grapevine powdery mildew by manipulating these two defence pathways in susceptible grapevine cultivars.

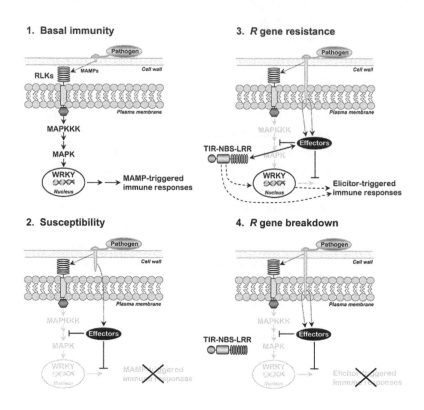

FIGURE 1: Model for the evolution of pathogen resistance in plants. This figure has been adapted from the model of Bent and Mackey (2007) using powdery mildew as the invading pathogen. Stage 1: Recognition of microbial-associated molecular patterns (MAMP; such as fungal chitin) by extracellular receptor-like kinases (RLKs) triggers basal immunity against non-adapted powdery mildew species, which includes signalling through MAP kinase cascades and transcriptional reprogramming mediated by plant WRKY transcription factors (proteins containing one or two highly conserved domains characterised by the heptapeptide WRYKGQK). Stage 2: Adapted powdery mildew species suppress basal immune responses possibly via release of effector molecules, allowing penetration of the cell wall. Stage 3: Plant resistance proteins (R gene products, such as TIR-NBS-LRR proteins: TIR, Toll/Interleukin-1 receptor domain; NBS, nucleotide-binding site domain; LRR, leucine-rich repeat domain) recognise effector activity and restore resistance through effector-triggered immune responses. Stage 4: Pathogen avoids R gene-mediated defences by modifying or eliminating the effector(s) that trigger those defences.

10.3 IDENTIFYING POWDERY MILDEW RESISTANCE GENES IN A WILD NORTH AMERICAN GRAPEVINE

10.3.1 GENETIC MAPPING AND POSITIONAL CLONING

The wild grape species *Muscadinia rotundifolia*, a native to the southeastern USA, is highly resistant to a number of pathogens known to affect cultivated grapevines, including powdery mildew, downy mildew, phylloxera and nematodes (Olmo 1986). *M. rotundifolia* is taxonomically separated from *Euvitis* species by anatomical and morphological characteristics (Planchon 1887) and a difference in chromosome number (Vitis 2n = 38, Muscadinia 2n = 40; Branas 1932). Not surprisingly, therefore, the first attempts to produce inter-specific crosses between *M. rotundifolia* and *V. vinifera* in the mid 1800s, were of only limited success. However, in 1919, L.R. Detjen, a grape breeder from North Carolina was successful in producing the first authentic hybrids between *V. vinifera* and *M. rotundifolia* using the *V. vinifera* cultivar Malaga as the female parent (Detjen 1919). One of the progeny, NC6-15, was used as the resistant parent in a series of pseudo-backcrosses with *V. vinifera* cultivars that led to the identification of a single, dominant locus designated *Run1* (Resistance to *Uncinula necator* 1) that provided complete resistance to powdery mildew (Bouquet 1986, Pauquet et al. 2001). Analysis of the *Run1*-mediated resistance response indicates that it involves induction of programmed cell death, specifically within the penetrated epidermal cell, approximately 24–48 h following powdery mildew infection (Figure 2). Later studies with the same populations also identified a second resistance locus, designated *Rpv1* (for Resistance to *Plasmopora viticola* 1) that provided partial resistance to downy mildew (Merdinoglu et al. 2003). Interestingly, *Run1* and *Rpv1* were found to co-segregate making it feasible to identify both resistance genes in a single map-based cloning effort.

Three back-cross populations Mtp3294 (BC4:VRH3082-1-42 × Cabernet Sauvignon), Mtp3328 (BC4:VRH3082-1-49 × Marselan) and Mtp3322 (BC5: VRH3176-21-11 × Cabernet Sauvignon) were used in genetic mapping studies to identify markers linked to the resistance loci (Pauquet et al. 2001, Barker et al. 2005). Using a bulked segregant analysis approach (Michelmore et al. 1991), a number of amplified fragment length polymorphism (AFLP) markers (Pauquet et al. 2001) and resistance gene analogue markers (RGAs; Donald et al. 2002) were identified that are linked to the *Run1/Rpv1* locus. RGA markers are based on the knowledge obtained from the cloning of a large number of plant disease resistance genes from plants, which indicate a high degree of structural conservation independent of the pathogen they confer resistance to. Highly conserved motifs within the NBS domain of R proteins (Takken et al. 2006) have been used to design degenerate primers capable of amplifying novel resistance gene analogues (RGAs) many of which have been found to map to resistance loci in a range of plant species (Aarts et al. 1998, Collins et al. 1998, Shen et al. 1998). Di Gaspero and Cipriani (2002) used a similar approach to clone a set of RGA sequences from the grape species *Vitis amurensis* and *Vitis riparia* which are known to show resistance to downy mildew and demonstrated that one of these RGA clones, rgVrip064, was linked to downy mildew resistance in segregating hybrid populations. In all a total of 14 AFLP, RGA and simple sequence repeat (SSR) markers were identified that showed linkage to the *Run1/Rpv1* resistance locus in the Mtp3294 and Mtp3328 populations (Barker et al. 2005). Based on the closest flanking SSR markers, VMC4f3.1 and VMC8g9, *Run1/Rpv1* is located on chromosome 12.

These genetic markers were used to screen a bacterial artificial chromosome (BAC) library constructed using genomic DNA of a single powdery mildew-resistant plant from the Mtp3294 mapping population (Barker et al. 2005). Assembly of the contigs between VMC4f3.1 and VMC8g9 indicated that the physical distance between these markers was much greater than initially suggested by the genetic mapping, necessitating the identification of more recombinants to further narrow the genomic region containing the *Run1/Rpv1* locus. To achieve this, approximately 3300 new progeny of the Mtp3294 and Mtp3322 populations were screened for recombination between VMC4f3.1 and VMC8g9 and the recombinants phenotyped for powdery and downy mildew resistance (C. Anderson, unpublished). This genetic analysis narrowed the *Run1/Rpv1* locus to a ~1 Mbp region of introgressed DNA. Sequencing of this region revealed the presence of a cluster of RGAs encoding TIR-NBS-LRR type resistance proteins. No other genes predicted to be involved in resistance against fungal pathogens were identified within this region.

FIGURE 2: Powdery mildew development on leaves of susceptible (V. vinifera cv. Cabernet Sauvignon; panels A and C) and resistant (Run1-containing BC4 progeny (see text for details); panels B and D) grapevines. Panels A and B: glasshouse vines 2 weeks after inoculation with powdery mildew. Panels C and D: detached leaves sampled 48 h post-inoculation and stained with Coomassie Blue to visualise fungal structures. Arrows indicate individual epidermal cells penetrated by powdery mildew that have undergone programmed cell death in leaves containing the Run1 locus. (c) conidium, (a) appressorium, (h) hypha.

In total, 11 RGAs have been identified at the *Run1/Rpv1* locus (Figure 3). Seven of the RGAs encode full-length TIR-NBS-LRR resistance proteins; the remaining genes are truncated and lack either the TIR or the NBS-LRR domain. Three of the RGAs (RGA-9, RGA-10 and RGA-11) were originally recovered as only partial length genomic clones associated with one of a number of 'gap' regions within the *Run1* BAC library for which overlapping BAC clones could not be found. Full-length genomic clones of these RGAs were obtained by genome walking techniques. The full-length RGAs are highly homologous apart from a variable number of LRR domains. Expression analysis indicates that all of the full-length RGA candidates are expressed in powdery mildew-resistant progeny, indicating that any one of these genes are candidates for conferring powdery mildew (and/or downy mildew) resistance. Significantly, the only other powdery mildew resistance genes which have been cloned to date are the Mla locus from barley (Zhou et al. 2001) and the *Pm3* locus from wheat (Srichumpa et al. 2005) which also encode NBS-LRR type proteins but possess a CC domain at the N-terminus instead of a TIR domain.

It is interesting to note that having screened over 5000 backcross progeny in the course of this mapping project, we have not identified any plants which segregate differently for powdery mildew (*Run1*) or downy mildew (*Rpv1*) resistance. This suggests that either both mildew resistances are encoded by the same resistance gene or different members of the same resistance gene cluster. An NBS-LRR resistance protein that confers resistance to two completely different pathogens, i.e. the ascomycete fungus *E. necator* and the oomycete *P. viticola*, would be highly unusual, but not unique. The *Mi* gene from tomato, for example, confers resistance against both root-knot nematodes and potato aphids (Rossi et al. 1998).

10.3.2 FUNCTIONAL TESTING OF CANDIDATE RESISTANCE GENES

NBS-LRR resistance genes show 'restricted taxonomic functionality' which generally means they are only functional within species of the same taxonomic family (Michelmore 2003). Furthermore, biotrophic pathogens such as *E. necator* and *P. viticola* are specifically adapted to the grapevine host and cannot infect other plant species. This means that whereas most genes isolated from grapevine can be functionally evaluated in readily transformable model species such as Arabidopsis, tobacco or tomato, candidate grapevine powdery mildew resistance genes are unlikely to reveal function unless experiments are carried out in transgenic grapevines, the generation of which remains a time-consuming and inefficient process. Some progress has been made in the development of transient assay systems in grapevine (Santos-Rosa et al. 2008), but whether these systems are efficient enough to enable testing of resistance to biotrophic pathogens, such as powdery mildew, remains to be seen.

Another consideration when evaluating R gene function in transgenic plants is the issue of promoter selection. The use of generic expression cassettes involving highly active constitutive promoters, such as 35S, may lead to autoactivity of the R proteins resulting in necrosis in the absence of the pathogen (Oldroyd and Staskawicz 1998, Tao et al. 2000, Bendahmane et al. 2002). This is thought to be either because of the titration of trans-acting repressors or an increase, beyond an activity threshold, of the steady-state level of R protein molecules, a proportion of which may exist in a spontaneously active state (Tao et al. 2000).

With these various considerations in mind, we are currently introducing individual genomic fragments (~12–14 kb) encompassing the endogenous promoter-RGA coding-endogenous terminator regions of each of the seven RGA candidates into powdery (and downy) mildew susceptible *V. vinifera* cultivars. This will enable us to determine not only which of the RGA candidates function as powdery and/or downy mildew resistance genes, but also test each of the candidate genes against different powdery and downy mildew isolates to determine if they are isolate-specific.

10.3.3 CONSIDERATIONS IN THE DEPLOYMENT OF QUALITATIVE POWDERY MILDEW RESISTANCE GENES IN THE VINEYARD

Durable resistance can be considered as resistance that remains effective when used in a large growing area, over a long period of time, under environments favourable to disease development (Leach et al. 2001). Perennial crops, like grapevines, provide particular challenges when considering the issue of durability. Most grapegrowers would have an expectation that individual vines would remain in the vineyard for at least 20 years and do not have the capacity to rapidly introduce new cultivars with different *R* genes, as is the case for annual crops such as cereals, should existing *R* genes fail.

Testing of *Run1* backcross lines against powdery mildew isolates in France and Australia has shown the *Run1* locus to be effective against all isolates tested to date. However, this may be because of the combined action of more than one *R* gene located at the *Run1* locus. The *M. rotundifolia* cultivar G52 from which the *Run1* locus was originally obtained is thought to have arisen from a cross between the two *M. rotundifolia* cultivars Thomas and Hope (Detjen 1919). Even if both cultivars have contributed different *R* genes to the *Run1* locus, it remains to be determined whether they would provide resistance to different powdery mildew isolates.

On the other hand, if it is determined that powdery mildew resistance at the *Run1* locus is only conferred by a single *R* gene then this has important implications for the deployment of this resistance gene within the vineyard. To date, there are few examples of single dominant *R* genes that have proven to be durable in the field. As outlined in Figure 1, interactions leading to host plant resistance involving these *R* genes can be lost through modification or elimination of pathogen effector(s) that trigger the resistance response. Even the barley *Rpg1* gene which had provided effective control for almost 50 years in North America against stem rust caused by *Puccinia graminis f. sp. tritici* was finally compromised by the evolution of a new virulent pathotype (Steffenson 1992).

One approach which may improve the durability of resistance in the field, is to stack or pyramid multiple *R* genes within a single cultivar such that, for a given pathogen, reproduction will be restricted even if individuals within the pathogen population are present that have lost, or modified, the interacting effector molecule for one of the *R* genes. At least two other sources of resistance (or at least reduced susceptibility) to grapevine powdery mildew have been reported that appear to map to a different locus to *Run1*. A major QTL originating from an undetermined North American *Vitis* species has been mapped to chromosome 15 (Akkurt et al. 2006, Welter et al. 2007). Hoffmann et al. (2008) also recently reported on a dominant locus for powdery mildew resistance (designated *REN1*) from the central Asian *V. vinifera* cultivar Kishmish vatkana which maps to a 10-cM region on chromosome 13. Interestingly, a family of NBS-LRR genes also maps to this same region raising the possibility that *REN1* is also an NBS-LRR gene. A number of groups have already commenced breeding programs to combine *Run1* resistance with these other powdery mildew resistance loci using marker assisted evaluation to follow the inheritance of both resistance loci (Eibach et al. 2007, Molnar et al. 2007)

The identification of dominant resistance to *E. necator* in *V. vinifera* cv. Kishmish vatkana, would appear to be at odds somewhat, with the model of evolution of plant resistance described earlier (Figure 1), which proposes that *R* gene mediated resistance is the result of a significant period of co-evolution between the host and an adapted pathogen. In addressing this conundrum, Hoffmann et al. (2008) proposed that wild *V. vinifera* populations in the Central Asia, unlike the clonally propagated cultivated grapevines in Europe, are undergoing sexual propagation, and may have evolved resistance since the arrival of *E. necator* from North America. An alternative explanation may be that *E. necator* has actually been in Central Asia for a much longer period than first thought. The existence of a large number of wild Chinese *Vitis* species which also show significant resistance to *E. necator* (Wan et al. 2007) also lends support for this hypothesis, although the actual mechanism of resistance in these Chinese species has not been reported.

10.4 AN ALTERNATIVE STRATEGY TO POWDERY MILDEW RESISTANCE: RE-ESTABLISHING BASAL IMMUNITY AGAINST *E. NECATOR*

According to the model outlined in Figure 1, powdery mildew species which have become adapted on particular plant hosts, are able to suppress basal immune responses to allow penetration, whereas non-adapted species are unable to do this and fail to enter the epidermal cell. For example, in the case of *V. vinifera* cv. Cabernet Sauvignon, inoculation with the adapted powdery mildew species *E. necator* results in rates of successful penetration (as determined by the presence of haustoria in epidermal cells) of over 90% within 48 h, whereas inoculation with the non-adapted cucurbit powdery mildew species *E. cichoracearum* under the same conditions, results in rates of successful penetration of <15% (A. Feechan, unpublished). It is still not understood how adapted powdery mildew species are able to suppress basal immunity to enable such high rates of successful penetration on host species. However, evidence obtained from studies on naturally occurring mutants, in which the basal immune response against a previously adapted powdery mildew species has been re-established, may offer some insights.

10.4.1 A ROLE FOR MLO IN SUPPRESSING BASAL IMMUNITY AGAINST POWDERY MILDEW?

German expeditions to Ethiopia in 1937–1938 collected seed samples of barley landraces which were found to exhibit broad-spectrum resistance against all isolates of barley powdery mildew (*Blumeria graminis f. sp. hordei*; Peterhänsel and Lahaye 2005). However, what was even more surprising was that this spontaneously occurring resistance allele, designated mlo (powdery mildew resistance gene o), was inherited in a recessive manner. The wild type *HvMLO* gene was cloned from barley in 1997 and found to encode a plant-specific 7 transmembrane domain protein that resides in the plasma membrane (Büschges et al. 1997, Devoto et al. 1999). Since then, a naturally occurring allele (ol-2), derived from a Central American tomato (*Solanum lycospersicum*) accession, that confers broad-spectrum and recessively inherited resistance to tomato powdery mildew (*Oidium neolycopersici*), has been demonstrated to be due to loss of function of the *SlMLO1* gene (Bai et al. 2008). Finally, a homozygous T-DNA insertion line in the Arabidopsis *AtMLO2* gene, has been shown to have significantly reduced susceptibility to *Erysiphe orontii* (Consonni et al. 2006), with mutation of two other closely related genes, *AtMLO6* and *AtMLO12* also required to achieve complete powdery mildew resistance.

One of the characteristics of the resistance mediated by *mlo* is the failure of the adapted powdery mildew species to successfully penetrate epidermal cells, which is reminiscent of the phenotype observed following inoculation with a non-adapted powdery mildew species. It has also been shown that *mlo* resistance in barley is dependent on the presence of *ROR2* (Collins et al. 2003), an orthologue of the Arabidopsis *PEN1* gene shown to be involved in basal immunity against non-adapted powdery mildews (see previous section). Furthermore, studies have revealed that, like the PEN1 and ROR2 proteins, wild-type MLO becomes polarised at the site of attempted powdery mildew ingress (Bhat et al. 2005). The shared histological and phytopathological characteristics, together with the conserved requirement for *PEN1/ROR2* in both basal immunity against non-adapted powdery mildew species, and MLO-mediated resistance against adapted powdery mildew species, strongly suggests a common mechanism of resistance (Humphry et al. 2006). In other words, mutation of the wild-type *MLO* gene effectively re-establishes basal immunity against the adapted powdery mildew species.

The mechanism by which wild-type MLO protein might interfere with or suppress PEN1-mediated basal immunity is yet to be determined. Meyer et al. (2009) have postulated that the co-localisation of MLO and PEN1/ROR2 within the plasma membrane, beneath the site of attempted penetration, might directly interfere or antagonise the PEN1/ROR2-mediated processes required for basal immunity against powdery mildew. If this were the case, it also implies that adapted powdery mildew species are able to hijack MLO to suppress basal immunity whereas the non-adapted species cannot.

10.4.2 FUNCTIONAL CHARACTERISATION OF THE MLO GENE FAMILY IN GRAPEVINE

Independent of the mechanism involved, the work described previously with barley, tomato and Arabidopsis clearly demonstrates that it is feasible to re-establish basal immunity against an adapted powdery mildew species by targeted mutation or silencing of host *MLO* gene(s). However, the challenge is to determine which *MLO* gene(s) to target. For example, *A. thaliana* contains 15 *MLO* genes, but only three (*AtMLO2*, *AtMLO6* and *AtMLO12*) appear to have a role in powdery mildew susceptibility (Consonni et al. 2006). Analysis of the recently released *V. vinifera* genome

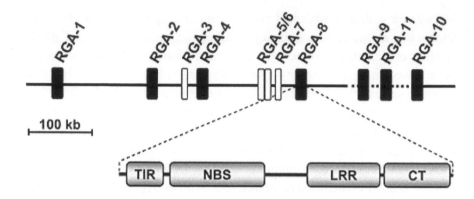

FIGURE 3: Arrangement and structure of resistance gene analogue (*RGA*) genes clustered at the Run locus. Black boxes represent full-length genes, open boxes are truncated genes. The dotted line represents regions of the genomic DNA for which overlapping clones could not be obtained from the *Run1* BAC library. Translation of the *RGA* genes at the Run1 locus reveals the presence of four domains characteristic of TIR-NBS-LRR resistance genes: TIR, Toll/Interleukin-1 receptor domain; NBS, nucleotide-binding site domain; LRR, leucine-rich repeat domain; CT, C-terminal domain of undetermined function.

indicates there may be as many as 17 *VvMLO* genes in grapevine (Feechan et al. 2008, Winterhagen et al. 2008).

As a first step, we undertook phylogenetic analysis of the translated products of the *VvMLO* gene family together with other known plant MLO sequences. Of the 17 *VvMLO* genes analysed, six (*VvMLO3, VvMLO4, VvMLO6, VvMLO9, VvMLO13* and *VvMLO17*) clustered within the same clade as the Arabidopsis and tomato *MLO* genes that have been demonstrated to be required for powdery mildew susceptibility (Feechan et al. 2008). Of these six genes, three (*VvMLO3, VvMLO4* and *VvMLO17*) were found to be induced significantly in grape leaves within 8 h of *E. necator* inoculation, which coincided with the commencement of fungal penetration (Figure 4). Transcript abundance of *VvMLO4* and *VvMLO17* then decreased, while *VvMLO3* expression remained relatively constant during the period 8–20 h post-inoculation. *VvMLO9* was also found to exhibit increased transcript levels following *E. necator* inoculation, but the timing was delayed relative to the other powdery mildew-induced *VvMLO* genes,

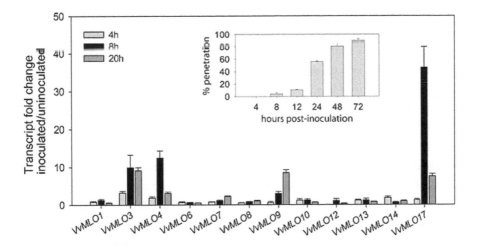

FIGURE 4: Transcriptional response of grapevine *VvMLO* genes to powdery mildew infection. Detached *V. vinifera* cv. Cabernet Sauvignon leaves were heavily inoculated with *E. necator* spores and sampled at 4, 8, 12, 24, 48 and 72 for scoring penetration efficiency and 4, 8 and 20 h for gene expression analysis. Inset: The incidence of epidermal host cell penetration by individual germinated conidia, scored via the presence of a haustorium. Each data point represents the scoring of a minimum of 100 germinated conidia with two replicates. The mean is shown and error bars are standard errors. Main graph: Quantitative RT-PCR analysis of *VvMLO* gene expression. Control (un-inoculated) detached leaves were incubated and sampled as for the inoculated leaves. Transcript levels of individual *VvMLO* genes were normalised against the transcript levels of *EF1* in each cDNA sample. Results are plotted as a mean ratio of the normalised transcript level of each *VvMLO* gene in inoculated leaves versus the normalised transcript level of each *VvMLO* gene in control leaves. Means are totals calculated from triplicate technical qRT-PCR measurements within three biological replicates and the data shown are representative of the results obtained in the three independent experiments. Error bars are standard errors. This figure has been modified from Functional Plant Biology 35(12) 1255–1266, http://www.publish.csiro.au/paper/FP08173.htm.

whereas none of the other eight *VvMLO* genes examined were induced greater than twofold in response to powdery mildew inoculation. The transient transcriptional response of *VvMLO17*, *VvMLO3* and *VvMLO4* to *E. necator* penetration is consistent with that previously reported for *HvMLO* in response to *B. graminis* infection (Piffanelli et al. 2002).

The similarities in both sequence homology and transcriptional regulation suggests that *VvMLO3*, *VvMLO4* and *VvMLO17*, may have functionally similar roles to those *MLO* genes from barley, Arabidopsis and tomato (described previously) which have been shown to modulate powdery mildew susceptibility. It may also indicate some level of functional redundancy between these three *VvMLO* genes as is the case in Arabidopsis. Experiments are now underway in our laboratory to try and establish if selective silencing of any of these *VvMLO* genes individually, or in combination, has any impact on powdery mildew susceptibility. The outcome of these experiments will have important implications for any future strategies to engineer powdery mildew resistance based on *MLO* suppression. If resistance in grapevine can be achieved through the silencing of a single *VvMLO* gene, as in tomato (Bai et al. 2008), it would then be feasible to initiate a search for naturally occurring mutant alleles within *V. vinifera* germplasm collections that could be used in marker-assisted selection to generate progeny that are homozygous recessive at this locus. If, on the other hand, complete powdery mildew resistance in grapevine is only achieved through the silencing of more than one *VvMLO* gene, then this is realistically only amenable to transgenic approaches using constructs designed to simultaneously silence multiple *VvMLO* genes. Such an approach has already been shown to be feasible in Arabidopsis (Consonni et al. 2006).

10.4.3 ONTOGENIC RESISTANCE: THE RETURN OF BASAL IMMUNITY?

Ontogenic or age-related resistance describes an increase in the ability of whole plants or plant tissues to resist pathogen infection as they age or mature. It is exhibited towards viral, bacterial, oomycete and fungal pathogens (Panter and Jones 2002, Develey-Rivière and Galiana 2007). Grapevines show ontogenic resistance to several fungal pathogens including powdery mildew (Gadoury et al. 2003), downy mildew (Kennelly et al. 2005), and black rot (Hoffman et al. 2002).

Delp (1954) originally reported field and laboratory observations that immature grapes were susceptible to infection, but that no new infections occurred once the concentration of soluble solids exceeded 8°Brix. Chellemi and Marois (1992) also reported that grapes became resistant to powdery mildew infection above 7°Brix. As the apparent decrease in powdery mildew susceptibility of grape berries post-veraison, coincided with a marked increase in the accumulation of pathogenesis-related proteins, it was postulated that these proteins may have a role in ontogenic resistance (Robinson et al. 1997, Tattersall et al. 1997). However, subsequent studies showed that ontogenic resistance to powdery mildew is actually expressed at a much earlier stage of berry development, with berries of several cultivars of *V. vinifera* and *V. labruscana* immune to infection within 3–5 weeks after bloom (Ficke et al. 2003, Gadoury et al. 2003).

Although the mechanism of ontogenic resistance in grape berries is not fully understood, it appears to operate at the level of the epidermal cell wall resulting in an inhibition of penetration (Ficke et al. 2003). This increased penetration resistance could not be correlated with cuticular or cell wall thickness or the production of antimicrobial phenolics, leading Ficke et al. (2004) to postulate that ontogenic resistance in grape berries is mediated by the development of either (i) a physical or biochemical barrier near the cuticle surface or (ii) an inducible system for the rapid synthesis of an antifungal compound.

It is interesting to note that the increased penetration resistance observed in developing grape berries against the adapted powdery mildew species *E. necator*, is comparable to the observed response of grape leaves to attempted penetration by a non-adapted powdery mildew species such as *E. cichoracearum* (see previous discussion). This raises the question as to whether the development of ontogenic resistance reflects a decrease in the capacity of *E. necator* to continue to suppress basal immunity. We are currently investigating components of the basal immunity pathway in grapevine in order to test this hypothesis.

If we could determine the molecular/genetic basis of ontogenic resistance to powdery mildew in grape berries, it may be possible to use this information to manipulate broad-spectrum, durable resistance to powdery mildew in other highly susceptible grapevine tissues. Recent work by Gee et al. (2008) has identified a strategy via which the determination of the mechanism of ontogenic resistance might be achieved. They screened a diverse collection of *Vitis* species and interspecific hybrids for the development of ontogenic resistance in berries under field conditions. Of the 79 genotypes investigated, 50 showed little or no sign of disease when inoculated, irrespective of the stage of berry development at inoculation, 24 exhibited a significant gain of resistance as berries aged and four genotypes exhibited no statistically significant pattern. However, more importantly, one genotype (*V. rupestris*'R-65-44') remained susceptible past the onset of ripening, over 1 month later than reported previously for *V. vinifera*. The discovery of significant genotypic variation in ontogenic resistance now offers the possibility of undertaking genetic studies to investigate the inheritance and molecular basis of this powdery mildew resistance character.

10.4.4 FINAL CONSIDERATIONS

We are employing a number of molecular approaches in our quest to try and introduce significant and durable resistance to powdery mildew into *V. vinifera*. Of the three approaches described, integration of the *Run1* locus from *M. rotundifolia* into *V. vinifera* has clear potential for generating new powdery mildew-resistant winegrape cultivars. Our research is focussed on identifying which of the individual resistance gene candidates, located at the *Run1* locus, contributes to powdery (and downy) mildew resistance. Functional demonstration of mildew resistance mediated by a single *R* gene in a *V. vinifera* genetic background, would open up the possibility of introducing this gene into existing elite cultivars with minimal impacts on wine quality. However, as work with other pathosystems has shown, the use of single dominant resistance genes is not without potential problems in terms of durability in the field. It will be important to obtain information regarding the existence and distribution of any *E. necator* isolates, most probably within North America, which can overcome *Run1*-mediated resistance. Where possible, new winegrape cultivars should also contain resistance genes from multiple sources, to reduce the risk of generating new compatible isolates. However, this clearly adds another level of complexity in terms of plant breeding or genetic transformation.

While as yet unproven in grapevine, the *MLO* mutation approach offers some advantages over resistance based on *Run1*, because of its potential to deliver broad-spectrum, durable resistance to powdery mildew. If a member of the *VvMLO* gene family is demonstrated to act as a powdery mildew susceptibility gene in grapevine, the most likely mode of delivery is via genetic modification of existing cultivars. However, powdery mildew resistance obtained through *MLO* mutation may not come without a price. Barley and Arabidopsis *mlo* mutants have been found to show enhanced susceptibility to some hemibiotrophic and necrotrophic pathogens (Jarosch et al. 1999, Kumar et al. 2001, Consonni et al. 2006). As grapevines are exposed to a broad range of pathogens within the different regions of the world in which they are cultivated, deployment of grapevine *MLO* mutants, would need to be preceded by field trials designed to evaluate the impact of this mutation on susceptibility to pathogens other than powdery mildew. It also remains to be seen if silencing of *VvMLO* gene(s) will have any pleiotropic effects on grapevine growth and development. *MLO* mutants of barley and Arabidopsis display a range of developmentally regulated pleiotropic effects, including the spontaneous deposition of callose-containing cell wall appositions and the premature onset of leaf senescence (Peterhänsel et al. 1997, Piffanelli et al. 2002, Consonni et al. 2006). However, similar developmentally regulated pleiotropic effects were not observed on powdery mildew resistant tomato mlo mutants (Bai et al. 2008). The impaired leaf physiology of barley *mlo* plants results in some reduced grain yield compared with wild-type plants and this initially hampered the use this germplasm despite its broad-spectrum resistance to powdery mildew. The possible appearance of similar pleiotropic effects on grapevine leaves is unlikely to cause significant issues for grapegrowers because photosynthetic capacity is not normally a limiting factor in grape production. Indeed, it is likely that most growers would even tolerate some loss in maximum cropping capacity in exchange for durable resistance to powdery mildew. However, it will be important to confirm that the mutation of specific *VvMLO* gene(s) in grapevine does not lead to any deleterious effects on berry yield or quality.

An over-arching issue across all of these considerations is the continuing uncertainty regarding the adoption of transgenic vines in the major wine-making countries and whether this could be used by some countries as a trade barrier. In comparison with classical breeding techniques, genetic transformation offers the capacity to move resistance genes between different grapevine species that cannot be readily hybridised by traditional breeding techniques, or selectively silence host genes which pre-dispose plants to infection. It also retains the unique genetic make-up of our elite wine cultivars, which may be lost through the use of classical breeding techniques because of the high level of heterozygosity of the grape genome. Indeed, despite the best efforts of grapevine breeders, not one new powdery mildew-resistant cultivar has been developed, since the invasion of *E. necator* into Europe in the 1840s, that has been adopted to any significant extent by growers. However, with the development of genetic markers linked to both qualitative and quantitative disease resistance loci and the employment of marker assisted selection strategies, the breeding of new disease resistant winegrape cultivars may be about to undergo a renaissance.

REFERENCES

1. Aarts, M.G.M., Te Lintel Hekkert, B., Holub, E.B., Beynon, J.L., Stiekema, W.J. and Pereira, A. (1998) Identification of r-gene homologous DNA fragments genetically linked to disease resistance loci in Arabidopsis thaliana. Molecular Plant Microbe Interactactions 11, 251–258.

2. Akkurt, M., Welter, L., Maul, E., Töpfer, R. and Zyprian, E. (2006) Development of SCAR markers linked to powdery mildew (Uncinula necator) resistance in grapevine (Vitis vinifera L. and Vitis sp). Molecular Breeding 19, 103–111.

3. Assaad, F.F., Qiu, J-L., Youngs, H., Ehrhardt, D., Zimmerli, L., Kalde, M., Wanner, G., Peck, S.C., Edwards, H., Ramonell, K., Somerville, C.R. and Thordal-Christensen, H. (2004) The PEN1 syntaxin defines a novel cellular compartment upon fungal attack and is required for the timely assembly of papillae. Molecular Biology of the Cell 15, 5118–5129.

4. Bai, Y., Pavan, S., Zheng, Z., Zappel, N.F., Reinstädler, A., Lotti, C., De Giovanni, C., Ricciardi, L., Lindhout, P., Visser, R., Theres, K. and Panstruga, R. (2008) Naturally occurring broad-spectrum powdery mildew resistance in a Central American tomato accession is caused by loss of Mlo function. Molecular Plant Microbe Interactions 21, 30–39.

5. Barker, C.L., Donald, T., Pauquet, J., Ratnaparkhe, M.B., Bouquet, A., Adam-Blondon, A-F., Thomas, M.R. and Dry, I. (2005) Genetic and physical mapping of the grapevine powdery mildew resistance gene, Run1, using a bacterial artificial chromosome library. Theoretical and Applied Genetics 111, 370–377.

6. Baudoin, A., Olaya, G., Delmotte, F., Colcol, J.F. and Sierotzki, H. (2008) QoI resistance of Plasmopara viticola and Erysiphe necator in the mid-Atlantic United States. Online. Plant Health Progress, Doi: 10.1094/PHP-2008-0211-02-RS.

7. Bendahmane, A., Farnham, G., Moffett, P. and Baulcombe, D.C. (2002) Constitutive gain-of-function mutants in a nucleotide binding site leucine rich repeat protein encoded at the Rx locus of potato. The Plant Journal 32, 195–204.

8. Bent, A.F. and Mackey, D. (2007) Elicitors, effectors, and R genes: the new paradigm and a lifetime supply of questions. Annual Review of Phytopathology 45, 399–436.

9. Bhat, R.A., Miklis, M., Schmelzer, E., Schulze-Lefert, P. and Panstruga, R. (2005) Recruitment and interaction dynamics of plant penetration resistance components in a plasma membrane microdomain. Proceedings of National Academy of Sciences U.S.A. 102, 3135–3140.

10. Bittel, P. and Robatzek, S. (2007) Microbe-associated molecular patterns (MAMPs) probe plant immunity. Current Opinion in Plant Biology 10, 335–341.

11. Bouquet, A. (1986) Introduction dans l'espèce Vitis vinifera L. d'un caractère de résistance à l'oidium (Uncinula necator Schw. Burr.) issu de l'espèce Muscadinia rotundifolia (Michx.) Small. Vignevini 12 (Suppl), 141–146.

12. Branas, M.M. (1932) Sur la caryologie des Ampélidées. C R Acad Sci Paris 194, 121–123.

13. Büschges, R., Hollricher, K., Panstruga, R., Simons, G., Wolter, M., Frijters, A., Van Daelen, R., Van Der Lee, T., Diergaarde, P., Groenendijk, J., Topsch, S., Vos, P., Salamini, F. and Schulze-Lefert, P. (1997) The barley Mlo gene: a novel control element of plant pathogen resistance. Cell 88, 695–705.

14. Calonnec, A., Cartolaro, P., Poupot, C., Dubourdieu, D. and Darriet, P. (2004) Effects of Uncinula necator on the yield and quality of grapes (Vitis vinifera) and wine. Plant Pathology 53, 434–445.

15. Chellemi, D.O. and Marois, J.J. (1992) Influence of leaf removal, fungicide applications, and fruit maturity on incidence and severity of grape powdery mildew. American Journal of Enology and Viticulture 43, 53–57.

16. Collins, N.C., Thordal-Christensen, H., Lipka, V., Bau, S., Kombrink, E., Qiu, J.L., Hückelhoven, R., Stein, M., Freialdenhoven, A., Somerville, S.C. and Schulze-Lefert, P. (2003) SNARE-protein-mediated disease resistance at the plant cell wall. Nature 425, 973–977.

17. Collins, N.C., Webb, C.A., Seah, S., Ellis, J.G., Hulbert, S.H. and Pryor, A. (1998) The isolation and mapping of disease resistance gene analogues in maize. Molecular Plant Microbe Interactions 11, 968–978.

18. Consonni, C., Humphry, M.E., Hartmann, H.A., Livaja, M., Durner, J., Westphal, L., Vogel, J., Lipka, V., Kemmerling, B., Schulze-Lefert, P., Somerville, S.C. and Panstruga, R. (2006) Conserved requirement for a plant host cell protein in powdery mildew pathogenesis. Nature Genetics 38, 716–720.

19. Delp, C.J. (1954) Effect of temperature and humidity on the grape powdery mildew fungus. Phytopathology 44, 615–626.

20. Detjen, L.R. (1919) The limits in hybridisation of Vitis rotundifolia with related species and genera. North Carolina Agricultural Experiment Station Bulletin 17, 409–429.

21. Develey-Rivière, M-P. and Galiana, E. (2007) Resistance to pathogens and host developmental stage: a multifaceted relationship within the plant kingdom. New Phytologist 175, 405–416.

22. Devoto, A., Piffanelli, P., Nilsson, I., Wallin, E., Panstruga, R., Von Heijne, G. and Schulze-Lefert, P. (1999) Topology, subcellular localization, and sequence diversity of the MLO family in plants. Journal of Biological Chemistry 274, 34993–35004.

23. Di Gaspero, G. and Cipriani, G. (2002) Resistance gene analogs are candidate markers for disease- resistance genes in grape (Vitis spp.). Theoretical and Applied Genetics 106, 163–172.

24. Donald, T.M., Pellerone, F., Adam-Blondon, A-F., Bouquet, A., Thomas, M.R. and Dry, I. (2002) Identification of resistance gene analogs linked to a powdery mildew resistance locus in grapevine. Theoretical and Applied Genetics 104, 610–618.

25. Eibach, R., Zyprian, E., Welter, L. and Topfer, R. (2007) The use of molecular markers for pyramiding resistance genes in grapevine breeding. Vitis 46, 120–124.

26. Erickson, E.O. and Wilcox, W.F. (1997) Distributions of sensitivities to three sterol demethylation inhibitor fungicides among populations of Uncinula necator sensitive and resistant to triadimefon. Phytopathology 87, 784–791.

27. Feechan, A., Jermakow, A.M., Torregrosa, L., Panstruga, R. and Dry, I.B. (2008) Identification of grapevine MLO gene candidates involved in susceptibility to powdery mildew. Functional Plant Biology 35, 1255–1266.

28. Ficke, A., Gadoury, D.M., Godfrey, D. and Dry, I.B. (2004) Host barriers and responses to Uncinula necator in developing grape berries. Phytopathology 94, 438–445.

29. Ficke, A., Gadoury, D.M., Seem, R.C. and Dry, I.B. (2003) Effects of ontogenic resistance upon establishment and growth of Uncinula necator on grape berries. Phytopathology 93, 556–563.

30. Gadoury, D.M., Seem, R.C., Ficke, A. and Wilcox, W.F (2003) Ontogenic resistance to powdery mildew in grape berries. Phytopathology 93, 547–555.

31. Gadoury, D.M., Seem, R.C., Pearson, R.C., Wilcox, W.F. and Dunst, R.M. (2001) Effects of powdery mildew on vine growth, yield, and quality of Concord grapes. Plant Disease 85, 137–140.

32. Gadoury, D.M., Seem, R.C., Wilcox, W.F., Henick-Kling, T., Conterno, L., Day, A. and Ficke, A. (2007) Effects of diffuse colonization of grape berries by Uncinula necator on bunch rots, berry microflora, and juice and wine quality. Phytopathology 97, 1356–1365.

33. Gee, C.T., Gadoury, D.M. and Cadle-Davidson, L. (2008) Ontogenic resistance to Uncinula necator varies by genotype and tissue type in a diverse collection of Vitis spp. Plant Disease 92, 1067–1073.

34. Glawe, D.A. (2008) The Powdery Mildews: a review of the world's most familiar (yet poorly known) plant pathogens. Annual Review of Phytopathology 46, 27–51.

35. Heintz, C. and Blaich, R. (1990) Ultrastructural and histochemical studies on interactions between Vitis vinifera L. and Uncinula necator (Schw.) Burr. New Phytologist 115, 107–117.

36. Hoffman, L.E., Wilcox, W.F., Gadoury, D.M. and Seem, R.C. (2002) Influence of grape berry age on susceptibility to Guignardia bidwelli and its incubation period length. Phytopathology 92, 1068–1076.

37. Hoffmann, S., Di Gaspero, G., Kovács, L., Howard, S., Kiss, E., Galbács, Z., Testolin, R. and Kozma, P. (2008) Resistance to Erysiphe necator in the grapevine 'Kish-

mish vatkana' is controlled by a single locus through restriction of hyphal growth. Theoretical and Applied Genetics 116, 427–438.

38. Humphry, M., Consonni, C. and Panstruga, R. (2006) mlo-based powdery mildew immunity: silver bullet or simply non-host resistance? Molecular Plant Pathology 7, 605–610.

39. Jarosch, B., Kogel, K.-H. and Schaffrath, U. (1999) The ambivalence of the barley Mlo locus: mutations conferring resistance against powdery mildew (Blumeria graminis f. sp. hordei) enhance susceptibility to the rice blast fungus Magnaporthe grisea. Molecular Plant Microbe Interactions 12, 508–514.

40. Kennelly, M.M., Gadoury, D.M., Wilcox, W.F., Magarey, P.A. and Seem, R.C. (2005) Seasonal development of ontogenic resistance to downy mildew in grape berries and rachises. Phytopathology 95, 1445–1452.

41. Kumar, J., Hückelhoven, R., Beckhove, U., Nagarajan, S. and Kogel, K.-H. (2001) A compromised Mlo pathway affects the response of barley to the necrotrophic fungus Bipolaris sorokiniana (Teleomorph: Cochliobolus sativus) and its toxins. Phytopathology 91, 127–133.

42. Leach, J.E., Cruz, V., Casiana, M., Bai, J. and Leung, H. (2001) Pathogen fitness penalty as a predictor of durability of disease resistance genes. Annual Review of Phytopathology 39, 187–224.

43. Lipka, V., Dittgen, J., Bednarek, P., Bhat, R., Wiermer, M., Stein, M., Landtag, J., Brandt, W., Rosahl, S., Scheel, D., Llorente, F., Molina, A., Parker, J., Somerville, S. and Schulze-Lefert, P. (2005) Pre- and postinvasion defences both contribute to nonhost resistance in Arabidopsis. Science 310, 1180–1183.

44. Merdinoglu, D., Wiedeman-Merdinoglu, S., Coste, P., Dumas, V., Haetty, S., Butterlin, G. and Greif, C. (2003) Genetic analysis of downy mildew resistance derived from Muscadinia rotundifolia. Acta Horticulturae 603, 451–456.

45. Meyer, D., Pajonk, S., Micali, C., O'Connell, R. and Schulze-Lefert, P. (2009) Extracellular transport and integration of plant secretory proteins into pathogen-induced cell wall compartments. The Plant Journal 57, 986–999.

46. Michelmore, R.W. (2003) The impact zone: genomics and breeding for durable resistance. Current Opinion in Plant Biology 6, 397–404.

47. Michelmore, R.W., Paran, I. and Kesseli, R.V. (1991) Identification of markers linked to disease-resistance genes by Bulked Segregant Analysis: a rapid method to detect markers in specific genomic regions by using segregating populations. Proceedings of National Academy of Sciences U.S.A. 88, 9828–9832.

48. Molnar, S., Galbacs, Z., Halasz, G., Hoffmann, S., Kiss, E., Kozma, P., Veres, A., Galli, Z., Szoke, A. and Heszky, L. (2007) Marker assisted selection (MAS) for powdery mildew resistance in a grapevine hybrid family. Vitis 46, 212–213.

49. Mur, L.A.J., Kenton, P., Lloyd, A.J., Ougham, H. and Prats, E. (2008) The hypersensitive response; the centenary is upon us but how much do we know? Journal of Experimental Botany 59, 501–520.

50. Oldroyd, G.E.D. and Staskawicz, B.J. (1998) Genetically engineered broad-spectrum disease resistance in tomato. Proceedings of National Academy of Sciences U.S.A. 95, 10300–10305.

51. Olmo, H.P. (1986) The potential role of (vinifera x rotundifolia) hybrids in grape variety improvement. Experientia 42, 921–926.

52. Panter, S.N. and Jones, D.A. (2002) Age-related resistance to plant pathogens. Advances in Botanical Research 38, 251–280.
53. Pauquet, J., Bouquet, A., This, P. and Adam-Blondon, A-F. (2001) Establishment of a local map of AFLP markers around the powdery mildew resistance gene Run1 in grapevine and assessment of their usefulness for marker assisted selection. Theoretical and Applied Genetics 103, 1201–1210.
54. Pearson, R.C. and Gadoury, D.M. (1992) Grape powdery mildew. In: Plant diseases of international importance, Vol. III, Diseases of fruit crops. Eds. J.Kumar, H.S.Chaube, U.S.Singh and A.N.Mukhopadhyay (Prentice Hall: Upper Saddle River, NJ) p. 456.
55. Pesticides Safety Directorate (2008) Pesticides Safety Directorate website. (UK Government)   http://www.pesticides.gov.uk/uploadedfiles/Web_Assets/PSD/Impact_report_76_final_(May_2008).pdf[accessed 29/10/2009].
56. Peterhänsel, C. and Lahaye, T. (2005) Be fruitful and multiply: gene amplification inducing pathogen resistance. Trends in Plant Science 10, 257–260.
57. Peterhänsel, C., Freialdenhoven, A., Kurth, J., Kolsch, R. and Schulze-Lefert, P. (1997) Interaction analyses of genes required for resistance responses to powdery mildew in barley reveal distinct pathways leading to leaf cell death. The Plant Cell 9, 1397–1409.
58. Piffanelli, P., Zhou, F.S., Casais, C., Orme, J., Jarosch, B., Schaffrath, U., Collins, N.C., Panstruga, R. and Schulze-Lefert, P. (2002) The barley MLO modulator of defense and cell death is responsive to biotic and abiotic stress stimuli. Plant Physiology 129, 1076–1085.
59. Planchon, J.E. (1887) Monographie des Ampelidae vraies. Monographia Phanerogamarum A et C de Candolle 5, 305–364.
60. Pollefeys, P. and Bousquet, J. (2003) Molecular genetic diversity of the French-American grapevine hybrids cultivated in North America. Genome 46, 1037–1048.
61. Ridout, J., Skamnioti, P., Porritt, O., Sacristan, S., Jones, J.D.G. and Brown, J.K.M. (2006) Multiple avirulence paralogues in cereal powdery mildew fungi may contribute to parasite fitness and defeat of plant resistance. The Plant Cell 18, 2402–2414.
62. Robinson, S.P., Jacobs, A.K. and Dry, I.B. (1997) A class IV chitinase is highly expressed in grape berries during ripening. Plant Physiology 114, 771–778.
63. Rossi, M., Goggin, F.L., Milligan, S.B., Kaloshian, I., Ullman, D.E. and Williamson, V.M. (1998) The nematode resistance gene Mi of tomato confers resistance against the potato aphid. Proceedings of National Academy of Sciences U.S.A. 95, 9750–9754.
64. Santos-Rosa, M., Poutaraud, A., Merdinoglu, D. and Mestre, P. (2008) Developement of a transient expression system in grapevine via agro-infiltration. Plant Cell Reports 27, 1053–1063.
65. Savocchia, S., Stummer, B.E., Wicks, T.J., Van Heeswijck, R. and Scott, E.S. (2004) Reduced sensitivity of Uncinula necator to sterol demethylation inhibiting fungicides in southern Australian vineyards. Australasian Plant Pathology 33, 465–473.
66. Shen, K.A., Meyers, B.C., Islam-Faridi, M.N., Chin, D.B., Stelly, D.M. and Michelmore, R.W. (1998) Resistance gene candidates identified by PCR with degenerate oligonucleotide primers map to clusters of resistance genes in lettuce. Molecular Plant Microbe Interactions 11, 815–823.

67. Srichumpa, P., Brunner, S., Keller, B. and Yahiaoui, N. (2005) Allelic series of four powdery mildew resistance genes at the Pm3 locus in hexaploid bread wheat. Plant Physiology 139, 885–895.
68. Steffenson, B.J. (1992) Analysis of durable resistance to stem rust in barley. Euphytica 63, 153–167.
69. Stein, M., Dittgen, J., Sánchez-Rodriquez, C., Hou, B-H., Molina, A., Schulze-Lefert, P., Lipka, V. and Somerville, S. (2006) Arabidopsis PEN3/PDR8, an ATP binding cassette transporter, contributes to nonhost resistance to inappropriate pathogens that enter by direct penetration. The Plant Cell 18, 731–746.
70. Stummer, B.E., Francis, I.L., Zanker, T., Lattey, K.A. and Scott, E.S. (2005) Effects of powdery mildew on the sensory properties and composition of Chardonnay juice and wine when grape sugar ripeness is standardised. Australian Journal of Grape and Wine Research 11, 66–76.
71. Takken, F.L.W., Albrecht, M. and Tameling, W.I.L. (2006) Resistance proteins: molecular switches of plant defence. Current Opinions in Plant Biology 9, 383–390.
72. Tao, Y., Yuan, F., Leister, R.T., Ausubel, F.M. and Katagiri, F. (2000) Mutational analysis of the Arabidopsis nucleotide binding site-leucine rich repeat resistance gene RPS2. The Plant Cell 12, 2541–2554.
73. Tattersall, D.B., Van Heeswijck, R. and Hoj, P.B. (1997) Identification and characterization of a fruit-specific, thaumatin-like protein that accumulates at very high levels in conjunction with the onset of sugar accumulation and berry softening in grapes. Plant Physiology 114, 759–769.
74. This, P., Lacombe, T. and Thomas, M.R. (2006) Historical origins and genetic diversity of wine grapes. Trends in Genetics 22, 511–519.
75. Wan, Y.Z., Schwaninger, H., He, P.C. and Wang, Y.J. (2007) Comparison of resistance to powdery mildew and downy mildew in Chinese wild grapes. Vitis 46, 132–136.
76. Welter, L.J., Göktürk-Baydar, N., Akkurt, M., Maul, E., Eibach, R., Töpfer, R. and Zyprian, E.M. (2007) Genetic mapping and localization of quantitative trait loci affecting fungal disease resistance and leaf morphology in grapevine (Vitis vinifera L). Molecular Breeding 20, 359–374.
77. Winterhagen, P., Howard, S.F., Qui, W. and Kovács, L.G. (2008) Transcriptional upregulation of grapevine MLO genes in response to powdery mildew infection. American Journal of Enology and Viticulture 59, 159–168.
78. Zhou, F., Kurth, J., Wei, F., Elliott, C., Valè, G., Yahiaoui, N., Keller, B., Somerville, S., Wise, R. and Schulze-Lefert, P. (2001) Cell-autonomous expression of barley Mla1 confers race-specific resistance to the powdery mildew fungus via a Rar1-independent signaling pathway. The Plant Cell 13, 337–350.

PART III

WHAT ROLE DO CONSUMERS PLAY IN SUSTAINABLE VITICULTURE?

CHAPTER 11

BIBERE VINUM SUAE REGIONIS: WHY WHIAN WHIAN[1] WINE

MOYA COSTELLO AND STEVE EVANS

11.1 INTRODUCTION

Bibere vinum suae regionis [2], to drink wine from one's own region, is an attempt to match the neologism 'locavore', local eater/local eating, with one for local drinker/drinking. In 2005, Olivia Wu (2005), staff writer for the *San Francisco Chronicle*, reported that three women had begun calling themselves locavores. Locavore was the *New American Oxford Dictionary*'s word of the year for 2007, ascribing invention of the term to Jessica Prentice (OUP Blog, 2006–2013). Locavore comes from the Latin roots of local (*locus*) and eating (*vorare*). But food enthusiasts don't always include wine in their sense of the local. Below, we discuss an Australian example of this. To drink is *bibere*; I drink *bibo*. *Bibendum* is the gerund, drinking. We could suggest *locabibo, locabibere* or *locabiber* for local drinker/drinking.

One illustration of the erasure of the wine locabiber is local produce dinners hosted by two restaurants in two towns in a single Australian region, the Northern Rivers [3] on the far north coast of New South Wales (NSW),

Costello M and Evans S. Bibere Vinum Suae Regionis: Why Whian Whian Wine. Locale: The Australasian-Pacific Journal of Regional Food Studies, 3 (2013). Reprinted with permission from the authors and *Locale: The Pacific Journal of Regional Food Studies*, a free-access, peer-refereed online journal, published by Southern Cross University's Regional Food Network initiative. http://www.localejournal.org

one of seven Australian states and territories (which include South Australia [SA] and Queensland [QLD]). The thirty-mile dinner at Lismore's Tommy's Bar (Destination Food, 2010) served SA McLaren Vale Region Maxwell wines. Byron at Byron used winemaker James Evers from SA's Barossa Zone ('A Delicious Gourmet Weekend', 2011: 54). However, a third restaurant in a third local town, Finns, Kingscliff, organising 'Finns 100 Mile Dinner' as part of the 2011 Crave Sydney '100 Mile Meal', did serve wines from two QLD Granite Belt Region wineries—Sirromet and Symphony Hill (p.c. October 2011)—which, we argue below, could be classified as wines local to the Northern Rivers. But just to note here, this lack of recognition of wines local to the Northern Rivers could be indicative of the difficulty of conceiving of the region as having any 'local' wines at all (despite the fact that the Northern Rivers is a classified wine zone; however, that zone does refer to a cluster of wineries lower down on the North Coast). Following, we contrast the ease of and restrictions on conceiving of and drinking 'local wine' in the two Australian locales from which the Australian regional restaurants drew the wines for their local produce dinners. We discuss how wine can be as equally important as food to the locavore/locabiber. Factors that support the concept and consumption of local produce are varied and complex. As other scholars have similarly discussed (see, for example DuPuis and Gillon, 2009; Garbutt, 2011; Stoneman, 2010; Heldke, 2007), identifying the local isn't always a simple equation, and local isn't always good or best.

11.2 CONCEIVING OF 'LOCAL WINE' IN TWO AUSTRALIAN LOCALES

The first locale is the Adelaide Superzone in SA [4], and the second is the Northern Rivers Zone and its nearby zones in northern NSW and southeast QLD (Beeston, 2002). [5] The Adelaide Superzone includes Mount Lofty, Fleurieu, Barossa and more distantly the Limestone Coast Zones that, in turn, include the Adelaide Hills and Clare Valley, McLaren Vale, and Coonawarra Regions. The Northern Rivers Zone is the classification for wineries on the mid-north coast of NSW, in the Hastings Valley Region (Beeston, 2002: 561). Nearby the Northern Rivers, on the far north coast, are the Northern Slopes that include Tenterfield and New England, and QLD Zones that include the Granite Belt Region and the Gold Coast Hinterland (Beeston, 2002: 568– 579; Wine Australia, nd).

The two locales are where the two authors of this article live or have lived; one of the authors, Moya Costello, has moved to the Northern Rivers. We already know from personal experience that Adelaide Superzone wines are more abundant, readily available and, on the whole, less expensive than those we are thinking of as local to the Northern Rivers. As experienced, but nonprofessional, wine drinkers in Adelaide, South Australia, Australia's 'wine capital' (Government of South Australia, 2011), we think of winemaking as an art form (discussed in more detail below). The Adelaide Hills, McLaren Vale, Coonawarra, Barossa and Clare Valleys produce among the best Australian, and international, wine. When Costello moved to the far NSW North Coast, she attempted to redeploy her Adelaide-based habit of drinking locally to a new region. The Adelaide Superzone has an international reputation (discussed below). The Northern Rivers is subtropical and not a diverse wine-growing area due to the high rainfall and humidity that produce mildew/fungus, destroying grape vines. Figure 1 (see Appendices at the end of this paper) aims to illustrate that one needs to be an avid detective to purchase local wine in the Northern Rivers, pursuing it with some personal cost to the wine consumer in terms of time, energy and finances, to know where to go, other than the cellar door or online, for what is a limited number of outlets selling a limited selection of local wines, given what is available at the cellar door or online. [6]

Because of contrasts in the two locales, (which we detail further below), they form an appropriate case study to query the idea of the local. As wine lovers from the two locales—as well as wine writers (see, for example, Costello 2010)—we are thinking of ourselves as the stand-in models for local wine drinkers. [7] We also want to note here that, in relation to food studies, which is interdisciplinary (Miller and Deutsch, 2009: 4–7), our discipline is not sociology, philosophy, ethnography, anthropology, history, or travel and tourism, but creative writing, close to cultural studies which is one of the 'major broad methodological baskets of food studies' (Miller and Deutsch, 2009: 6). Creative writing makes for an acute angle or refracted view on the field. Having addressed why the specific locales and who the drinkers are, following we consider why wine, why Australian wine, and why local. In this discussion we draw on industry sources (alongside scholarly ones), because they are among voices that act as influential forces in the formation of the local. This article represents a nascent case study, primarily experiential, and does not draw on travel and tourism marketing studies. This would be an appropriate component for a further research.

11.3 WHY WINE AND WHY AUSTRALIAN WINE

Generally speaking, drinking wine, whether local or not, is, like eating food, a regular part of our practice of the everyday. The symbiosis of food and drink with our lives "unfold[s] the realities in which we live" (Dolphijn, 2004: 9). However, obviously, wine is not as vital to survival or health as food is. Concerning health, although the Australian Wine Research Institute "suggests that the regular and moderate consumption of … wine, may reduce your risk of diseases, such as coronary artery disease … stroke and heart failure", the Institute also notes that "[t]he consumption of … wine above this moderate amount will, conversely, increase your risk of … diseases" (Stockley, 2009: 5).

Max Allen, an Australian wine critic, believes "we carry within us an archetypal idea of wine as a natural product of the earth" and "a remnant awareness of wine's ancient cultural and spiritual significance" (2010: 3). Wine, like food, is associated with affect: sensations and emotions. Winemaking as art generates and intensifies "sensation" which impacts "on bodies, nervous systems, organs" (Grosz, 2008: 16). Wine can set the dreaming, remembering mind in action, and thence links can emerge to contemporary issues in the practice of the everyday; just the aroma and appearance of wine can act as triggers which open doors onto our individual pasts, and, thence, develop identity (Costello, 2010). We want to drink, in this case specifically wine, locally, because wine (as well as other forms of drink), like food, can make up both our quotidian realities and our metaphysical capacities for identity and belonging. Choosing wine from a particular region brings the drinker closer to their sense of place and amplifies that feeling of belonging (see for example, Costello, 2010).

In relation to winemaking as an artform, Eric Rolls calls blending grape varieties to make wine the great art" (1997: 117), James Halliday, another Australian wine critic, has posed the question: "is winemaking an art?" (2010: 31) His answer—"most would say so" (ibid)—is guarded but gratifying for those who think it is. Mary Douglas and Baron Isherwood write that food and drink "are no less carriers of meaning than ballet and poetry" (1996: 49). Art is important culturally because it enables us to understand others, and ourselves; and to imagine possibilities related to problem-solving innovation and survival (see, for example, Buell, 2005; Potter, 2005).

As to the quality of Australian wine, Andrew Jefford states that "[n]o other country in the southern hemisphere has made an impact to rival that of Australia on the world winemaking scene over the last two decades" (2006: 65). The Australian Government Department of Foreign Affairs and Trade states that "Australia is consistently one of the top 10 wine-producing countries in the world", "one of the few wine producers to make every one of the major wine styles", and, in 2006–7, "Australian wine exports were worth $2.87 billion" with the top destinations being the United Kingdom, United States, Canada, Germany and New Zealand (2008: np). Australia has a strong culture of professional tertiary education in wine that includes Technical and Further Education institutions and at least nine universities, with more than that number of courses—Adelaide, Charles Sturt, Curtin, La Trobe, Melbourne, Southern Cross, Southern Queensland and Western Australia universities deliver Food Studies and Writing, Wine Appreciation, Oenology, Viticulture, Wine Production, Wine Science and Technology. The quality of Australian wine is acknowledged even in popular culture. In an episode of the American comedy television series, Frasier, the two protagonists, professionals who pride themselves on exquisite taste and highly skilled aesthetic judgment, compete to be corkmaster of their wine club in a blind fine-wine tasting which includes an Australian shiraz (Hartley, 2001). This is indicative of international recognition of quality.

South Australian wine in particular has a good reputation nationally and internationally. Jefford states that SA is Australia's "leading wine state in terms of both volume and quality" (2006: 65). Specific wines have notable claims about national and international reputations. For example, Charles Melton's Rose of Virginia was described as "the best ... in Australia" (Charles Melton Fine Barossa Red Wines, nd) in London's *Observer* newspaper. Online, Primo Estate claim that their La Biondina is "the world's premier Colombard based wine" (nd) and Allen deems it "the best example in the country" (2010: 64). (In relation to our experience of drinking these wines, see, as an example, Costello, 2012.)

The Northern Rivers Zone and its nearby zones are not nearly as well known as SA wine. For example, 2011 was the first time Queensland wine was available in the UK (Wines Unfurled, 2011). But there is evidence of quality for wines produced in the Northern Rivers and nearby zones. In 2011, Granite Belt's Sirromet won gold medals in European wine shows (Halliday, 2011: 33). The Granite Belt's Boireann and Symphony Hill and Gold Coast Hinterland's Witches Falls supply wine to the finedining Brisbane restaurant, Aria. Aria Sydney has a Two-Chef's-Hats' status (a restaurant rating compiled by the *Sydney Morning Herald*'s 'Good Food Guide'), won the *Sydney Morning Herald*'s 'Good Food Guide' Wine List of the Year Award in 2009, and was inducted into the Hall of Fame as part of the Australian Wine List of the Year Awards in 2008. Symphony Hill's 2003 Reserve Shiraz was Queensland's first gold medal winner at the 2005 Sydney Royal Wine Show. Wines from the area have won medals in other national competitions such as the Australian Small Winemakers Show and Australian Alternative Varieties Wine Show. (See Costello, 2012, for one of the author's experiences of drinking these 'local', alternate-variety wines.)

The wineries in this northeastern Australian locale are fewer in number, smaller in size, primarily boutique and therefore the wine often higher in bottle price. In contrast, short distances from Adelaide to local wineries, and the quality, quantity, variety and price, make the availability of local wine unproblematic in the locale of Adelaide, and SA wine is present in most, and most probably all bottle shops and restaurants in the city and its suburbs. [8] Moreover, it is fair to say the latter is true of the Northern Rivers, as is possibly the case in the whole of Australia. The Granite Belt Wine Country advertises more than thirty wineries, and there are about six in the Hastings Valley Region (Vinodiversity, 2011). But within the Adelaide Superzone, there are about seventy cellar doors in McLaren Vale and over eighty in the Barossa Valley alone (South Australian Tourist Commission, 2009). In the whole of Queensland there are "1,500 hectares under vine", but the figure is 5,000 in SA's McLaren Vale alone (Madigan, 2008: 4). Queensland produces approximately 0.25% of Australia's wine grapes, and Boireann, for example, a Granite Belt Winery, is only 1.5 hectares in size. Tonnes produced per hectare in the Northern Rivers and nearby zones are in the two-three figure range, but in the Adelaide Superzone they can be four to five (see Beeston, 2002). In relation to pricing, we note what Master of Wine David Stevens says: "No matter how skilled winemakers may be they are still at the mercy of nature's elements and in recent times the retail markets" (quoted in Geddes 2007: 8). Wines from the Northern Rivers Zone and surrounds usually have a starting price of around AU $15 and are most often in the AU $20–$40 or more range, while a good-drinking and alternative variety such as Dopff au Moulin Alsace Pinot Blanc is around AU $12 from the Australian large-chain retailer, Dan Murphy's, despite it coming all the way from France. The Vermentino from the Granite Belt's Golden Grove Estate was first in production in 2011 with 150 bottles at AU $26 each, while the SA Yalumba Vermentino, in its keenly priced, high-volume but good-quality Y series, is around AU $12.

Low and Vogel (2011) found that productive climate and topography, good transportation and information, and proximity to markets favour higher levels of direct-to-consumer sales of local produce. So, in summary, it is more challenging to drink what could be seen as constituting wine local to the Northern Rivers Region and its surrounds than it is in the Adelaide Superzone, because of price and the complexities of procurement, which include winery size, location/distance, distribution and representation/marketing, and reputation/identity. South Australian wineries have the advantage of being more numerous, larger, relatively closely grouped, generally highly regarded, and also near a capital city, compared to their counterparts in the Northern Rivers, all of which facilitate consumer access and effective umbrella marketing programs.

11.4 WHY LOCAL AND WHY *TERROIR*

What might local mean? In terms of distance alone, the details vary. In *The 100-Mile Diet*, Alisa Smith and J B MacKinnon found one hundred miles was the distance from where they lived within which they could readily access food and wine (2007: 10; 215). Crave Sydney said of its International Food Festival's '100 Mile Meal' that "[a]l ingredients ... come from just down the road or a nearby paddock—within a 160km radius". David Goodman has specified '1500 miles' of travel from produce growth 'to your table' (2011: np). As a further example, Clare Hinrichs found that, in Iowa, local has "shifted from signifying food grown within a county or a neighbouring one to food grown anywhere in the state" (quoted in DeWeerdt, 2012: np).

Distance is complicated by changes in climate and geography. For what could be defined as local wines in the Northern Rivers in terms of distance alone, grapes are grown in distinctly, even radically different climates, vegetation and soils (a much more recognisable and significant difference than in the Adelaide Superzone). Problems of identifying a wine as local may include these climactic, geographical and geological differences. Mt Tamborine, the Granite Belt, Tenterfield and New England are all higher above sea level than towns and villages in the Northern Rivers, with a consequent colder climate, different vegetation, geography, etc. (see Figure 3). [9] Such differences are not always the case in SA where, for example, the Fleurieu Peninsula, south of Adelaide, is very similar to the coastal area of suburban Adelaide. Halliday (2011: 33) notes that regions such as the Granite Belt can produce good wine only because of their altitude. Grapes are grown and wine is made in Whian Whian, a village outside of subtropical Lismore, in the Northern Rivers; however it specialises most successfully in Chambourcin (as do wineries further south in the Northern Rivers Zone), a tough-skinned grape that will resist mildew and fungus.

Some potential wineries local to the Northern Rivers, like the Granite Belt, are in another state, Queensland. Australia has a long history of state rights and different state cultures. Anything from the ticketing system on public transport to vocabularyuse can be different in another state. Perhaps one of the best places to discover national attitudes, or predilections, is in literature. In her short fiction, 'The Bangalow Story' (2008), Barbara Brooks likens the northerly move from the temperate to the subtropical zone on Australia's east coast as crossing "a kind of Mason-Dixon line" (2008: 22). Moving in the opposite direction, Andrew McGahan's protagonist in his novel *Last Drinks*, trying to escape Queensland, can only get as far as "a few miles from the border" (2000:11; 60) in a mountainous, cold town of the Border Ranges. So issues of specific identity and place are further complicated for the formation of the concept of wines local to the Northern Rivers Zone because of cross-border and borderland issues. The borderland space was ever ambiguous and unruly. Further, notions of *terroir* contribute to highly specified identities; *terroir* is, for some, a key definer of locality and identity.

What constitutes *terroir* and what are its implications in relation to defining the local and its production of identity? According to respected Australian agricultural scientist and terroir specialist John Gladstones, the original meaning of *terroir* is:

the vine's whole natural environment, the combination of climate, topography, geology and soil that bears on its growth and the characteristics of its grapes and wines [as well as] local yeasts and microflora. (2011: 2)

This outlook is echoed by Bruno Prats, the former owner of Chateau Cos d'Estournel in the Médoc: "[t]he terroir is the coming together of the climate, the soil, and the landscape" with variations in temperature, rain, light, slope and drainage (in Halliday & Johnson, 1994: 19; Geddes, 2007: 25). Gladstones sees *terroir* as allowing the consumer to predict the style, if not so much the quality, of the wine they are about to drink with a fair degree of reliability (2011: 2). For Thomas Girgensohn, "Experienced and educated palates can detect the difference" of *terroir*, (2011: 37). While a sommelier's rigorous professional training may make for a highly sophisticated palate, a broad recognition of *terroir* may also be available to an experienced but untrained and amateur palate.

Terroir is also argued to be culture as well as nature. "[W]inemaking techniques that enhance and complement the raw material flavor are part of the terroir—because terroir includes the culture and practices of the region", says Master of Wine Andrew Corrigan (2010: 37). Halliday and Hugh Johnson note that while:

[c]haracter ... is determined by terroir; quality is largely determined by man [sic] ... Incompetent winemaking can destroy the

potential of a given site to produce wine of great character (first) and quality (second). (1994: 20)

But they add: "the role of the winemaker, although critical, is nevertheless, limited" (ibid: 22). However, Australian wine critic Philip White (2011) believes that while you can (and must) have the right conditions to make wine, the critical element is always the maker who sees how best to capitalise on them. His notion is that *terroir* is ultimately about the winemaker. In his own provocative style, White asserts:

Humans are the single biggest aspect of terroir. All that dandy fluff about landscape, geology, climate and aspect provides an obvious mass to the old French theory of terroir, but whatever you think, the human intervention factor is the biggest when it comes to tipping somebody's works into one's own personal body. ... Only great winemakers can really influence terroir for the betterment of their wines ... those sufficiently sympathetic and sensitive to their piece of country. (2011: np)

White also argues that there is more to take into account than *who* does the work; it is *how* they do it. He contrasts "sugar mining by industrial grape farming" with more sensitive attention "to the whims, folds and crannies of the land they farm, and the life abundant in it" (ibid).

But *terroir* and the local can be highly contentious. For example, arguments about who is inside or outside SA's Coonawarra Region testify to the importance of marketing *terroir* (Port, 2010: 36–39), and to the nuances of the local. As DuPuis and Gillon state, "market boundaries" are set, determining "who can participate in the market and who cannot" (2009: 10). Robert Geddes stresses that:

[t]he dark side of terroir lies when successful marketing creates recognition and riches that blunt the winemaker's desire to keep improving a wine because the vineyard or appellation is widely recognised and guarantees high prices based solely on the reputation of the land rather than what is in the bottle. (2011: 44)

It is worth noting in relation to *terroir*, Australian wine, and formation of the local, that Allen (2010: 19) says *terroir* has come late to its
wine industry (see also Pinney and Goldberg, 2006: 476–477; and Gladstones, Smart and Lindley, 2006: 694–695, on New-World wine making).
Australian winemakers have commonly blended grapes from different
regions, with 'the winemaker more important than the place'. More recently, a number of "quality-focused smaller winemakers" are producing
"more single-vineyard, single-site, terroir-driven wines" (Allen, 2010:
20). Given the climatic difficulties of growing wine grapes notably in the
Northern Rivers Zone, an interaction of factors, rather than singularity,
marks some winemaking. [10] The Hastings Valley Region's Cassegrain
takes some Chardonnay, Semillon and grapes for its Rosé from the local
area, and some, along with Viognier, Verdelho, Sauvignon Blanc, Shiraz,
Merlot and Pinot Noir, from elsewhere, including New England. Witches
Falls grows its Fiano and Viognier nearby in Boyland, but uses grapes for
its red wine and some of its white from the Granite Belt Region. Viticulturist Mark Kirkby of Topper's Mountain, New England, uses winemaker
Mike Hayes of Symphony Hill in the Granite Belt Region. Jefford rightly
notes that the viticulturist is "just as important as the winemaker, and quite
possibly more so" (2006: 73). SA's Yalumba team, from the Eden Valley
Region, has influenced Hayes (specifically in the use of wild yeast) (Stelzer, 2008: 21). Here is the mix of *terroir*, viticulturist and winemaker.

We suspect that if we were to put these experts together (Gladstones,
Corrigan, Halliday, White, Allen, Jefford et al.), they would agree about
the necessary plural ingredients in the recipe for a specificity of locale.
Knowledge about locale may make local wine consumption preferable,
since it permits the local wine drinker to judge quickly and confidently,
assuming that: the local product is known to the local wine drinker, the
language associated with its consumption a knowledge already acquired,
and the local product is accessible, affordable and of quality. The narrower
focus may allow an easier command of detail. The local maker and drinker
may have a better chance of actually meeting. For example, as a new local
resident, Costello picked the Chambourcin grapes in the 2011 and 2012
vintage at Imogen's Farm, Whian Whian, after receiving an invitation to
do so on a visit to her new medical practitioner, a relation of the wine-

maker. This anecdote conveys the characteristics of the local in relation to communication, knowledge, and a shared culture. For the French, "taste … is a form of local knowledge", writes Amy B Trubek: "[l]ocal taste, or *goût du terroir*, is … evoked when an individual wants to remember an experience, explain a memory, or express a sense of identity" (2005: 268–9).

While local consumption is about identity, it is also, more finely, about ethical and environmental issues. In 2011 the U.S. Department of Agriculture reported that sales of 'local foods' were on the increase (Suhr, 2011). Local, in this report, meant roadside stands, farmers' markets, grocers and restaurants. The ethics of consumption of local food and drink include taking responsibility for personal health, where freshness, consequently coupled with closeness, is considered to have a positive influence—however, this latter point is clearly an issue with food and not always with wine which can improve with age in some cases. An ethics of local consumption is also about supporting the local economy to sustain one's immediate social and cultural networks. As Talpalaru notes, "Local eating … exemplifies efforts to de-link from various large-scale forms of globalization and of getting away from the grasp of corporate bio-power" (2010: np). Further, an ethics of local consumption includes lessening the adverse impact on the global environment from unsustainable energy usage and pollution or greenhouse-creating/global warming carbon emissions in transportation. Clive Hamilton writes:

> *[t]he reluctant conclusion of the most eminent climate scientists is that the world is now on a path to a very unpleasant future and it is too late to stop it … global warming … will this century bring about a radically transformed world that is much more hostile to the survival and flourishing of life … climate change does not mean we should do nothing. Cutting global emissions … can at least delay some of the worst effects of warming. (2010: vii–xi)*

In turn, "Locavores aim to lower their ecological footprint and honour their environment by eating … food … produced within a limited radius" (Talpalaru, 2010: np). As well, the rising cost of oil and gas focuses attention on the local product (Suhr, 2011).

11.5 CONCLUSION

Adelaide wine drinkers have the easier and more environmentally friendly task, compared with those in the Northern Rivers, in drinking locally simply from the perspective of driving distance to the cellar door. Further, orders from the Northern Rivers for "local" wine may go to a capital city such as Brisbane or Sydney first, as central distribution points, before it comes to the region. (Of course, this happens with Adelaide wines consumed in Adelaide too, in distribution to wine clubs, big retailers, and online purchases.) But winemakers and wine consumers in both SA and the Northern Rivers, however, can sensibly focus on the produce being linked to identification with regional character and a feeling of ownership. Moreover, the growing concern—for environmental, health, economic and cultural reasons—about food miles might help 'local' wineries to be perceived of as such and be accessed.

Philip White says: "I tend to favour great wines which remind me fondly of their maker. I can see the face" (2011: np). While this comment is not necessarily generated by or exclusively concerned with the idea of the local, from the perspective of the local perhaps we can also see something of our own faces in the wines that come from the locale we inhabit. Local wine that is well made ought to win the sentiments of a local consumer, be the bottle from Wirra Wirra or Whian Whian. [11]

ENDNOTES

1. Pronounced 'wine wine'.
2. With thanks for this Latin phrase to Dr Jacqueline Clarke, University of Adelaide.
3. It is not unusual to use 'Northern Rivers' as a catchment term for all the townships on the far North Coast of NSW.
4. Adelaide is the capital of the state of South Australia
5. Beeston lists the official Australian zones and regions.
6. See Figure 1 for availability/accessibility of 'local' wine in Northern Rivers/Lismore.
7. As continuing academics, we currently have relatively good incomes. We note what Guthman (Stoneman 2010: online) says, among other things, about an income to consume locally: '[Y]ou have to question localization as an ethical (or coherent)

response. At the very least it is ironic that re-localization efforts have gained traction in some of the most well-off regions in the world...'

8. See Figure 2 for contrasts in road/travel distances to wineries from a central site in the specific locales.
9. See Figure 3 for height above sea level contrasts among wine-growing areas that could be considered 'local' to the Northern Rivers.
10. Queensland Wine (online) also details Brisbane and Scenic Rim and Somerset Valley as wine areas which are around 143–168kms/89–104miles in distance from Lismore, and Toowoomba/Darling Downs (216–234/135) is marked by Wine Australia (nd).
11. Wirra Wirra is a small to mid-size winery in the Adelaide Superzone.

BIBLIOGRAPHY

1. 'A Delicious Gourmet Weekend' (2011) The Northern Star 12 November: 54
2. Allen, Max (2010) The Future Makers: Australian Wines for the 21st Century, Victoria: Hardie Grant
3. Australian Government, Dept of Foreign Affairs and Trade (2008) 'Australian Wine Industry', online at: http://www.dfat.gov.au/facts/wine.html (accessed October
4. 2011)
5. Beeston, John (2002) The Wine Regions of Australia, Crows Nest: Allen & Unwin
6. Bonzle.com: Bonzle Digital Atlas of Australia (2011) online at: http://www.bonzle.com/c/a (accessed 12 October 2011)
7. Brooks, Barbara (2008) 'The Bangalow Story', Leaving Queensland, Sydney: Sea Cruise Books
8. Buell, Lawrence (2005) The Future of Environmental Criticism: Environmental Crisis and Literary Imagination, Malden, MA: Blackwell
9. Charles Melton Fine Barossa Red Wines (nd) 'History', online at: http://www.charlesmeltonwines.com.au/about-us/history (accessed 6 October 2011)
10. Corrigan, Andrew (2010) 'The Trouble with Terroir', Winestate (November/December) v33i6: 34–37
11. Costello, Moya (2010) 'The Drink Has Called It into Being: A Year in a Wine Column', Text, Special Issue n9, October, online at: http://www.textjournal.com.au/speciss/issue9/content.htm (accessed October 2011)
12. Costello, Moya (2012) 'Living Australian Wine', China-Australia Entrepreneurs, v2, 70–71, online at: http://www.caemedia.com/En_Book2012210125243/index.html (accessed February 2013)
13. Crave Sydney International Food Festival, '100 Mile Meals', online at: http://cravesydney.com/events (accessed 5 October 2011)
14. Destination Food (2010) 'Tommy's 30 Mile Dinner', online at: http://www.destinationfood.com.au/2011/03/tommys-30–mile-dinner/ (accessed October 2011)
15. DeWeerdt, Sarah (2012) 'Is Local Food Better?', Worldwatch Institute: Visions for a Sustainable World, online at: http://www.worldwatch.org/node/6064 (accessed January 2013)

16. Dolphijn, Rick (2004) Food Scapes: Towards a Deleuzian Ethics of Consumption, Delft: Eburton Publishers
17. Douglas, Mary and Isherwood, B (2006) The World of Goods: Towards an Anthropology of Consumption, London: Routledge
18. DuPuis, Melanie E and Gillon, Sean (2009) 'Alternative Modes of Governance: Organic as Civic Engagement', Agriculture and Human Values v26n1: 43–56, online at: https://www.researchgate.net/publication/225869314_Alternative_modes_of_gove rnance_organic_as_civic_engagement (accessed January 2013)
19. Garbutt, Rob (2011) The Locals: Identity, Place and Belonging in Australia and Beyond, Oxford: Peter Lang
20. Geddes, Robert (2007) A Good Nose and Great Legs, Millers Point: Murdoch Books
21. Girgensen, Thomas (2011) 'Subtleties of the Barossa Terroirs', Winestate (March/April) v34i2: 36–39
22. Gladstones, John (2011) Wine, Terroir and Climate Change, Adelaide: Wakefield Press
23. Gladstones, John, Smart, Richard and Lindley, Dennis (2006) in Janice Robinson (ed.) The Oxford Companion to Wine, third edition, Oxford: Oxford University Press: 694–695
24. GlobeFeed.Com (2009) 'Distance Calculator', online at: http://distancecalculator. globefeed.com/Country_Distance_Calculator.asp (accessed November 2011)
25. Goodman, David (2010) 'Winter Solstice Letter' 21 December, in NetAge Endless Knots: David Goodman on Food, Water, Energy, Health & Wellness, Governance & Public Policy, 4 January 4 2011, online at: http://endlessknots.netage.com/endlessknots/2011/01/each-year-my-friendantioch-alum-too-we-met-in-college-david-goodman-sends-a-letter-to-friends-withhis-thoughts-of-days.html (accessed January 2013)
26. Government of South Australia (2011) 'Wine Regions', South Australia, online at: http://www.southaustralia.com/food-and-wine/wine-regions.aspx (accessed September 2011)
27. Grosz, Elizabeth (2008) 'Chaos, Territory, Art. Deleuze and the Framing of the Earth', New York: Rutgers University: 15–28
28. Halliday, James & Hugh Johnson (1994) The Art & Science of Wine, Prahran: Hardie Grant
29. Halliday, James & Hugh Johnson (2010) 'An Artist's Spirit', News Articles 14 February, online at: http://www.winecompanion.com.au/News/News-Articles/2010/2/An-artistsspirit (accessed September 2013)
30. Halliday, James & Hugh Johnson (2011) 'Small Is Beautiful', The Weekend Australian Magazine J4–5 June: 33
31. Hamilton, Clive (2010) Requiem for a Species, Crows Nest: Allen & Unwin
32. Hartley, Nick (2001) '[17.7] Whine Club', ReoCities, online at: http://reocities.com/Hollywood/derby/3267/717.html (accessed September 2011)
33. Heldke, Lisa (2007) 'Staying Home for Dinner: Ruminations on Local Foods in a Cosmopolitan Society', online at: http://liberalarts.iupui.edu/mpsg/Staying%20 Home%20for%20Dinner (accessed 17 January 2013)
34. Imogen's Farm, Imogen's Farm Winery and Farmstay, online at: http://imogensfarm. com/wine/ (accessed February 2011)

35. International Food Festival, '100 Mile Meals', online at: http://cravesydney.com/events.php?intcategoryid=85&inteventcategoryid=&intlocation=&intcuisine=&intday=&strrestaurantname=&strkeyword=&intvegetarian=&intchildren=&intwheelchair= (accessed October 2011)

36. Jefford, Andrew (2006) The World of Wine, New York: Ryland Peters and Small

37. Low, Sarah A and Vogel, Stephen (2011) Direct and Intermediated Marketing of Local Foods in the United States, Economic Research Report (ERR-128), November, Department of Agriculture, United States, online at: http://www.ers.usda.gov/publications/err128/ (accessed November 2011)

38. Madigan, Anthony (2008) 'Queensland Wine Front Page News', WBM: Australia's Wine Business Magazine 4 July, online at: http://www.symphonyhill.com.au/wbm.pdf (accessed January 2012)

39. McGahan, Andrew (2000) Last Drinks, St Leonards: Allen & Unwin

40. Miller, Jeff and Deutsch, Jonathan (2009) Food Studies: An Introduction to Research Methods, Oxford: Berg

41. OUP Blog (2006–2013) 'Oxford Word of the Year: Locavore', 3 January and 20 November, online at: http://blog.oup.com/2007/11/locavore/ and http://blog.oup.com/2007/11/prentice/ (accessed February 2013)

42. Pinney, Thomas and Goldberg, Howard (2006) in Janice Robinson (ed.) The Oxford Companion to Wine, third edition, Oxford: Oxford University Press: 476–477

43. Port, Jenny (2010) 'Search for an Identity', Winestate, July/August v33i4: 36–39

44. Potter, Emily (2005) 'Ecological Consciousness in Australian Literature: Outside the Limits of Environmental Crisis', Hawke Research Institute for Sustainable Societies, Working Papers, 29, online at: http://www.unisa.edu.au/hawkeinstitute/publications/downloads/wp29.pdf (accessed January 2012)

45. Primo Estate, 'Primo Estate La Biondina Colombard 2011', online at: http://www.primoestate.com.au/index.php?option=com_content&view=article&id=21&Itemid=51 (accessed October 2011)

46. Queensland Wine (nd), online at: http://www.queenslandwine.com.au/index.php?MMID=123 (accessed November 2011)

47. Rolls, Eric (1997) A Celebration of Food and Wine, St Lucia: University of Queensland Press

48. Smith, Alisa and MacKinnon, J B (2007) The 100–Mile Diet, Canada: Random House

49. South Australian Tourist Commission (2009) 'McLaren Vale Wine Region', online at: http://www.southaustralia.com/media/documents/food-and-wine/mclaren-valewine-region-map-2010.pdf (accessed November 2011)

50. Stelzer, Tyson (2008) 'Ewen's Day in the Sun', WBM: Australia's Wine Business Magazine, July: 18–22, online at: http://www.symphonyhill.com.au/wbm.pdf (accessed January 2012)

51. Stockley, Creina (2009) Wine and Health Information, Adelaide: Australian Wine Research Institute

52. Stoneman, Scott (2010) 'Julie Guthman, on Globalization, Neoliberalism, Obesity, Local Food and Education', Politics and Culture n2, online at: http://www.politicsandculture.org/2010/10/27/an-interview-with-julie-guthman/ (accessed January 2013)

53. Strong, Jeremy (ed.) (2011), 'Introduction', Educated Tastes: Food, Drink, and Connoisseur Culture, Nebraska: University of Nebraska Press: ix-xii

54. Suhr, Jim (2011), '"Locally Grown" Food Now a $4.8 Billion Business, Says USDA Report', Huff Post, 17 November, online at: http://www.huffingtonpost.com/2011/11/14/locally-grownfood_n_1092146.html?ref=fb (accessed November 2011)

55. Talpalaru, Margrit (2010) 'Lines of Flight: Does the Locavore Movement Offer an Alternative to Corporatism?', Rhizomes: Boundaries of Publication: Posthumography 20, summer, online at: http://www.rhizomes.net/issue20/talpalaru.html (accessed November 2011)

56. Trubek, Amy B (2005) 'Place Matters', in Korsmeyer, Carolyn (ed.), The Taste Culture Reader: Experiencing Food and Drink, Oxford: Berg: 260–271

57. Vinodiversity (2011) 'Hastings Valley Wine Region', online at: http://www.vinodiversity.com/hastings-river-wine-region.html (accessed November 2011)

58. White, Philip (2011) 'Bukowski & Gladstones Have a Write: Putting the Terror Back into Terroir Dr John's Second Epistle No Emo in This Chill Memo', 22 June, online at: http://drinkster.blogspot.com/2011/06/bukowski-gladstones-have-write.htm (accessed July 2011)

59. Wine Australia (nd) 'Australian Wine Regions', online at: http://wineaustralia.com/australia/Default.aspx?tabid=5462 (accessed September 2011)

60. Wines Unfurled (2011) 'Latest News from Wines Unfurled', online at: http://www.winesunfurled.co.uk/news/13/112/Wines-from-Queensland-s-Granite-Belt-coming-to-the-UK (accessed November 2011)

61. Witches Falls Winery (2011) '2011 Granite Belt Co-inoculated Sauvignon Blanc', Witches Falls, online at: http://www.witchesfalls.com.au/ourwines/whiterange/2011_granite_belt_coinoculated_sauvignon_blanc (accessed October 2011)

62. Wu, Olivia (2005) 'Environment in Focus: Diet for a Sustainable Planet: The Challenge: Eat locally for a Month (You Can Start Practicing Now)', SF Gate, online at: http://www.sfgate.com/green/article/ENVIRONMENT-IN-FOCUS-Diet-for-asustainable-2630631.php#ixzz2JtJtgTSM (accessed February 2013).

CHAPTER 12

SENSORY DESCRIPTORS, HEDONIC PERCEPTION AND CONSUMER'S ATTITUDES TO SANGIOVESE RED WINE DERIVING FROM ORGANICALLY AND CONVENTIONALLY GROWN GRAPES

ELLA PAGLIARINI, MONICA LAUREATI, AND DAVIDE GAETA

In recent years, produce obtained from organic farming methods (i.e., a system that minimizes pollution and avoids the use of synthetic fertilizers and pesticides) has rapidly increased in developed countries. This may be explained by the fact that organic food meets the standard requirements for quality and healthiness. Among organic products, wine has greatly attracted the interest of the consumers. In the present study, trained assessors and regular wine consumers were respectively required to identify the sensory properties (e.g., odor, taste, flavor, and mouthfeel sensations) and to evaluate the hedonic dimension of red wines deriving from organically and conventionally grown grapes. Results showed differences related mainly to taste (sour and bitter) and mouthfeel (astringent) sensations, with odor and flavor playing a minor role. However, these differences did not influence liking, as organic and conventional wines were hedonically comparable. Interestingly, 61% of respondents would be willing to pay more for organically produced wines, which suggests that environmentally sustainable practices related to wine quality have good market prospects.

12.1 INTRODUCTION

The sensory analysis of wine has always given rise to interest both in the scientific community and among consumers. Wine is tightly tied to psychological aspects besides being purely sensory. There have been many studies carried out on different aspects connected with wine tasting such as the cognitive and perceptual processes that characterize wine expertise. Wine-tasting expertise involves advanced discriminative and descriptive abilities with respect to wine. While the basis of wine expertise remains unknown, differences in performance between experts and novices are relatively clear (Lawless, 1984; Noble et al., 1987; Solomon, 1990; Hughson and Boakes, 2002; Zucco et al., 2011). Wine-tasting experts such as sommeliers have obviously a greater sensory ability than inexperienced novices, but their knowledge of wine may sometimes lead them to misperception of the product (Pangborn et al., 1963; Morrot et al., 2001). Pangborn et al. (1963) and Morrot et al. (2001) carried out experiments in which white wines were colored to obtain rosé and red wines, respectively. Pangborn et al. (1963) found that such a modification led wine experts but not novices to judge the product as sweeter than colorless controls. Similarly, Morrot et al. (2001) showed that wine experts described the white wine with the characteristics of a red wine.

While there are several studies on wine perception, little is known about sensory characteristics of wines deriving from organically and conventionally grown grapes. Organic agriculture is a production management system that promotes and enhances biodiversity, biological cycles, and soil biological activity. The primary goal of organic agriculture is to minimize all forms of pollution and to avoid the use of synthetic fertilizers and pesticides, thus optimizing the health and productivity of soil, plants, animals, and humans.

In recent years, consumers have become increasingly concerned by the effects of conventional agricultural production practices on both human and environmental health. As a consequence, production obtained from organic farming methods has been rapidly growing in developed countries. This may be explained as organic food adequately meets all requirements for quality, genuineness, and healthiness (Forbes et al., 2009). Recent evidence has also shown an increase of the related literature, even though studies are still few in number. The studies comparing foods derived from organic and conventional growing systems focused mainly on three topics: nutritional value, sensory quality, and food safety (Bourn and Prescott, 2002).

Relative to the nutritional value of wine, its antioxidant activity and benefit on health were addressed (Renaud and De Lorgeril, 1992), showing that phenolic compounds are natural anti-inflammatory and efficient scavengers of free radicals (Akçay et al., 2004).

As to the sensory quality of food products, reports indicate that organic and conventional fruits and vegetables may differ on a variety of sensory aspects; however, findings are inconsistent (Bourn and Prescott, 2002). Therefore, the assumption of organic food having a better taste may be explained by the consumer's expectation of a healthier and safer product evoked by the label "organic food" (Deliza and MacFie, 1996). Indeed, expectations greatly influence subject responses (see e.g., Dalton et al., 1997).

Few studies compared sensory properties of wines derived from organically and conventionally grown grapes. Moyano et al. (2009) for instance, examined the aroma profile of sherry wines that had been cultivated conventionally and organically and found that organic wines had a sensory profile similar to that of the conventional ones, but lower odor intensity. The same findings were reported by Dupin et al. (2000), who examined German wines and found that organic products tended to be less aromatic than conventional ones.

"Sangiovese" (*Vitis vinifera* L.) is the most widely consumed Italian wine. It is used to produce prestigious Tuscan wines such as Chianti and Brunello di Montalcino. To our knowledge no studies are available on Sangiovese red wine sensory quality. Thus, the main aim of this work is to identify and describe the sensory properties, such as odor, taste, flavor, and mouthfeel sensations, that characterize organically and traditionally grown Romagna Sangiovese red wines. Also, as sensory properties greatly influence food preference, the hedonic dimension of organic and conventional wines was investigated.

12.2 MATERIALS AND METHODS

12.2.1 WINES

The red wines evaluated in the present study were produced from ripe grapes from *Vitis Vinifera* Sangiovese harvested in September 2007 and 2008 in the region of Faenza (Italy). The grapes were derived from two different farms located in adjacent areas and subjected to similar environmental conditions. For both vintages, one farm produced grapes according to organic techniques whereas the other adopted conventional agricultural techniques. At variance from conventionally cultivated grapes neither insecticides nor synthetic fertilizers were used in organic agriculture during the growth.

All wines were produced following the same process according to PDO (Protected Designation of Origin) specifications. Wines were analyzed 6 months after they were bottled. Three bottles from the organic and three from the traditional production of vintage 2007 were randomly selected to be used for sensory analysis and the same procedure was used for vintage 2008.

12.3 SENSORY ANALYSIS

12.3.1 PARTICIPANTS

Descriptive analysis of wines: 12 assessors (seven women and five men) aged on average $27.0 \pm$ (SD) 3.5 years (range 23–35 years) were selected. They were trained to evaluate organic and conventional wines from vintages 2007 and 2008.

Hedonic test of wines: a second group of 100 (50 women and 50 men) regular red wine consumers (inexpert individuals with no formal wine training) aged on average 32.1 ± (SD) 9.6 years (range, 20–60 years) participated.

The participants were students and employees of the University of Milan, who reported liking red wine and consuming it more than twice a month. None of the participants had previous or present taste or smell disorders. The study was in accordance with the Declaration of Helsinki. The protocol was approved by the Institutional Ethics Committee at the study site. Informed consent was obtained from all subjects.

12.3.2 DESCRIPTIVE ANALYSIS

Descriptive analysis (Lawless and Heymann, 1998; ISO International Organization for Standardization, 2003) was used to identify and quantify the sensory properties of organic and conventional wines from two successive vintages.

Training phase: subjects were trained over a period of 2 months. During the first part of the training, assessors tasted Romagna Sangiovese wines and set up a list of descriptors that characterized the wines. To do so, assessors wrote down as many terms as they could to describe the sensory characteristics fully. Assessors agreed through panel discussion on what terms were relevant, and arrived at definitions for each term. At this stage, a reference product was provided in order to help the assessors to understand each term.

Evaluation phase: after training was completed, the panel evaluated the two wines (organic vs. conventional) in triplicate. Judges were instructed to drink and swallow each sample and rate the intensity of each attribute using a nine-point scale (1 = absence of the sensation and 9 = maximum intensity). The sessions were performed on the same day (with a minimum 2-h break between the sessions) at the sensory laboratory of the Department of Food, Environmental and Nutritional Sciences (DeFENS, Università degli Studi di Milano) designed in accordance with ISO guidelines (ISO International Organization for Standardization, 2007). Data acquisition was done using Fizz v2.31 software (Biosystèmes, Couternon,

France). Assessors were asked not to smoke, eat or drink anything, except water, at least 1 h before the tasting sessions. For each sample, judges received a 30 ml sample served in glasses coded with a three-digit number and covered with a Petri dish to avoid the escape of volatile components. Participants were provided with mineral water and unsalted crackers to clean their mouth between tastings. Wines were served at $18 \pm 1°C$. Presentation orders were systematically varied over assessors and replicates in order to balance the effects of serving order and carryover (MacFie et al., 1989).

12.3.3 CONSUMER'S PREFERENCE AND ATTITUDE TOWARD WINE CONSUMPTION

Since the sensory properties of a food are among the primary determinants of food preference and choice, we also investigated the hedonic qualities of organic and conventional Romagna Sangiovese wines. For this purpose, the two wines under study, organic and conventional from vintage 2008, were evaluated along with four other Romagna Sangiovese wines from the same vintage produced according to conventional agriculture techniques, which were purchased in local wineries and were comparable for price category to those under study. Due to practical constraints (i.e., no availability of wine), the wines from vintage 2007 were not included in the hedonic evaluation.

Consumers were invited to take part in a hedonic test carried out at the DeFENS sensory laboratory. Each participant received a series of six wines (20 ml for each product) served in glasses coded with three-digit numbers and covered with Petri dishes. For each sample, participants were instructed to drink and swallow the wine and rate the degree of liking using a seven-point hedonic scale (with 1 = extremely disliked and 7 = extremely liked; Lawless and Heymann, 1998). Consumers were asked to drink mineral water and to eat a piece of unsalted cracker to clean their mouth between tastings. Also, they were asked not to smoke, eat or drink anything, except water, 1 h before the tasting session. Data were collected using Fizz v2.31g software program (Biosystemes, Couternon, France). Wines were evaluated under standard light conditions at a temperature of

$18 \pm 1°C$. In order to balance the effects of serving order and carryover, the presentation order of the wines was randomized. After the liking test, the subjects were asked a few questions about their wine consumption habit and organic wine purchase likelihood.

12.4 RESULTS

12.4.1 DESCRIPTIVE ANALYSIS

The panel generated a total of 12 descriptors that characterize the sensory profile of the wines: four odor descriptors (fruity, spicy, woody, and vanilla), two taste descriptors (sour and bitter), three flavor descriptors (fruity, spicy, and woody) and three mouthfeel sensations (astringent, alcohol, and body). Complete definitions and standard products for all descriptors are listed in Table 1.

Mean intensity ratings of organic and conventional wines are reported in Figures 1 and 2. Intensity data for each sensory descriptor from the two vintages were analyzed separately through ANOVA with Wines (organic vs. conventional), Judges, Replicates (rep 1 vs. rep 2 vs. rep3) as factors. Relative to vintage 2007, Wines were significantly different for sour taste ($F = 10.31$, $p < 0.01$), bitter taste ($F = 8.87$, $p < 0.05$) and astringency ($F = 51.13$, $p < 0.001$). Post-hoc comparison using the Bonferroni test ($p < 0.05$) showed that organic wine was perceived as having a higher intensity of sour taste, and astringent sensation but lower bitter taste. Differences between the two wines from vintage 2008 concerned only astringency ($F = 13.66$, $p < 0.01$), with organic wine having a higher intensity. The effect of Judges was significant ($p < 0.05$), which is expected because individuals can of course have different sensitivities to the different descriptors. This effect can seldom be changed by training (Lea et al., 1997). Also, data analysis showed that F values for Replicates and interactions between Wines and Judges, Judges and Replicates and Wines and Replicates were not significant ($p < 0.05$) for nearly all the attributes. These results indicated that the mean scores for each wine given by the assessors for each attribute could be assumed to be satisfactory estimates of the sensory profile of the samples (i.e., good panel reliability).

TABLE 1: List of the 12 sensory descriptors of Romagna Sangiovese PDO wines with their relevant definitions and reference standards.

Descriptor	Definition	Reference standard
	Odor	
Fruity	Characteristic odor of a combination of blueberry, raspberry, and blackberry perceived by means of the sense of smell (orthonasal perception)	Infusion (24h, 4°C) of 12 blueberries, two raspberries, and one blackberry in 0.5 l of red table wine
Spicy	Characteristic odor of a combination of spices (cinnamon and clove) perceived by means of the sense of smell (orthonasal perception)	Infusion (24h, 4°C) of 16 cloves and one cinnamon stick in 0.5 l of red table wine
Vanilla	Characteristic odor of vanilla perceived by means of the sense of smell (orthonasal perception)	Commercial liquid vanilla odorant (2 ml) dissolved in 0.5 l of red table wine
Woody	Characteristic odor of toasted wood perceived by means of the sense of smell (orthonasal perception)	Guaiacol in red table wine (2 ppb)
	Taste	
Sour	One of the basic tastes, caused by solution of acidic compounds perceived in the oral cavity	Anhydrous citric acid (2 g) in 0.7 l of red table wine
Bitter	One of the basic tastes, caused by solution of bitter compounds perceived in the oral cavity	Caffeine (0.8 g) in 0.5 l of red table wine
Flavor		
Fruity	Characteristic odor of a combination of blueberry, raspberry, and blackberry perceived by means of the snese of smell during swallowing (retronasal perception)	Infusion (24h, 4°C) of 12 blueberries, two raspberries, and one blackberry in 0.5 l of red table wine
Spicy	Characteristic odor of a combination of spices (cinnamon and clove) perceived by means of the sense of smell during swallowing (retronasal perception)	Infusion (24h, 4°C) of 16 cloves and one cinnamon stick in 0.5 l of red table wine
Woody	Characteristic odor of toasted wood perceived by means of the sense of smell during swallowing (retronasal perception)	Guaiacol in red table wine (2 ppb)
	Mouthfeel	
Astringent	Mouth dryness caused by tannins and perceived in the oral cavity	Dissolve 1.5 g of tannin in 750 ml of red table wine
Alcohol	Characteristic heat/burning sensation perceived in the oral cavity	Mix 40 ml of 95% ethyl alcohol with 500 ml of red table wine
Body	Characteristic perceived in the oral cavity, due to the friction among the molecules in a liquid, that gives to it a limited fluidity and mobility.	Mix 6 ml of glycerol with 1 l of red table wine.

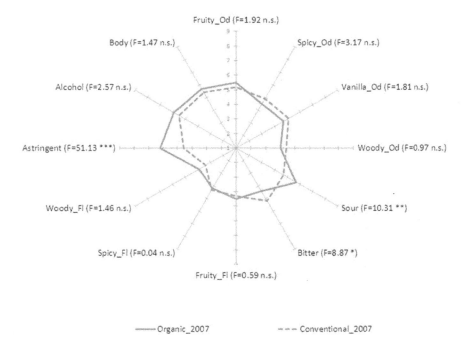

FIGURE 1: Descriptive analysis results: mean values for each sensory descriptor by method of production (organic vs. conventional) for vintage 2007. For each descriptor the relevant significance is reported (***p < 0.001, **p < 0.01, *p < 0.05).

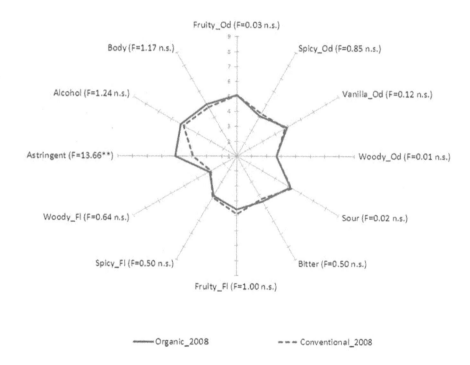

FIGURE 2: Descriptive analysis results: mean values for each sensory descriptor by method of production (organic vs. conventional) for vintage 2008. For each descriptor the relevant significance is reported (**p < 0.01).

12.4.2 STUDY OF CONSUMER PREFERENCE AND ATTITUDE TOWARD WINE CONSUMPTION

Mean hedonic ratings and standard errors for organic and conventional Romagna Sangiovese wines are reported in Table 2. Data analysis by means of one-way ANOVA showed significant differences ($F = 2.42$, $p < 0.05$) between wines for liking ratings. Post-hoc comparison using the Bonferroni test ($p < 0.05$) showed that organic and conventional wines from vintage 2008 were not significantly different and showed liking ratings comparable to other commercial wines (Sangiovese A, B, and C).

TABLE 2: Mean hedonic ratings (±STDERR) for organic and conventional Romagna Sangiovese wines from vintage 2008 and other four commercial Romagna Sangiovese wines from conventional agricultural techniques (Sangiove A–D).

Wines ($F = 2.42$; $p < 0.05$)	Hedonic rating
Sangiovese A	$4.2^a \pm 0.3$
Conventional 2008	$4.4^a \pm 0.3$
Organic 2008	$4.5^a \pm 0.3$
Sangiovese B	$4.8^{a,b} \pm 0.3$
Sangiovese C	$4.9^{a,b} + 0.3$
Sangiovese D	$5.3^b \pm 0.3$

Mean hedonic ratins with different superscripts are significantly different according to Bonferroni test ($p < 0.005$).

The same subjects involved in the hedonic study were also asked to answer a few questions about their attitude toward wine consumption (see, Table 3). About 59% of the subjects were habitual red wine consumers. The largest part (85%) of the wine used was mostly for home consumption. Wine is purchased at retail shops (59%) and most of the consumers are used to spending no more than 7 euros for a bottle of wine. Finally, it is interesting to note that when asked about the purchase of organically produced wine, 61% of them declared they would be willing to pay more for such product.

TABLE 3: Results from the questionnaire related to wine consumption habit and organic wine purchase intention.

Question	Answer (%)	Items
How would you define yourself?	59	Habitual wine consumer (2 or more times a month)
	41	Occasional wine consumer (less than twice a week)
Wine purchase is mainly destined to...	85	Home consumption
	15	Restaurant consumption
Where do you usually buy wine?	12	Wine shops
	59	Retail shops
	29	Wineries
How much do you usually pay for a bottle of wine?	3	Less than 3 euros
	19	Between 3 and 5 euros
	49	Between 5 and 7 euros
	28	Between 7 and 10 euros
	2	More than 10 euros
Would you be willing to pay an extra charge for an organically produced wine?	23	Yes, less than 10%
	34	Yes, between 10 and 20%
	4	Yes, between 20 and 30%
	0	Yes, more than 30%
	39	No

12.5 DISCUSSION

The present study investigated the sensory and hedonic qualities of red wines derived from organically and conventionally grown grapes. The examined wines were Romagna Sangiovese red wines. The descriptive analysis identified specific olfactory properties that characterize these wines, namely fruity, spicy, vanilla, and woody odors and flavors. Odor is a relevant sensory attribute of food, as well as of wines, which lead consumer's preference and choice. Also, the quality and specificity of each wine are associated in most cases with a specific odorant.

This study has shown that the organic and conventional wines differed marginally in the intensity of sensory descriptors. Only the properties of taste and mouthfeel sensations distinguished the two types of wine, whereas odor and flavor seemed to play a minor role. Organic wine from vintage 2007 was perceived as more sour and astringent but less bitter than its conventional counterpart, whereas differences between wines from vintage 2008 concerned only astringency.

In addition, the differences between wines did not influence liking, as organic and conventional wines were hedonically comparable. This means that consumers are not able to discriminate among organic and conventional wines from a hedonic point of view. One reason relates to their lack of formal training in sensory evaluation, which leads them only to detect major differences among products with less sensitivity to more subtle differences. It may be assumed that differences in liking could have been perceived between organic and conventional wines from vintage 2007, which showed larger differences in the intensity of some sensory qualities (i.e., bitter taste, sour taste and astringency) than wines from vintage 2008. Unfortunately, this hypothesis could not be verified, as wines from vintage 2007 were not included in the hedonic comparison. Nevertheless, self-reported comments by the participants suggest that even though the organic wine from vintage 2007 showed a high intensity of sourness and astringency, it was judged equally liked as its conventional counterpart.

The issue of comparing the hedonic qualities of organically and conventionally produced food has been tackled by various authors with respect to different food products, e.g., yogurt (Laureati et al., 2013), cheese (Napolitano et al., 2010a), meat (Napolitano et al., 2010b), and beer (Caporale and Monteleone, 2004). Interestingly, in these studies the liking of organic and conventional products has been evaluated under different information conditions: the blind condition (i.e., consumers taste and judge the product without any kind of information); the expected condition (i.e., consumers do not taste the product and judge it only on the basis of written or visual information); and the informed condition (i.e., consumers taste and judge the product after having read written information and/or seen an image). The main outcome of these studies is that organic products are liked more than their conventional counterparts but only in informed conditions, namely when consumers knew that they were to taste an organic food. Thus, it would seem that organic products are liked more because of the "healthier" connotation they have in the consumer's mind rather than for an actual preference based on perceptual attributes. Also, the influence of information about organic production on consumers' food preferences and expectations is especially evident in the case of consumers who are more interested in and proactive for "sustainable" products (Laureati et al., 2013). This suggests that expectation plays an important role for food consumption, since it may improve or degrade the perception of a product, even before it is tasted (Deliza and MacFie, 1996; Dalton et al., 1997). In this respect, it should be pointed out that the Sangiovese wines used in the present study were evaluated under blind conditions, without any information concerning production method. Thus, consumers' liking derives mainly from the mere sensory perception of the wines without any preconceived ideas due to their knowledge about the product.

Finally, an interesting result is that most of the consumers declared themselves willing to pay more for organically produced wines. This result is in line with the finding of a recent study by Lockshin and Corsi (2012) who reported that consumers in European countries as well as in the United States, New Zealand and Australia are willing to pay more for organic wines mainly for health and environmental reasons but also because consumers are interested in helping producers who adopt these innovations. Of course cognitive factors as personal expectancies—addressed above—have room. Therefore, a greater predisposition to pay an additional charge for organic wine may be due to specific consumer's attitude and involvement in sustainability issues.

In conclusion, the present study evidenced the sensory properties that characterize red wines from organically and conventionally grown grapes. The differences detected from a quantitative point of view are only marginal, and do not seem to have an impact on consumer's hedonic perception. A limitation of this study may be that only two vintages of one grape variety of organic and conventional wines were considered. Further research is needed to clarify this aspect. In this context, future perspectives of study should deal with the study of sensory and hedonic qualities of wine, which are undoubtedly the strongest determinants of consumer's expectations and play a key role in consumer's purchase attitude. This aspect seems to be particularly relevant for wines deriving from organically and conventionally grapes since environmentally sustainable practices related to wine quality seem to have good market prospects.

REFERENCES

1. Akçay, Y. D., Yıldırım, H. K., Güvenç, U., and Sözmen, E. Y. (2004). The effects of consumption of organic and nonorganic red wine on low-density lipoprotein oxidation and antioxidant capacity in humans. Nutr. Res. 24, 541–554. doi:10.1016/j.nutres.2004.04.004

2. Bourn, D., and Prescott, J. (2002). A comparison of the nutritional value, sensory qualities, and food safety of organically and conventionally produced foods. Crit. Rev. Food Sci. 42, 1–34. doi: 10.1080/10408690290825439

3. Caporale, G., and Monteleone, E. (2004). Influence of information about manufacturing process on beer acceptability. Food Qual. Pref. 15, 271–278. doi: 10.1016/S0950-3293(03)00067-3

4. Dalton, P., Wysocki, C. J., Brody, M. J., and Lawley, H. J. (1997). The influence of cognitive bias on the perceived odor, irritation and health symptoms from chemical exposure. Int. Arch. Occup. Environ. Health 69, 407–417. doi: 10.1007/s004200050168

5. Deliza, R., and MacFie, H. J. H. (1996). The generation of sensory expectation by external cues and its effect on sensory perception and hedonic ratings: a review. J. Sens. Stud. 7, 253–277. doi: 10.1111/j.1745-459X.1996.tb00036.x

6. Dupin, I., Schlich, P., and Fischer, U. (2000). "Differentiation of wines produced by organic or conventional viticulture according to their sensory profiles and aroma composition," in Proceedings of Proceedings 6th International Congress on Organic Viticulture, eds H. Willer and U. Meier (Bad Dürkheim, D: Print-Online), 245–251.

7. Forbes, S. L., Cohen, D. A., Cullen, R., Wratten, S. D., and Fountain, J. (2009). Consumer attitudes regarding environmentally sustainable wine: an exploratory study of the New Zealand marketplace. J. Clean. Prod. 17, 1195–1199. doi: 10.1016/j.jclepro.2009.04.008

8. Hughson, L., and Boakes, R. (2002). The knowing nose: the role of knowledge in wine expertise. Food Qual. Pref. 13, 463–472. doi: 10.1016/S0950-3293(02)00051-4

9. ISO International Organization for Standardization. (2003). Sensory Analysis – Methodology – General Guidance for Establishing a Sensory Profile (ISO 13299:2003). Geneva, Switzerland.

10. ISO International Organization for Standardization. (2007). Sensory Analysis – General Guidance for the Design of Test Rooms (ISO 8589:2007). Geneva, Switzerland

11. Laureati, M., Jabes, D., Russo, V., and Pagliarini, E. (2013). Sustainability and organic production: how information influences consumer's expectation and preference for yogurt. Food Qual. Pref. 30, 1–8. doi: 10.1016/j.foodqual.2013.04.002

12. Lawless, H. T. (1984). Flavor description of white wine by expert and non-expert wine consumers. J. Food Sci. 49, 120–123. doi: 10.1111/j.1365-2621.1984.tb13686.x

13. Lawless, H. T., and Heymann, H. (1998). "Descriptive analysis," in Sensory Evaluation of Food: Principles and Practices, eds H. T. Lawless and H. Heymann (New York: Chapman & Hall), 341–372.

14. Lea, P., Næs, T., and Rødbotten, M. (1997). "Further aspects of design and modeling," in Analysis of Variance for Sensory Data (Chichester: John Wiley & Sons Ltd), 20–21.

15. Lockshin, L., and Corsi, A. M. (2012). Consumer behavior for wine 2.0: a review since 2003 and future directions. Wine Econ. Policy 1, 2–23. doi: 10.1016/j.wep.2012.11.003

16. MacFie, H. J. H., Bratchell, N., Greenhoff, K., and Vallis, L. V. (1989). Designs to balance the effect of order of presentation and first order carry-over effects in hall tests. J. Sens. Stud. 4, 129–148. doi: 10.1111/j.1745-459X.1989.tb00463.x

17. Morrot, G., Brochet, F., and Dubourdieu, D. (2001). The color of odors. Brain Lang. 79, 309–320. doi: 10.1006/brln.2001.2493

18. Moyano, L., Zea, L., Villafuerte, L., and Medina, M. (2009). Comparison of odor-active compounds in sherry wines processed from ecologically and conventionally grown pedroximenez grapes. J. Agric. Food Chem. 57, 968–973. doi: 10.1021/jf802252u

19. Napolitano, F., Braghieri, A., Piasentier, E., Favotto, S., Naspetti, S., and Zanoli, R. (2010a). Cheese liking and consumer willingness to pay as affected by information about organic production. Food Qual. Pref. 18, 280–286. doi: 10.1017/S0022029910000130. Epub 2010 Mar 3.

20. Napolitano, F., Braghieri, A., Piasentier, E., Favotto, S., Naspetti, S., and Zanoli, R. (2010b). Effect of information about organic production on beef liking and consumer willingness to pay. Food Qual. Pref. 21, 207–212. doi: 10.1016/j.foodqual.2009.08.007

21. Noble, A., Arnold, R. A., Buechsenstein, J., Leach, E. J., Schmidt, J. O., and Stern, P. M. (1987). Modification of a standardized system of wine aroma terminology. Am. J. Enol. Vitic. 38, 143–146.

22. Pangborn, R., Berg, H., and Hansen, B. (1963). The influence of color on discrimination of sweetness in dry table-wine. Am. J. Psychol. 76, 492–495. doi: 10.2307/1419795

23. Renaud, S., and De Lorgeril, M. (1992). Wine, alcohol, platelets, and the French paradox for coronary heart disease. Lancet 339, 1523–1526. doi: 10.1016/0140-6736(92)91277-F

24. Solomon, G. (1990). Psychology of novice and expert wine talk. Am. J. Psychol. 105, 495–517. doi: 10.2307/1423321

25. Zucco, G., Carassai, A., Baroni, M. R., and Stevenson, R. J. (2011). Labeling, identification, and recognition of wine-relevant odorants in expert sommeliers, intermediates, and untrained wine drinkers. Perception 40, 598–607. doi: 10.1068/p6972

PART IV

HOW DOES VITICULTURE INTERACT WITH OTHER ENVIRONMENTAL ISSUES?

CHAPTER 13

AVIAN CONSERVATION PRACTICES STRENGTHEN ECOSYSTEM SERVICES IN CALIFORNIA VINEYARDS

JULIE A. JEDLICKA, RUSSELL GREENBERG, AND DEBORAH K. LETOURNEAU

13.1 INTRODUCTION

Ecosystem services such as pest control and pollination are functions provided by biological diversity that are critical to human societies and their agricultural production [1], [2]. Nevertheless, agriculture often generates environmental pollution, contributes to habitat loss and, hence, decreases biodiversity [3], [4]. Environmentally sustainable farming practices are designed to foster biodiversity and ecosystem services. For example, bird-friendly® coffee systems are well-known for their conservation value, particularly in providing habitat for insectivorous migrant bird species [5], [6]. Studies comparing insect herbivore abundance with and without net caging over plants (exclosures) suggest that insectivorous birds significantly reduce both herbivorous arthropod abundance and plant damage in agricultural and natural systems [7], [8]. As a result, conservation of birds in agricultural landscapes may benefit growers through the provision of pest control services. For example, outside exclosures avian predation of

Avian Conservation Practices Strengthen Ecosystem Services in California Vineyards. © Jedlicka JA, Greenberg R, Letourneau DK. PLoS ONE 6,11 (2011), doi:10.1371/journal.pone.0027347. This work is made available under the Creative Commons 1.0 Universal Public Domain Dedication, http://creativecommons.org/publicdomain/zero/1.0/.

insect pests increased quantities of marketable fruit and raised farmer income in apple [9], [10] and coffee [11], [12] production systems.

Experimental methods for quantifying ecosystem services are fraught with complications, because in situ manipulations (e.g. predator exclosures) can have hidden or confounding effects [13]. An alternative methodology to quantify avian predation in agroecosystems combines the manipulation of specific predator populations via the establishment of nest boxes with a sentinel prey experiment that controls for density dependent population effects. Sentinel prey studies, which monitor removal rates of immobilized, tethered, or frozen prey in the field are common in the entomology literature for comparing relative predation pressure under different conditions e.g. [14], [15], [16]. Often sentinel prey experiments are used in concert with predator abundance data to test the effects of management practices (mulching, crop diversification, plant density) on biological control by predators and parasitoids e.g. [17], [18], [19], [20] or to measure behavioral responses of natural enemies [21], [22], [23]. We know of only one experiment, however, that uses sentinel prey to quantify the activity of vertebrate predators. Perfecto et al. compared net differences in removal rates of sentinel prey (outside versus inside exclosures) in two coffee agroecosystems and found that the farm with relatively greater structural diversity had a significantly higher removal rate of prey [24]. Using vineyards as a model system, we tested for an increase in regulating services (pest removal) in agriculture by measuring sentinel prey removal with and without avian predator augmentation through the provision of nest boxes.

In California (CA), USA, grapes are the second most economically important agricultural commodity, generating over $3.2 billion US dollars in 2009 [25]. Since 1950, the expansion of vineyards has contributed to the conversion of over 1,000,000 acres of CA oak woodlands and savannas to agricultural and urban land [26], [27]. Recently the American Bird Conservancy included CA oak savannas on their list of the 20 most threatened bird habitats in the United States [28] due to the rapid conversion of breeding habitat and loss of nesting sites [29]. However, erecting nest boxes in vineyards may provide compensatory resources for Western Bluebirds (*Sialia mexicana*) [30].

The Western Bluebird, hereafter simply bluebird, is one of the species that nest in natural oak cavities and the primary occupant of vineyard nest boxes in the North and Central Coast of CA [30], serving as the focal predator species of this study. Western Bluebirds forage by perching in low vegetation and striking arthropods on the ground, air, or vegetation [31], and potentially serve as an important natural predator to many vineyard insect pest species [32]. They produce one or two broods per year between April and July and clutches usually contain four to six eggs [31]. The average energy requirement for a nine to twelve day old bluebird nestling is approximately 65 kJ per day [33]. Consequently for broods of five nestlings, about 78 g of arthropods per day must be delivered to the nest to maintain growth and development of chicks, in addition to the 23 g of arthropods per day necessary to sustain each adult bird [34].

To determine if conserving insectivorous avian predators results in increased pest control services in vineyards, we enhanced nesting opportunities for local songbird communities by establishing nest boxes in one half of two CA vineyards. By mimicking a pest outbreak in the vineyards, we investigated the response of the predator concentration treatment and control to such a perturbation. The study was designed to address the following questions: (1) How do vineyard nest boxes affect local avian abundance and composition? (2) Is avian activity restricted to the immediate location of occupied bluebird nest boxes? And (3) does the establishment of vineyard nest boxes result in increased insect pest mortality as indicated by removal rates of sentinel prey?

13.2 RESULTS

13.2.1 NEST BOX OCCUPANCY

In 2009, three avian species were the predominant occupants of vineyard nest boxes: Western Bluebirds (76.1% of box pairs), Tree Swallows, and Violet-green Swallows (*Tachycineta bicolor* and *Tachycineta thalassina* respectively, 17.4% of box pairs combined). Ash-throated Flycatchers (*Myiarchus cinerascens*) built one nest and were the only other species occupying vineyard boxes. All four species are predominately insectivorous during the breeding season.

TABLE 1: Total number of bird sightings by species in nest box treatments and control areas of vineyards.

Species	Latin Name	Guild	Nest Box	Control
Western Bluebird	*Sialia mexicana*	I	313	39
Chipping Sparrow	*Spizella passerina*	I	132	100
Tree Swallow	*Tachycineta bicolor*	I	4	0
Bullock's Oriole	*Icterus bullockii*	I	0	2
Ash-throated Flycatcher	*Myiarchus cinerascens*	I	1	1
Norther Flicker	*Colaptes auratus*	I	1	1
Black Phoebe	*Sayornis nigricans*	I	1	0
Nutall's Woodpecker	*Picoides nuttallii*	I	1	0
Yellow-rumped Warbler	*Dendroica coronata*	I	0	1
Orange-crowned Warbler	*Vermivora celata*	I	1	0
Western Tanager	*Piranga ludoviciana*	I	1	0
European Starling	*Sturnus vulgaris*	O	3	22
Brewer's Blackbird	*Euphagus cyanocephalus*	O	5	8
American Robin	*Turdus migratorius*	O	3	4
Lark Sparrow	*Chondestes grammacus*	O	1	2
American Crow	*Corvus brachyrhynchos*	O	0	2
Steller's Jay	*Cyanocitta stelleri*	O	2	0
Brown-headed Cowbird	*Molothrus ater*	O	1	0
Dark-eyed Junco	*Junco hyemalis*	O	0	1
American Goldfinch	*Spinus tristis*	G	81	150
House Finch	*Carpodacus mexicanus*	G	67	81
Wild Turkey	*Meleagris gallopavo*	G	28	21
Lesser Goldfinch	*Carduelis psaltria*	G	10	17
Mourning Dove	*Zenaida macroura*	G	4	5
California Towhee	*Pipilo crissalis*	G	0	5

Species were categorized into guilds based on the Birds of North America reference collection, where I = mostly insectivore, O = omnivore, and G = Granivore.

Eggs were laid in the earliest bluebird nests in mid-April at both sites. Over the breeding season, pooling both sites, 44 bluebird nesting attempts were made. On average, each bluebird nest contained almost five eggs (mean = 4.91, SE = 0.13). Bluebird nests fledged between mid-May and late July.

13.2.2 AVIAN SPECIES RICHNESS

A total of 1122 birds representing 25 species were observed at the vineyard sites (Table 1). The most common insectivorous species observed in the vineyards were Western Bluebird and Chipping Sparrows (*Spizella passerina*). Both species are associated with woodlands and savannas. Whereas bluebirds are a cavity-nesting species, Chipping Sparrows build open cup nests in vegetation, including grapevines (Jedlicka pers. obs).

Mean avian species richness did not differ significantly, but the species richness of insectivorous birds was over 50% greater in nest box treatments than in control areas of vineyards (Table 2). This increase in the average number of insectivores per observation was due to the higher frequency of bluebird sightings and, to a lesser extent, Chipping Sparrow and Tree Swallow (Table 1).

TABLE 2: Mean (± SE) avian species richness observed or heard over the 30-minute observations and average avian abundance per 5-minute observation interval for nest box treatments and control areas.

Parameter	Nest Box	Control	P
Avian Species Richness	4.23 ± 0.39	3.67 ± 0.19	0.104
Insectivore Richness	2.01 ± 0.07	1.21 ± 0.25	0.002
Total Avian Abundance	3.71 ± 0.43	2.09 ± 0.33	0.003
Western Bluebird Abundance	1.82 ± 0.14	0.18 ± 0.05	<0.001
Non-bluebird Insectivore Abundance	0.84 ± 0.11	0.47 ± 0.15	0.119
Omnivore Abundance	0.14 ± 0.09	0.18 ± 0.02	0.307
Granivore Abundance	1.20 ± 0.31	1.23 ± 0.11	0.454

Treatment means and standard errors were calculated from both sites over early, middle, and late time periods. Estimated P-values are from bootstrap resampling (see Methods).

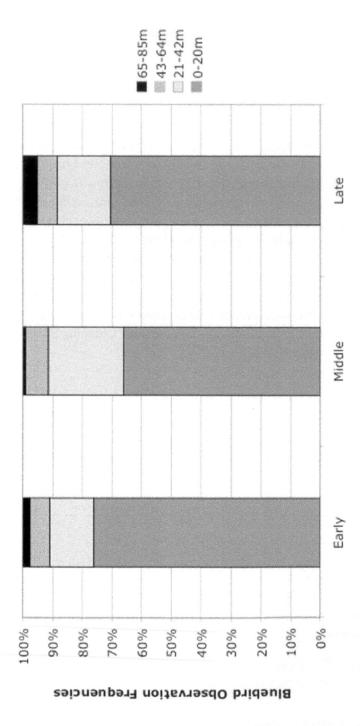

FIGURE 1: Frequency of Western Bluebird observations categorized as distance (in m) from active nest box locations during the breeding season (x-axis).

13.2.3 AVIAN ABUNDANCE

Total avian abundance doubled in nest box treatments early in the season and experienced a 2.6 factor increase late in the breeding season when fledglings were seen foraging with adults throughout the vineyard. Across all time periods, nest box treatments contained significantly higher avian abundances than control areas (P = 0.003). The increase in avian abundance in nest box treatments was driven by a single species. Western Bluebird abundance was an order of magnitude greater in nest box treatments than in control areas without nest boxes, averaging 1.8 individuals surveyed every 5 minutes compared to 0.18 individuals in control areas (P<0.001, Table 2). Total insectivore abundance excluding bluebirds was not significantly different across treatments (P = 0.119). Likewise, the abundance of both omnivores and granivores showed no consistent pattern by treatment (Table 2).

In nest box areas, the majority of the bluebirds were observed foraging near active nests (Fig. 1) and 1–5% of observations recorded bluebirds at distances over 65 m away. The number of bird detections varied over time (n = 122 early, n = 130 middle, n = 61 late season) likely because of a decreased detectability late in the breeding season due to fewer vocalizations. Bluebirds were found disproportionally closer to nest boxes early in the season, corresponding to nest building, egg laying, and incubation (Fig. 1). During the middle of the season (when first broods fledged but other nests contained eggs), bluebirds were increasingly observed at intermediate distances (21–42 m) from active nests. Late in the season proportionally more bluebirds were observed over 65 m away from active nests when bluebird adults were often seen foraging with fledglings (young that recently left the nest) in small flocks of three to five individuals.

13.2.4 SENTINEL PREY EXPERIMENTS

The number of sentinel larvae removed varied by treatment (control vs. nest box treatment vs. near occupied nest boxes, df = 2, X^2 = 16.6, P<0.001). Those treatment effects did not vary between the vineyards (df = 1, X^2 = 0.5, P = 0.48) nor was there an interaction effect between treatment and site (df = 2, X^2 = 1.2, P = 0.54). Pooled removal rates of sentinel larvae were 2.4 times greater in the nest box treatment than in the control half of the vineyard (Fig. 2, n = 10 transects, mean$_{trmt}$ = 2.9 ± 0.6SE vs. mean$_{control}$ = 1.2 ± 1.0SE, z = 3.4, P = 0.002). The highest average removal rate of sentinel larvae occurred on transects placed within 25 m of the seven remaining active bluebird nest boxes (Fig. 2, n = 7 transects, mean = 4.14 ± 0.6 SE larvae removed out of 5), indicating that beneficial effects of avian foraging in these vineyards can be enhanced significantly when nest boxes are occupied (larvae removed near active nests vs. control, z = 4.8 P<0.001). Removal rates by active nests were also higher than removal from transects placed randomly in the nest box treatments (z = 2.2, P = 0.066).

13.3 DISCUSSION

Providing songbird nest boxes in vineyards nearly quadrupled the abundance of insectivorous birds, most notably the Western Bluebird whose density increased tenfold. Nest boxes were placed in the vineyard just over one year prior to the study, however bluebirds occupied over 75% of all box pairs. Occupancy rates may further increase over time as bird populations become aware of nest box locations. Establishing nest box treatments created significant differences in avian predator densities that allowed for comparisons to baseline predator levels. Such experimental designs are advantageous because they allow for precise quantification of predator effects without the potential distortions that may be associated with exclosure methodologies [13]. One potential disadvantage of exclosures is that arthropod movement in and out of the exclosure may equalize the effects of predation pressure between experimental and control plants. This could take the form of an "osmotic effect" if protection from avian predation inside exclosures increases prey density resulting in increased prey dispersal rates away from exclosures. Structurally, the exclosure may serve to attract organisms such as web-building spiders, unnaturally increasing predation levels on other taxa within the exclosure. For-example, in a recent meta-analysis Mooney et al. found that arachnid abundance was over two times higher inside predator exclosures [8]. Each of these factors may cause an underestimation of the effects of bird predation. Furthermore, overestimates may result from a faster reproduction rate of prey inside the mesh, especially if mates are easier to find or if microclimatic conditions are favorable. Finally, exclosure studies compare presence and absence of a suite of vertebrate predators, including bats [35] and lizards [36], which makes it difficult to assess the predation effect of a particular species, or even class of predator [37].

The CA winegrape growing season overlaps with the migratory bird breeding season when, due to the energetic demands of reproductive activities, the strongest predatory pressures occur [38]. A bluebird pair with five nestlings requires 124 g of arthropods daily [34]. They produce one or two broods per year between April and July and clutches usually contain four to six eggs [31]. Data from the sentinel pest experiment during the breeding season showed a greater predation rate of larvae in the nest box treatment compared to vineyard control areas with no nest boxes. Moreover, removal rates near active nest boxes were nearly 3.5 times greater than the control. Such high predation of grapevine pests is likely a significant ecosystem service the birds provide to winegrape growers.

Bluebirds are generalist arthropod predators, preying upon insects in a range of different orders such as Lepidoptera, Orthoptera, Hemiptera, and Coleoptera [31]. As a result, bluebird presence may help provide resilience against novel pest outbreaks. The presence of new, exotic, and economically important insect pest species in United States vineyards is increasing with the notable discoveries of European grapevine moth (*Lobesia botrana, Lepidoptera: Tortricidae*) in Napa County in 2009, light brown apple moth (*Epiphyas postvittana, Lepidoptera: Tortricidae*) on the California North Coast in 2007, and glassy-winged sharpshooter (*Homalodisca vitripennis, Hemiptera: Cicadellidae*) identified in 1989 in California. Maintaining an abundant and diverse community of generalist insectivores may provide local protection against current and future pest challenges [39].

As generalist predators, bluebirds consume spiders and other arthropod enemies of herbivorous pests [31], acting as intraguild predators with uncertain net top-down trophic effects on pest levels and plant biomass [39]. Although birds may play conflicting roles as primary and secondary predators, a recent meta-analysis of exclosure studies by Mooney et al. suggests that despite their reduction of intermediate predator densities, insectivorous birds still significantly lower arthropod herbivores resulting in increased plant biomass [8].

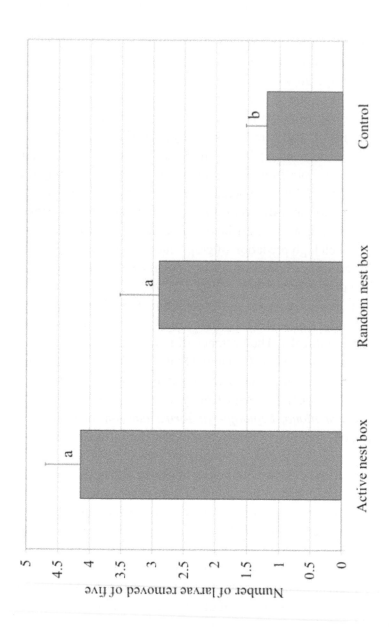

FIGURE 2: Mean number (± SE) of five lepidopteran larvae removed per transect in the pooled control (n = 10), nest box treatment (n = 10), and below active Western Bluebird nests (n = 7). Different letters indicate significant differences (P<0.05).

To minimize disturbance, sentinel prey were not observed during their six hours of exposure, consequently it was not possible to determine if bluebirds were responsible for removing all sentinel larvae. Mechanical removal did not occur because humans and machinery were prevented from entering vineyard sites during the experiment. Sentinel prey were only accessible in the morning and unavailable for nocturnal predators such as bats (Chiroptera), raccoons (Carnivora: Procyonidae: Procyon lotor), and mice (Rodentia: Muridae). Many diurnal predators large enough to remove fastened larvae (e.g. squirrels (Rodentia: Sciuridae)) did not frequent these groomed habitats that are subject to frequent tilling and spray applications. Some larvae may have been removed by other animals such as lizards (Squamata), frogs (Anura), or ants (Formicidae). No ant swarms or evidence of larva dissection were present upon collection of transects, and no lizards were seen at vineyard sites during the entire field season. Besides housing avian predators the presence of nest boxes is not likely to influence other explanatory factors causing larvae disappearance. Both controls and predator enhancement treatments were adjacent and equidistant from wooded riparian vegetation where higher predator abundance and diversity may exist. Nevertheless the removal rate of sentinel prey in the nest box treatment averaged nearly 2.5 times higher than control areas and targeted transects below active bluebird nests resulted in 3.5 times greater larval predation than no nest box areas.

The potential for enhancing the density of insectivorous birds locally through the establishment of nest boxes, possibly increasing their population size and pest control services, is not restricted to California vineyards. As urban and agricultural expansion takes place, the popularity of bluebird trails and citizen science programs such as NestWatch (an NSF funded program run by Cornell Lab of Ornithology) has grown and bluebirds across the United States have colonized artificial nesting sites [40]. The combined range of three different species of bluebirds extends throughout the continental USA: Western, Mountain (*Sialia currucoides*), and Eastern Bluebirds (*Sialia sialis*). Therefore, USA growers will likely be able to attract breeding bluebirds wherever there are suitable habitats, including annual row crops [41]. In southwestern Germany, the cavity-nesting and insectivorous Eurasian hoopoe (*Upupa epops*) experienced strong local population declines. Stange and Havelka [42] installed nest boxes throughout vineyards and, after nine years, one hoopoe population increased from three to twelve breeding pairs. The authors concluded that providing additional nest sites and reducing pesticide applications in the area contributed to the increased population size.

Wildlife-friendly viticulture practices may be necessary to maintain breeding populations of birds in vineyards. This study was performed in organic vineyards of the California North Coast where experimental nest boxes were rapidly inhabited by breeding birds. Remnant gallery forests along the Russian River may help maintain steady food resources for nest box occupants. Other vineyard landscapes and cultural practices may not be able to recruit such high bird abundances. Further research investigating how birds use novel vineyard habitat is urgently needed as vineyards increasingly compose greater proportions of the CA land-scape, often at the expense of oak savannas and woodlands. Nest boxes were placed throughout vineyard rows on existing trellises, supporting high densities of insectivorous birds. This nest box placement is suit-able for many winegrape growers in the region whose machinery is built to accommodate trellises and/or who employ workers to harvest crops. Nest box placement in vineyard rows may not be feasible for highly mechanized vineyard systems where suitable box placement may be lim-ited to the vineyard perimeter.

13.3.1 CONSERVATION IMPLICATIONS

In 2008, over 318,000 hectares in CA were devoted to grape cultivation [43]. Tremendous potential exists to expand avian conservation practices by increasing the numbers of songbird nest boxes in vineyards. Growers may benefit not only from the pest control services provided by breeding birds, but may also target their bird-friendly® wine to the growing organic and eco-friendly consumer markets [44]. Developing and marketing bird-friendly® wine could differentiate producers in the marketplace and em-power environmentally-conscious consumers to support more sustainable production systems.

In this study, we did not monitor the conservation impact of nest box placement, but rather documented how conservation practices benefit growers. Fiehler et al. demonstrated that California vineyard nest boxes provide compensatory breeding resources for bluebirds [30]. Bluebird clutch size was larger and nest initiation date earlier in vineyards compared to neighboring oak-savanna habitat. These findings offer promise, but studies that measure the population dynamics of birds across landscapes will be required to assess the conservation potential of vineyard nest boxes throughout the state. In particular, the reproductive success of vineyard box occupants must be greater than local replacement rates. If vineyards do not serve as 'sink' habitats and breeding populations are sustained year after year, then the practice of providing vineyard nest boxes may be a vital component of bird conservation efforts.

Research that broadens conservation biological control to include avian predators may appear to be a novel step for Integrated Pest Management. However, these investigations resurrect a former research focus within the US Department of Agriculture (USDA) before the advent of DDT and other cheaply produced materials for pest control. From 1885 to 1940 a division of the Bureau of Biological Survey (part of the USDA) called economic ornithology was devoted to researching avian biological control [45], [46]. Our study revitalizes economic ornithology in the context of ecosystem services, and shows that the conservation practice of providing nest boxes increases the abundance of mobile, recruiting, insectivorous predators that can rapidly consume sentinel pests in contemporary, high-value crop production systems.

13.4 MATERIALS AND METHODS

13.4.1 ETHICS STATEMENT

Vertebrate animals were approved for use in the study by the United States Geological Survey (Permit Number: 22665) and the University of California's Institutional Animal Care and Use Committee (Permit Number: Letod0705) and efforts were made to minimize animal suffering. Sentinel prey were approved for use by the United States Department of Agriculture's Animal and Plant Health Inspection Service (Permit Number: P526P-08-00396).

13.4.2 STUDY SITES

Vineyards chosen for this experiment were located 12 km from each other in Mendocino County, CA, USA: in Hopland (33.2 ha, 38°59′N, 123°06′W) and near Ukiah (51.4 ha, 39°04′N, 123°09′W). Both study sites were certified organic vineyards planted between 1985 and 1988. In addition to vineyards, forest remnants and wooded riparian vegetation are common landscape features in the county. Vineyard sites were both adjacent to the Russian River and managed identically by the same grower who is responsible for an additional 351.4 hectares of winegrape in the region. Chardonnay grapevines were grown on trellises forming rows. Tilling occurred in every other tractor row, alternating with cultivated cover crops—97% clover (*Trifolium spp*), and 3% Queen Anne's Lace *(Daucus carota)*. Grapevines were pruned to 6 buds per lineal foot of cordon with yields averaging 6 metric tons per acre [47]. Timing of the annual harvest is climate-dependent, but usually occurs in September and October.

13.4.3 NEST BOX MANAGEMENT

Each vineyard was divided in half, and randomly assigned either as a control or predator enhancement (nest box) treatment. A buffer of at least 250 m was left between the nest box treatment and control because nearest-neighbor distances of bluebird nests ranged from 120–240 m over a 5-year CA study [48], (Fig. 3). Nest boxes were constructed from redwood following recommendations of the North American Bluebird Society (13.9 cm by 10.2 cm by at least 23.8 cm tall with entrance hole opening of 3.8 cm diameter) [49]. Because swallow species occupy vineyard boxes and defend territories from conspecific pairs but not bluebirds, we erected boxes back-to-back in pairs within nest box treatments. Because no other bird species are common occupants of vineyard boxes, this design ensured unoccupied boxes throughout the vineyard would be available for bluebirds. In Jan 2008, nest box pairs were placed in predator treatments, spaced 85 m from each other based on nearest-neighbor distances measured by Dickinson & Leonard where a 68% nest box occupancy rate was achieved [48]. Twenty-three to 24 nest box pairs were established in a grid pattern in 5 to 6 rows (Fig. 3). Each row consisted of 3 to 6 pairs of boxes on 3.1 m t-posts placed 0.6 m into the ground along grapevine trellises. All nest boxes were cleaned of previous reproductive materials in February 2009 and checked weekly for nesting activity during the 2009 avian reproductive season from March through July. Once bluebird nests were found to contain eggs, Noel predator guards made of wire mesh hardware cloth were attached to the outside of the boxes to prevent predation by raccoons (*Procyon lotor*) or domestic cats (*Felis catus*) [50].

13.4.4 AVIAN OBSERVATIONS

Avian observations were performed at five locations in both the control and nest box treatments. In both treatments, observation points were only selected if they were located at least 85 m from each other and any vineyard edges (i.e. riparian habitat and roads). In the control, observation points were selected by arbitrarily placing a finger down on maps of control areas. In nest box treatments, active bluebird nest boxes were randomly selected as observation points. Nest boxes were monitored weekly to assess bluebird reproductive activity. A nest was defined as active if it contained eggs and/or live nestlings. Abandoned nests with eggs were no longer considered active if eggs had not hatched in three weeks and no adults appeared to be entering the box.

All observations were conducted on days without strong winds or rain. Observations began shortly after sunrise and continued until approximately 10 am when avian activity decreased. Treatments were sampled on consecutive days, weather permitting. Sampling occurred biweekly at each vineyard (alternating sites between weeks) from mid-April to mid-July. The sampling order of observation points was altered each week to avoid temporal biases in observations.

At observation points in both treatments, standard avian point count procedures [51] were modified as follows. Point counts were performed from a camouflaged ground hunting blind (Ameristep one-person chair blind #403580, gandermountain.com) by the same observer (JJ). Once in position, the observer waited five minutes before sampling to minimize human disturbance. All birds seen or heard on vineyard vegetation (not flying overhead) within an 85 m radius from the observation point were recorded for one minute. Samples were repeated at five-minute intervals for a 30-minute duration at each point. Once birds were located, their species identity and distance from the observation point were recorded. Because vineyards were established in a mechanized grid where all tractor rows were 3.05 m wide, distances were relatively easy to estimate.

13.4.5 AVIAN CLASSIFICATION AND JUSTIFICATION

In California vineyards, nine of the ten avian species that occupy nest boxes of the dimension used in this study are insectivorous, with House Sparrow (*Passer domesticus*; omnivore) being the one exception [52]. In this study we focused on bluebirds because of their high nest box occupancy rate and greater likelihood to forage on vineyard insect pests. For example, swallows forage upon aerial insects over great distances [53], [54] and are not likely to be consuming pest insects from vineyard vegetation.

Avian species were divided into three guilds (insectivores, omnivores, or granivores) according to their predominant diets during the breeding season based on the Birds of North America reference collection. For example, although Chipping Sparrows (*Spizella passerina*) regularly consume seeds, they are categorized as insectivores because stomach-content analyses show invertebrates (primarily insects) to comprise the majority of their diet during the breeding season [55]. The omnivorous guild includes partial frugivores, some of which consume ripe grapes. Avian species that opportunistically forage on grape crops include the granivorous House Finch (*Carpodacus mexicanus*) and several omnivorous species such as European Starling (*Sturnus vulgaris*), Brewer's Blackbird (*Euphagus cyanocephalus*), and American Robin (*Turdus migratorius*). These potential pest species did not occupy nest boxes during the duration of this study, as some are ground or open cup nesters and others (e.g. starlings) could not fit through the box entrance hole.

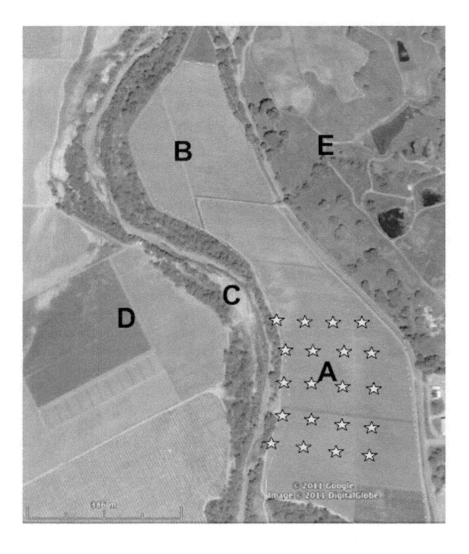

FIGURE 3: Aerial view of one vineyard site illustrating: (A) experimental treatment; (B) no nest box control; (C) wooded riparian zone; (D) surrounding vineyards; and (E) oak savannas. Within nest box treatment (A), each star indicates one pair of nest boxes mounted back-to-back 85 m from each other.

13.4.6 SENTINEL PREY EXPERIMENT

The University of California Division of Agriculture and Natural Resources recognizes many lepidopteran species, including beet armyworm (*Spodoptera exigua, Lepidoptera: Noctuidae*), as California vineyard pests [56]. *S. exigua* eggs are laid on vineyard weeds or cover crops and larvae may feed on ground vegetation or climb up grapevines producing plant damage [56]. Fifth instars of larvae (~12 mm long) were purchased from Bio-Serv and used for sentinel prey experiments at each vineyard site on consecutive days in June, 2009. *S. exigua* larvae were placed on the ground in transects containing five individuals pinned through their last abdominal segment to 10.2 cm^2 brown cardboard squares, restricting the movement but not killing the insect. Each larva was placed 5m apart with cardboard squares staked into the ground in vineyard tractor rows containing cover crops. Larvae were pinned directly before placement in transects, and all sentinel pests were set out before 7:00 am. One transect, consisting of five presentation stations, was established at 10 different locations in each vineyard: at five randomly selected points in the nest box treatment, and at the five randomly selected vineyard control points chosen for avian observations. In addition, all active Western Bluebird nest boxes located at least 85 m from the riparian edge were used to quantify the maximum predatory response to sentinel prey (n = 4 and 3 at each vineyard site near the end of the season when these trials were conducted). The first larva of each transect was placed in the tractor row adjacent to the occupied box such that the final larva was approximately 25 m from the active nest. All remaining larvae were recollected approximately 6 hours later the same day and each presentation station was recorded as either present (dead from sun exposure) or missing, signifying consumption from predators. No vineyard workers or machines were present within the duration of the experiment.

13.4.7 DATA ANALYSIS

To reflect changes in phenology, avian observations were categorized into one of three 4-week long time periods during the breeding bird season corresponding to early (22-Apr–22-May; birds finding territories, building nests, some with eggs), middle (23-May–20-Jun; first broods are fledging, other nests with eggs), and late (21-Jun–19 Jul; second broods fledging, less singing). From the avian observation data we calculated (1) the mean species richness (of all birds and strictly insectivorous birds) over the 30 min sample; and (2) the mean abundance of all birds, Western Bluebirds, and avian species divided into three guilds (insectivores, omnivores, granivores) per 5-minute observation interval. For the latter calculations, 5-minute observation means were averaged together to provide one representation of abundance per treatment at each site in early, middle and late time periods.

Avian observation data (either raw or transformed) did not meet ANOVA assumptions and were randomly resampled (with replacement) using bootstrap estimation. Means and standard deviations were calculated per time period (n = 3), treatment (n = 2) and site (n = 2). Consequently each treatment contained 6 replicates (3 time periods by 2 sites). In order to test each dependent variable against the null hypothesis of no difference between treatments, we pooled treatment means and randomly resampled 1000 means based on a sample size of six. The resampling was performed twice and the difference between these two samples was calculated to form a distribution of means representing the null hypothesis of no treatment effect. Actual differences in nest box and control means were compared to the null distribution of differences, enabling the estimation of an associated P-value.

In the sentinel prey experiment, number of larvae removed per transect ranged from zero to five and was analyzed with a generalized linear mixed model (GLMM) using a binomial distribution and logit link function. The full GLMM included treatment (active nests, random nest box, or control), site (n = 2), and treatment x site as fixed effects that were nested by spatial location in nest box or control areas of the vineyard (random effect). To test for effect, the full GLMM was compared to a null GLMM that was identical except that it excluded the fixed effect of interest. Full and null GLMMs were compared with an ANOVA. The GLMMs, ANOVA, and post-hoc contrasts were performed with R version 2.13 [57] and the lme4 package [58]. All other statistical tests were performed with Systat version 12.

REFERENCES

1. Daily GC, editor. (1997) Nature's services: societal dependence on natural ecosystems. Washington D.C.: Island Press. 392 p.
2. Daily GC, Matson PA (2008) Ecosystem services: From theory to implementation. Proc Natl Acad Sci U S A 105: 9455–9456.
3. Tilman D (1999) Global environmental impacts of agricultural expansion: The need for sustainable and efficient practices. Proc Natl Acad Sci U S A 96: 5995–6000.
4. Butler SJ, Vickery JA, Norris K (2007) Farmland biodiversity and the footprint of agriculture. Science 315: 381–384.
5. Greenberg R, Bichier P, Angon AC, Reitsma R (1997) Bird populations in shade and sun coffee plantations in central Guatemala. Conservation Biology 11: 448–459.
6. Perfecto I, Rice RA, Greenberg R, VanderVoort ME (1996) Shade coffee: A disappearing refuge for biodiversity. Bioscience 46: 598–608.
7. Van Bael SA, Philpott SM, Greenberg R, Bichier P, Barber NA, et al. (2008) Birds as predators in tropical agroforestry systems. Ecology 89: 928–934.
8. Mooney KA, Gruner DS, Barber NA, Van Bael SA, Philpott SM, et al. (2010) Interactions among predators and the cascading effects of vertebrate insectivores on arthropod communities and plants. Proc Natl Acad Sci U S A 107: 7335–7340.
9. Mols CMM, Visser ME (2002) Great tits can reduce caterpillar damage in apple orchards. Journal of Applied Ecology 39: 888–899.
10. Mols CMM, Visser ME (2007) Great Tits (Parus major) reduce caterpillar damage in commercial apple orchards. PLoS ONE 2(2): e202. doi:10.1371/journal. pone.0000202.
11. Kellermann JL, Johnson MD, Stercho AM, Hackett SC (2008) Ecological and economic services provided by birds on Jamaican blue mountain coffee farms. Conservation Biology 22: 1177–1185.
12. Johnson MD, Kellermann JL, Stercho AM (2010) Pest reduction services by birds in shade and sun coffee in Jamaica. Animal Conservation 13: 140–147.
13. Englund G (1997) Importance of spatial scale and prey movements in predator caging experiments. Ecology 78: 2316–2325.
14. Walker KR, Welter SC (2004) Biological control potential of Apanteles aristoteliae (Hymenoptera: Braconidae) on populations of Argyrotaenia citrana (Lepidoptera: Tortricidae) in California apple orchards. Environmental Entomology 33: 1327–1334.
15. Cossentine JE (2008) Testing the impact of laboratory reared indigenous leafroller (Lepidoptera: Tortricidae) parasitoids (Hymenoptera: Ichneumonidae, Braconidae) on sentinel hosts in controlled orchard releases. European Journal of Entomology 105: 241–248.
16. Hagler JR (2006) Development of an immunological technique for identifying multiple predator-prey interactions in a complex arthropod assemblage. Annals of Applied Biology 149: 153–165.
17. Prasifka JR, Schmidt NP, Kohler KA, O'Neal ME, Hellmich RL, et al. (2006) Effects of living mulches on predator abundance and sentinel prey in a corn-soybean-forage rotation. Environmental Entomology 35: 1423–1431.

18. Danne A, Thomson LJ, Sharley DJ, Penfold CM, Hoffmann AA (2010) Effects of native grass cover crops on beneficial and pest invertebrates in Australian vineyards. Environmental Entomology 39: 970–978.

19. Ehler LE (2007) Impact of native predators and parasites on Spodoptera exigua, an introduced pest of alfalfa hay in northern California. Biocontrol 52: 323–338.

20. Chang GC, Snyder WE (2004) The relationship between predator density, community composition, and field predation of Colorado potato beetle eggs. Biological Control 31: 453–461.

21. Koptur S (1984) Experimental evidence for defense of Inga (Mimosoideae) saplings by ants. Ecology 65: 1787–1793.

22. Letourneau DK (1983) Passive aggression: an alternative hypothesis for the Piper-Pheidole association. Oecologia 60: 122–126.

23. Pearce S, Zalucki MP (2006) Do predators aggregate in response to pest density in agroecosystems? Assessing within-field spatial patterns. Journal of Applied Ecology 43: 128–140.

24. Perfecto I, Vandermeer JH, Bautista GL, Nunez GI, Greenberg R, et al. (2004) Greater predation in shaded coffee farms: The role of resident Neotropical birds. Ecology 85: 2677–2681.

25. California Department of Food and Agriculture (2011) California Agricultural Resource Directory 2010–2011. Available: http://www.cdfa.ca.gov/Statistics/. Accessed 2011 May 6.

26. Merenlender AM, Crawford J (1998) Vineyards in an oak landscape: Exploring the physical, biological, and social benefits of maintaining and restoring native vegetation in and around the vineyard. 21577. University of California Division of Agriculture and Natural Resources Publication.

27. Heaton E, Merenlender AM (2000) Modeling vineyard expansion, potential habitat fragmentation. California Agriculture 54: 12–19.

28. American Bird Conservancy (2007) Top 20 most threatened bird habitats in the U.S. Available: http://www.abcbirds.org/newsandreports/releases/habitatreport.pdf. Accessed 16 Oct 2011.

29. California Partners in Flight (2002) Version 2.0. The oak woodland bird conservation plan: a strategy for protecting and managing oak woodland habitats and associated birds in California (S. Zack, lead author). Point Reyes Bird Observatory, Stinson Beach, CA. Available: http://www.prbo.org/calpif/plans.html. Accessed 2011 Oct 16.

30. Fiehler CM, Tietje WD, Fields WR (2006) Nesting success of Western Bluebirds (Sialia mexicana) using nest boxes in vineyard and oak-savannah habitats of California. Wilson Journal of Ornithology 118: 552–557.

31. Guinan JA, Gowaty PA, Eltzroth EK Poole A, editor. (2008) Western Bluebird (Sialia mexicana). The Birds of North America Online. Ithaca: Cornell Lab of Ornithology. Available: http://bna.birds.cornell.edu.bnaproxy.birds.cornell.edu/bna/species/510. Accessed 2011 Oct 16.

32. Martin AC, Nelson AL, Zim HS (1951) American wildlife & plants: A guide to wildlife food habits; the use of trees, shrubs, weeds, and herbs by birds and mammals of the United States. New York: Dover Publications. 500 p.

33. Mock PJ, Khubesrian M, Larcheveque DM (1991) Energetics of Growth and Maturation in Sympatric Passerines That Fledge at Different Ages. Auk 108: 34–41.
34. Mock PJ (1991) Daily Allocation of Time and Energy of Western Bluebirds Feeding Nestlings. Condor 93: 598–611.
35. Williams-Guillen K, Perfecto I, Vandermeer J (2008) Bats limit insects in a Neotropical agroforestry system. Science 320: 70–70.
36. Borkhataria RR, Collazo JA, Groom MJ (2006) Additive effects of vertebrate predators on insects in a Puerto Rican coffee plantation. Ecological Applications 16: 696–703.
37. Gradwohl J, Greenberg R (1982) The effect of a single species of avian predator on the arthropods of aerial leaf litter. Ecology 63: 581–583.
38. Holmes RT (1990) Ecological and evolutionary impacts of bird predation on forest insects: An overview. Studies in Avian Biology 13: 6–13.
39. Letourneau DK, Jedlicka JA, Bothwell SG, Moreno CR (2009) Effects of natural enemy biodiversity on the suppression of arthropod herbivores in terrestrial systems. Annual Review of Ecology, Evolution, and Systematics 40: 573–592.
40. Cornell Lab of Ornithology (2008) NestWatch. Available: http://watch.birds.cornell. edu/nest/home/index. Accessed 1 Mar 2011.
41. Jacobson SK, Sieving KE, Jones GA, Van Doorn A (2003) Assessment of farmer attitudes and behavioral intentions toward bird conservation on organic and conventional Florida farms. Conservation Biology 17: 595–606.
42. Stange C, Havelka P (2003) Breeding population, nest site competition, reproduction, and feeding ecology of Hoopoe Upupa epops in the upper Rhine valley, SW Germany. Vogelwelt. 124. : 25–34. (in German).
43. California Department of Food and Agriculture (2010) California Agricultural Production Statistics 2009–2010. Available: http://www.cdfa.ca.gov/Statistics. Accessed 2010 Nov 1.
44. Raynolds LT (2004) The globalization of organic agro-food networks. World Development 32: 725–743.
45. Kirk DA, Evenden MD, Mineau P (1996) Past and current attempts to evaluate the role of birds as predators of insect pests in temperate agriculture. Current Ornithology 13: 175–269.
46. Evenden MD (1995) The laborers of nature: Economic ornithology and the role of birds as agents of biological pest control in North American Agriculture, ca. 1880-1930. Forest and Conservation History 39: 172–183.
47. Koball D (2010) Fetzer/Bonterra Vineyard Manager. Hopland, pers. comm. 20 Jan.
48. Dickinson JL, Leonard ML (1996) Mate attendance and copulatory behaviour in western bluebirds: Evidence of mate guarding. Animal Behaviour 52: 981–992.
49. North American Bluebird Society (2008) Eastern or Western Bluebird Nestbox. Available: http://www.nabluebirdsociety.org/eastwestbox.htm. Accessed 2007 Nov 1.
50. Toops C (1994) Bluebirds Forever. Stillwater, MN: Voyageur Press. 144 p.
51. Ralph CJ, Geupel GR, Pyle P, Martin TE, DeSante DF (1993) Handbook of field methods for monitoring landbirds. Albany, CA: Pacific Southwest Research Station, Forest Service, U.S. Dept of Agriculture. 41 p.

52. Heaton E, Long R, Ingels C, Hoffman T Tietje W, editor. (2008) Songbird, bat, and owl boxes: Vineyard management with an eye toward wildlife. University of California Agriculture and Natural Resources Publication 21636.

53. Winkler , DW , Hallinger KK, Ardia DR, Robertson RJ, Stutchbury BJ, Cohen RR Poole A, editor. (2011) Tree Swallow (Tachycineta bicolor). The Birds of North America Online. Ithaca: Cornell Lab of Ornithology. Available: http://bna.birds.cornell.edu.bnaproxy.birds.cornell.edu/bna/species/011. Accessed 2011 Oct 16.

54. Brown CR, Knott AM, Damrose EJ Poole A, editor. (1992) Violet-green Swallow (Tachycineta thalassina). The Birds of North America Online. Ithaca: Cornell Lab of Ornithology. Available: http://bna.birds.cornell.edu.bnaproxy.birds.cornell.edu/bna/species/014. Accessed 2011 Oct 16.

55. Middleton AL Poole A, editor. (1998) Chipping Sparrow (Spizella passerina). The Birds of North America Online. Ithaca: Cornell Lab of Ornithology. Available: http://bna.birds.cornell.edu.bnaproxy.birds.cornell.edu/bna/species/334. Accessed 2011 Oct 16.

56. University of California Agriculture and Natural Resources (1992) Grape Pest Management 2nd ed. : University of California Agriculture and Natural Resources Publication 3343:

57. R Development Core Team (2011) R: A language and environment for statistical computing. R Foundation for Statistical Computing, Vienna, Austria. Available: http://www.R-project.org. Accessed 2011 Oct 16.

58. Bates D, Maechler M, Bolker B (2011) lme4: Linear mixed-effects models using S4 classes. R package version 0.999375-40. Available: http://CRAN.R-project.org/package=lme4. Accessed 2011 Oct 16.

AUTHOR NOTES

CHAPTER 1

Competing Interests
The authors declare that they have no competing interests.

Author Contributions
CS carried out the systematic review on wine and set the database. CS analyzed together with AC the dataset. In particular CS wrote the paragraphs entitled "Entrepreneurs and top management" and "Sustainability and strategy", "Research orientations: a selective systematic literature review". AC carried out the general review on sustainability and together with CS has performed the dataset analysis. AC contributed to write, more specifically the paragraph entitled "Institutions, associations, regulators and market demand": AC also contributed specifically to the development of the following paragraphs: "The role of Research" LC contributed together with the other authors to conclusions, discussion and introduction. All authors read and approved the final manuscript.

Acknowledgments
We would like to thank reviewers for the useful and accurate suggestions that have contributed to improve the final version of the paper.

CHAPTER 2

Acknowledgments
The present work has been conducted in the framework of the Italian project V.I.V.A. Sustainable Wine, supported by the Italian Ministry for the Environment, Land and Sea. The authors would also like to thank the program managers for their availability in providing all the necessary information in order to make this evaluation possible.

Author Contributions

Chiara Corbo has realized this work in the framework of her research project. Lucrezia Lamastra, as the scientific expert, has supervised the overall work. Ettore Capri, director of the research centre OPERA, has coordinated the initiative.

Conflict of Interest

The authors declare no conflict of interest.

CHAPTER 3

Acknowledgments

This research would not have been possible without the dedication and time of the main moderator of the focus groups sessions, Dudley Brown. The authors would like to express their gratitude to all focus group participants of the sustainability project for their generous acceptance to our invitation and time to be part of our research. We would also like to thank the organisations where the sessions were held and people involved in the grape growing industry that helped with introductions or interviews in Australia, Chile, New Zealand, South Africa and United States. We also gratefully acknowledge support and discussions with Gerardo Leal (main moderator in Chile) and Joanna Kenny (transcriptions) and funding from the University of Adelaide and The Grape and Wine Research and Development Corporation—GWRDC.

Author Contributions

Irina Santiago-Brown organized the focus groups, performed interviews, organised and interpreted data and wrote the manuscript. Andrew Metcalfe, Cate Jerram and Cassandra Collins supervised the development of work, helped in data interpretation, manuscript evaluation and edits.

Conflicts of Interest

The author(s) declared no potential conflicts of interest with respect to the research, authorship, and/or publication with the following exception: Irina Santiago-Brown has participated in research collaboration with some of the informants in this study and also manages and has developed the McLaren Vale Sustainable Winegrowing Australia program.

CHAPTER 4

Acknowledgments
The authors wish to thank the Lincoln University Internal Research Fund
for financial support of this research.

CHAPTER 5

Acknowledgments
Some of the studies reviewed were financed by the Spanish Ministry of Education and Research – projects BFU2005-03102/BFI 'Effects of drought
on photosynthesis and respiration: acclimation and recovery', BFU2008-
01072/BFI 'Regulation of mesophyll conductance to CO_2 in relation to
plant photosynthesis and respiration', AGL2005-06927-CO2-01/AGR
'Optimización del uso del agua en la vid: Regulación y control fisiológico y agronómico y efectos en la calidad del fruto' and AGL2008-4525
'Hydraulic conductivity, mesophyll conductance and control of vegetative development effects on the WUE in grapevine: environmental and
genetic variations'.

CHAPTER 6

Funding
This research was conducted in the framework of the "Biodivigna" project
funded by the Veneto Region (Mis. 124 PSR Veneto). The funders had no
role in study design, data collection and analysis, decision to publish, or
preparation of the manuscript.

Acknowledgments
We are grateful to Mariano Brentan (University of Padova) who helped in
the identification of some critical specimens, to Filippo Taglietti (Conegliano-Valdobbiadene DOCG Consortium) and Elisabetta Barizza (University of Padova) who provided technical support for site selection. We thank
the owners of the vineyards who kindly allowed us to work in their fields
and provided information on management practices.

Author Contributions

Conceived and designed the experiments: JN LM DI MZ. Performed the experiments: JN DI. Analyzed the data: JN LM. Contributed reagents/materials/analysis tools: MZ. Wrote the manuscript: JN LM.

CHAPTER 7

Acknowledgments

This work was funded by the Spanish Ministry of Science and Innovation (SAFESPRAY project AGL2010-22304-C04-03 and AGL2010-22304-C04-04, and OPTIDOSA project AGL2007-66093-C04-02 and AGL2007-66093-C04-03), and the ERDF (European Regional Development Fund).

Conflict of Interest

The authors declare no conflict of interest.

CHAPTER 8

Acknowledgments

We thank Prof. H.G. Jones, University of Dundee, Dundee, UK, Dr. Gaëtan Louarn, INRA Centre Poitou-Charentes, Lusignan, France and Dr. Joachim Schmid, Forschungsanstalt Geisenheim, Germany for providing published and unpublished material. Financial support was provided by the 'Forschungsschwerpunkt Umweltstress und nachhaltige Produktion', Forschungsanstalt Geisenheim, Germany.

CHAPTER 9

Acknowledgments

The present paper resulted from a keynote presentation given at the 8th International Symposium on Grapevine Physiology and Biotechnology in Adelaide, Australia. I thank the conference organisers for the invitation and travel funding. Work for this review was partly funded by the University of Washington Climate Impacts Group.

CHAPTER 10

Acknowledgments
This work was funded by the Grape and Wine Research Development Corporation of Australia. We thank Nicole Kempster, Karina Swann and Pat Corena for excellent technical assistance. This paper is dedicated to the memory of Dr Alain Bouquet who passed away suddenly on 11th May 2009. Alain was an enthusiastic, committed and extremely generous scientist whose vision and creativity has paved the way for many of the breakthroughs in grapevine genetics that are now being achieved by grape researchers throughout the world.

CHAPTER 12

Conflict of Interes
The authors declare that the research was conducted in the absence of any commercial or financial relationships that could be construed as a potential conflict of interest.

Acknowledgment
The authors would like to thank professor Zucco for comments and criticisms on an early draft.

CHAPTER 13

Funding
Funding for this research was provided by: The Organic Farming Research Foundation (http://ofrf.org); Animal Behavior Society (http://animalbehaviorsociety.org); Wilson Ornithological Society (http://www.wilsonsociety.org); Smithsonian Migratory Bird Center (http://nationalzoo.si.edu/scbi/migratorybirds/default.cfm); Annie's Sustainable Agriculture Graduate Scholarship (http://www.annies.com/sustainable_agriculture_scholarship); and the Environmental Studies Department at University of California Santa Cruz (http://envs.ucsc.edu). The funders had no role in study design, data collection and analysis, decision to publish, or preparation of the manuscript.

Competing Interests

The authors have declared that no competing interests exist.

Acknowledgments

We thank David Koball, Mary Fetzer and Bonterra/Fetzer Vineyards for their cooperation and access to study sites. D. Thayer and D. Smith at UCSC were instrumental in helping to construct 200 nest boxes. L. Barrow, M. Poonamallee, and C. Mueller provided invaluable field assistance. We thank E. Heaton and C. Moreno for advice; P. de Valpine, P. Raimondi, and J. Freiwald for statistical consultations; and the Hopland Research and Extension Center for housing and services. H. Briggs, S. Bothwell-Allen, T. Cornelisse, T. Krupnik, A. Merenlender, and anonymous reviewers provided helpful comments on earlier drafts of the manuscript.

Author Contributions

Conceived and designed the experiments: JAJ RG DKL. Performed the experiments: JAJ. Analyzed the data: JAJ. Contributed reagents/materials/analysis tools: JAJ RG. Wrote the paper: JAJ. Revised the article: RG DKL.

INDEX